GIScience and Geo-environmental Modelling

The "GIScience and Geo-environmental Modelling" book series seeks to publish a broad portfolio of scientific books addressing the interface between geography and the environment. The aim of the book series is to present geospatial technology approaches to data mining techniques, data analytics, modeling, risk assessment, visualization, and management strategies. The series includes peer-reviewed monographs, edited volumes, textbooks, and conference proceedings. The focus of Geo-environmental is on geospatial modelling in the frontier area of earth-environment related fields, such as urban and peri-urban environmental issues, ecology, hydrology, basin management, geohazards, estuarine-ecology, groundwater, agriculture, climate change, land-water, and forest resources, and related topics.

Geo-environmental modelling techniques have enjoyed an overwhelming interest in recent decades among the earth environmental and social sciences research communities for their powerful ability to solve and understand various complex problems and develop novel approaches toward sustainable earth and human society. Geo-environmental modelling using data mining, machine learning, and cloud computing technology is focused on spatiotemporal data analysis and modeling for sustainability in our environment.

Swapan Talukdar · Ranit Chatterjee ·
Somnath Bera · Atish Prashar ·
Rajib Shaw
Editors

Communication, Science, Technology, and Innovation in Disaster Risk Management

Recent Trends and Approaches in South Asia

Springer

Editors
Swapan Talukdar
Department of Geography
Asutosh College
University of Calcutta
Kolkata, India

Ranit Chatterjee
Graduate School of Media
and Governance
Keio University
Fujisawa, Kanagawa, Japan

Somnath Bera
Department of Geography
Central University of South Bihar
Gaya, Bihar, India

Atish Prashar
Department of Mass Communication
and Media
Central University of South Bihar
Gaya, India

Rajib Shaw
Graduate School of Media
and Governance
Keio University
Fujisawa, Kanagawa, Japan

ISSN 2730-7506 ISSN 2730-7514 (electronic)
GIScience and Geo-environmental Modelling
ISBN 978-3-031-77188-0 ISBN 978-3-031-77189-7 (eBook)
https://doi.org/10.1007/978-3-031-77189-7

This Springer imprint is published by the registered company Springer Nature Switzerland AG
The registered company address is: Gewerbestrasse 11, 6330 Cham, Switzerland

This book is dedicated with heartfelt gratitude to Dr. Sweta Singh, whose constant inspiration has made this journey happy and lovely

Preface

South Asia faces increasing challenges from disasters exacerbated by climate change. The region, home to a quarter of the global population, is highly vulnerable to hazards such as floods, cyclones, earthquakes, and landslides. Its dense population, varied geography, and significant socioeconomic disparities make disaster response and risk management complex. Given this vulnerability, effective disaster risk management is imperative.

This book, *Communication, Science, Technology, and Innovation in Disaster Risk Management: Recent Trends and Approaches in South Asia*, examines contemporary challenges and cutting-edge approaches in managing disaster risks in this region. It highlights how communication, technological advancements, and innovation are transforming preparedness, response, and recovery efforts.

Communication plays a critical role in disaster management, ensuring that accurate and timely information is disseminated to vulnerable populations. Technology offers new opportunities to improve disaster management capabilities. Innovations such as satellite remote sensing, geographic information systems (GIS), mobile applications, and social media platforms are enabling more efficient data collection, analysis, and decision-making in real-time disaster scenarios. Innovation in policy, practices, and technology remains key to improving resilience, promoting preparedness, and enabling rapid response.

This book is divided into four major parts:

Part I: Impact of Disaster and Risk Management in South Asia

This part presents case studies on disaster impacts and response strategies. Examples include post-earthquake housing reconstruction in Nepal, the combined challenges of flooding and the COVID-19 pandemic in India, the effects of climate variability on coastal fishermen's livelihoods, and drought-related agricultural risks in India.

Part II: Communication in Disaster Management

This part focuses on the role of communication in disaster preparedness and mitigation, with a particular emphasis on India. Case studies include the use of social media, artificial intelligence (AI), and machine learning to disseminate real-time disaster information.

Part III: Science and Technology in Disaster Risk Management

The third part explores the use of scientific and technological tools in disaster risk management. Case studies include the application of Geo-AI for landslide hazard modeling, geospatial assessments of cyclone impacts, and the use of Google Earth Engine for flood hazard analysis.

Part IV: Innovation in Disaster Risk Management

This final part explores new tools, methodologies, and policies that address the specific challenges faced by South Asian countries. It examines innovations such as drone technology, social innovation, and sustainable infrastructure design for disaster resilience.

The book highlights the intersection of communication, science, technology, and innovation as essential to enhancing disaster risk management efforts and building resilience in South Asia's disaster-prone areas.

Kolkata, India Dr. Swapan Talukdar
Fujisawa, Japan Dr. Ranit Chatterjee
Gaya, India Dr. Somnath Bera
Gaya, India Prof. Atish Prashar
Fujisawa, Japan Prof. Rajib Shaw

The original version of the book has been revised. A correction to this book can be found at https://doi.org/10.1007/978-3-031-77189-7_17

Acknowledgements

The successful completion of this volume owes much to the unwavering support and contributions of numerous individuals and organizations, to whom we extend our heartfelt gratitude. First, we express sincere appreciation to all the authors who embraced and supported the original idea behind this book. Your dedication and expertise have been instrumental in shaping its contents and advancing the field of disaster management. Your valuable contributions have significantly enriched the discourse and knowledge in this area, and we are deeply grateful for your involvement. We also thank the anonymous reviewers whose diligent efforts and insightful comments elevated the quality and rigor of the work presented in this book. Your scholarly critiques and thoughtful suggestions have been invaluable in refining the ideas and findings contained within these pages. We appreciate the time you devoted to reviewing the chapters and are truly grateful for your input. Furthermore, our heartfelt thanks go to Ms. Pritha Ghosh, Mr. Shivam Priyadarshi, and Md. Noor Ali Imam Hassan for their invaluable editorial assistance throughout the publishing process. Your meticulous editing and dedication to ensuring the clarity and coherence of the content have been indispensable in bringing this book to fruition. Our sincere appreciation also extends to the entire Springer team for their continuous support throughout this journey. Your timely provision of information, guidance, and encouragement at every stage of this project has been invaluable. We thank all the individuals who worked behind the scenes to make this publication possible, from the editorial and production teams to the marketing and distribution personnel. We extend our gratitude to all the contributors who demonstrated unwavering dedication in completing their papers promptly. Their scholarly efforts have significantly contributed to creating this enlightening and invaluable book. This multidisciplinary volume is designed to appeal to a wide range of readers, including academics, scientists, students, planners, disaster management professionals, remote sensing experts, and GIS specialists. With a wealth of cutting-edge research and practical insights, *Communication, Science, Technology, and Innovation in Disaster Risk Management* stands as a valuable resource for enhancing understanding of disaster risk and resilience. We extend our heartfelt gratitude to the esteemed Springer Publishing House, Prof. Prabhat Kumar Shit, Springer Series Editor, and Doris Bleier, along with her highly efficient team. Special thanks to Cecil Joselin Simon for his timely communication. Their professionalism and collaborative efforts have been crucial in ensuring the timely publication of this book, and we deeply appreciate their support and assistance.

Finally, we hope that the publication of this book will inspire students and researchers to engage in more detailed investigations and analyses of disaster risk management. Once again, we express our deepest gratitude to everyone who contributed to making this endeavor a reality. Your support and encouragement have been a source of inspiration, and we are sincerely thankful for your invaluable contributions.

About This Book

The risk of disasters is predominated in the South Asia. In the coming decades, the region will experience rapid economic development and population growth, which, combined with the effects of climate change, will increase the likelihood of being affected by disasters. As disasters can negatively impact every sector of the economy and undo years of economic progress, it is imperative that effective measures to reduce the risk of disasters are incorporated into development plans and poverty reduction strategies. At the same time, technological advances have opened up previously unimaginable opportunities to strengthen resilience and improve connectivity. Both in this region, it has been shown that the prevention of disasters and the development of emergency plans enabled by advances in communications and space technology can be much more successful and significantly less costly than in the past. The use of space technology in disaster monitoring, early warning, and emergency response has shown remarkable efficiency. New approaches to disaster risk management (DRM) are becoming more available due to recent advances in space and communications technology. The proposed book therefore presents the five pillars of DRM, taking into account the communication, innovation, science, and technology.

The book will be composed of four parts, including 16 chapters. Part I will present various case studies on the impact of disasters and risk management in different regions of South Asia. Part II will examine the role of communication in disaster management. Part III will cover disaster risk assessment based on recent spatial technologies and tools such as GIS, geo-statistics, geo-AI, and Google Earth Engine. Part IV will uncover innovations in disaster management from different perspectives. Overall, *Communication, Science, Technology, and Innovation in Disaster Risk Management—Recent Trends and Approaches in South Asia* is an informative book that will contribute to improve disaster resilience plan and policy in South Asia and beyond.

Contents

Editors and Contributors

About the Editors

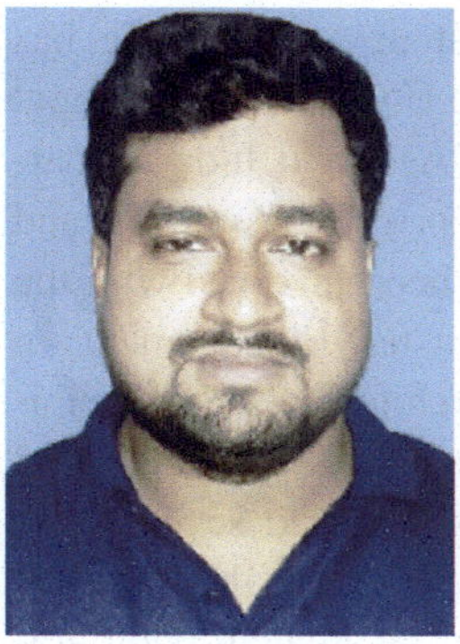

Dr. Swapan Talukdar is an assistant professor at Department of Geography, Asutosh College, University of Calcutta, Kolkata, India. He has worked as Dr. D. S. Kothari Post-doctoral fellow in the Department of Geography, Faculty of Natural Sciences, Jamia Millia Islamia, New Delhi, for the period of 20.09.2021–29.02.2024. He completed his Ph.D. from the Department of Geography, University of Gour Banga, West Bengal, India, in 2020. He served as the junior research fellow (JRF) during 2015–17 and senior research fellow (SRF) during 2017–19 in the Department of Geography, University of Gour Banga. He has completed two international projects (as Co-PI) in collaboration with the King Khalid University, Saudi Arabia. He has published more than 83 scientific research articles in journals of international repute such as *Journal of Cleaner Production*, *Environmental Pollution*, *Scientific Reports*, *Ecological Indicators*, and *Sustainable Cities and Society*. Also, one edited book has been accepted by Springer Nature. His main areas of research are urban environment, urban wetlands, natural hazards, remote sensing and AI, and their applications in environmental management. He has served as review editor for *Frontiers in Ecology and Evolution*, *Frontiers in Environmental Science*, *Frontiers in Earth Science*, and *Frontiers in Water*. He has served as a guest editor in *Sustainability* journal (MDPI). He has served as a reviewer for many journals of Springer, Elsevier, Taylor and Francis, MDPI, and Frontiers.

Google Scholar: https://scholar.google.co.in/citations?user=e9MgpZkAAAAJ&hl=en

Dr. Ranit Chatterjee is the visiting associate professor in Graduate School of Media and Governance at Keio University, Japan. He is also co-founder of a Delhi (India)-based social entrepreneur startup Resilience Innovation Knowledge Academy (RIKA) and RIKA Institute, a not-for-profit entity of RIKA. He is a Ph.D. in Environmental Management from Kyoto University, Japan. Trained as an Architect, he did his master's in Disaster Management from Tata Institute of Social Sciences, Mumbai. His work focuses mainly on disaster management while spanning across architecture and heritage, governance, private sector, and ecosystem services. He has worked previously with the UN agencies, national and local Governments in Asia and India in particular, local communities, private businesses, and NGOs. In addition, he has been involved with the National Institute of Disaster Management in training of engineers and architects. He is a recipient of IRDR Young Scientist Fellowship and CEM member of IUCN. He has authored several academic publications and reports on the topic of Disaster Management. He is an amateur photographer with a few publications in National Geographic.

Dr. Somnath Bera is currently an assistant professor in the Department of Geography at the Central University of South Bihar, India. He pursued his M.Phil. and Ph.D. in Disaster Studies at the Tata Institute of Social Sciences, Mumbai. Throughout his academic journey, he extensively explored disaster-affected regions across India, Nepal, China, and Japan enriching his research with firsthand experiences. Following his doctoral studies, he served as a research associate at the Department of Civil Engineering, Indian Institute of Technology Delhi. His scholarly pursuits revolve around disaster risk and resilience. His contributions to the field have earned him numerous fellowships, grants, and awards. He has authored a plethora of research articles published in esteemed journals like the *International Journal of Disaster Risk Reduction*, *Journal of Cleaner Production*, *Journal of Environmental Management*, *Remote Sensing Applications: Society and Environment*, *Natural Hazards*, *Bulletin of Engineering Geology*

and the Environment, and *Progress in Disaster Science*, among others. Additionally, he serves as a reviewer for prestigious international journals published by "Elsevier," "Springer," and "Taylor and Francis." Beyond academia, he finds joy in landscape painting.

https://scholar.google.com/citations?user=g0fSpo0AAAAJ&hl=en

Prof. Atish Prashar is a leading newspaper columnist and a well-known media educator with expertise in all media platforms. He is the Dean of the School of Media, Arts, and Aesthetics and works relentlessly as the head of the Department of Mass Communication (since 2016) at the Central University of South Bihar. He made his entry in academia in the year 2002 and since then has been associated with Delhi University, IP University, and many other institutions of excellence in the country.

He believes in maintaining high standards in media education which has led him make significant contributions in various universities and institutes as a selection committee member also he is contributing in Bihar State University service commission as a subject expert. He is well-known trainer and associated with Bihar Institute of Public Administration and Rural Development (BIPARD) and Bihar Police Academy. He has been training participants from Indian Administrative Services and Bihar Administrative Services.

He is a prolific writer with many reputed publications to his credit with an authorship of 5 books, 12 chapters, and 30 research papers. He is currently leading a major project as a Project Director at CUSB, funded by ICSSR. He has been closely worked with NIDM and organized many workshops on disaster management. Quite often his editorials appear in reputed newspapers and magazines. He is a member of various Board of Studies like Central University of Jharkhand, Assam University, CUSB, Patna University, Amity University, etc. He is also an academic council member at CUSB.

Prof. Rajib Shaw is a professor in Graduate School of Media and Governance of Keio University, Japan. He is the co-founder of a Delhi (India)-based social entrepreneur startup Resilience Innovation Knowledge Academy (RIKA), and RIKA Institute. He is the Co-chair of the United Nations Asia Pacific Science Technology Advisory Group (AP-STAG) and was CLA for IPCC's 6th Assessment Report. He is the recipient of prestigious "Pravasi Bharatiya Samman Award (PBSA)" in 2021 for his contribution in education sector. He also received United Nations Sasakawa Award in 2022 for his lifetime contributions in the field of disaster risk reduction. He published 64 books and over 450 research papers. More about his work can be found in: www.rajibshaw.org.

Contributors

Basanta Raj Adhikari U-Inspire Nepal, Kathmandu, Nepal; Institute of Himalayan Risk Reduction, Lalitpur, Nepal; Centre for Disaster Studies, Institute of Engineering, Tribhuvan University, Kathmandu, Nepal

Sohail Akhtar School of Social Work, Centre for Health and Mental Health, Tata Institute of Social Sciences, Mumbai, India

Guru Balamurugan Department of Geology, School of Earth Sciences, Central University of Tamil Nadu, Thiruvarur, India

Somnath Bera Department of Geography, Central University of South Bihar, Gaya, India

Subhas Bera Department of Remote Sensing and GIS Applications, Hemvati Nandan Bahuguna Garhwal University, Srinagar Garhwal, India

Ranit Chatterjee Graduate School of Media and Governance, Keio University, Fujisawa, Japan; RIKA Institute, New Delhi, India

Virajitha Chimalapati WRI India, Cuttack, India; University of York, York, United Kingdom

Sanchita Das Research and Development, Tata Institute of Social Sciences, Mumbai, India

K. Dharanirajan Department of Coastal Disaster Management, Pondicherry University, Puducherry, India

Suraj Gautam U-Inspire Nepal, Kathmandu, Nepal; Institute of Himalayan Risk Reduction, Lalitpur, Nepal; Nepal Flying Labs, Kathmandu, Nepal

Pritha Ghosh Department of Geography, Central University of South Bihar, Gaya, India

Balamurugan Guru Department of Geology, Central University of Tamil Nadu, Thiruvarur, India

N. A. Hareesh Centre for Geoinformatics, Tata Institute of Social Sciences, Mumbai, India

Chetna Jaiswal Department of Psychological Sciences, Central University of South Bihar, Gaya, Bihar, India

Toushif Jaman North Eastern Space Application Centre (NESAC), Department of Space, Government of India, Shillong, Meghalaya, India; Department of Coastal Disaster Management, Pondicherry University, Puducherry, India

Unnikannan P. Santosh Kumar Department of Psychological Sciences, Central University of South Bihar, Gaya, Bihar, India

Pravin Lamsal U-Inspire Nepal, Kathmandu, Nepal; Geovation Nepal, Kathmandu, Nepal

Shibu K. Mani Department of Civil Engineering, School of Engineering and Technology, Christ University, Mysore Road, Bengaluru, India

Prerna Vijaykumar Mehta WRI India, Cuttack, India; WRI India, Delhi, India

Niti Mishra Centre for Disaster Management, Jamsetji Tata School of Disaster Studies, Tata Institute of Social Sciences, Mumbai, India

Ramakanta Naik Jamsetji Tata School of Disaster Studies, Tata Institute of Social Sciences, Mumbai, India

Smitha S. Nair School of Social Work, Centre for Health and Mental Health, Tata Institute of Social Sciences, Mumbai, India

Peehu Pardeshi Tata Centre for Technology and Design, Indian Institute of Technology Bombay, Powai, Mumbai, India

Alisha Pathak Department of Mass Communication and Media, Central University of South Bihar, Gaya, India

Atish Prashar Department of Mass Media and Communication, Central University of South Bihar, Gaya, India

Shivam Priyadarshi Department of Geography, Central University of South Bihar, Gaya, India

Uttam Pudasaini U-Inspire Nepal, Kathmandu, Nepal; Nepal Flying Labs, Kathmandu, Nepal

Shailendra Rai Natural Resources Institute Finland (LUKE), Oulu, Finland; University of Jyväskylä, Kokkola University Consortium Chydenius, Kokkola, Finland

P. K. Ryngnga Department of Geography, North-Eastern Hill University, Shillong, India

Ranadeep Sarma Department of Geography, North-Eastern Hill University, Shillong, India

Md. Shahid Akhter Department of Humanities and Social Sciences, BITS, Hyderabad, Telengana, India

Rajib Shaw Graduate School of Media and Governance, Keio University, Fujisawa, Japan

S. V. Shivaprasad Sharma Disaster Management Support Division, National Remote Sensing Centre (ISRO), Hyderabad, India

Dharmendra Kumar Singh Department of Psychological Sciences, Central University of South Bihar, Gaya, Bihar, India

K. P. Soma Makaam (Mahila Kisan Adhikaar Manch), Secunderabad, India

Sandra Suresh Department of Psychological Sciences, Central University of South Bihar, Gaya, Bihar, India

Swapan Talukdar Department of Geography, Asutosh College, University of Calcutta, Kolkata, India

Part I
Impact of Disaster and Risk Management in South Asia

1 Disaster Risk Management in South Asia Through Innovations, Communication, and Technological Advances

Swapan Talukdar, Ranit Chatterjee, Somnath Bera, Atish Prashar, and Rajib Shaw

Abstract

Disasters continue to pose a significant threat to communities across South Asia and require effective disaster risk management strategies. This book addresses critical gaps in the understanding and implementation of disaster risk reduction measures through a multidisciplinary lens. The chapters in this book address various aspects of disaster management, including capacity development in the built environment, the impact of climate variability linked to extreme weather events on livelihoods, disaster communication strategies, and the role of science and technology in risk reduction. Each chapter contains unique insights from empirical research and case studies from different South Asian contexts. The need for this book arises from the increasing frequency and severity of disasters in the region, exacerbated by factors such as climate shocks, urbanization, and socio-economic vulnerabilities. By synthesizing research findings and practical experiences, this book aims to provide a comprehensive insight into the complexities of disaster risk management in South Asia, highlight innovative approaches and best practices to strengthen resilience and reduce vulnerability, and facilitate knowledge sharing and collaboration among stakeholders in the region. By fostering collaboration and innovation, this book seeks to empower communities to build resilience and adapt to the challenges of disasters in South Asia. We consider this book will be a valuable resource for disaster management practitioners, policymakers, researchers, and community leaders who want to improve their understanding of disaster risk dynamics and their preparedness and response capabilities. By systematically presenting research findings and practical insights, this book aims to inspire positive change and contribute to the development of sustainable strategies for disaster risk management in South Asia.

S. Talukdar
Department of Geography, Asutosh College, University of Calcutta, Kolkata 700026, India
e-mail: swapan.talukdar@asutoshcollege.in

R. Chatterjee · R. Shaw
Graduate School of Media and Governance, Shonan Fujisawa Campus, Keio University, Fujisawa, Japan

R. Chatterjee
RIKA Institute, New Delhi, India

S. Bera (✉)
Department of Geography, Central University of South Bihar, Gaya, India
e-mail: somnath@cusb.ac.in

A. Prashar
Department of Mass Communication and Media, Central University of South Bihar, Gaya, India

S. Talukdar et al. (eds.), *Communication, Science, Technology and Innovation in Disaster Risk Management*, GIScience and Geo-environmental Modelling, https://doi.org/10.1007/978-3-031-77189-7_1

Keywords

Disaster risk management · South Asia · Innovation · Resilience · Policy

1.1 Introduction

Disaster and risk management in South Asia is of paramount importance due to the region's high vulnerability to various natural and anthropogenic hazards: the region, which includes countries such as India, Pakistan, Bangladesh, Nepal, Bhutan, Sri Lanka, and the Maldives, is home to around 1.8 billion people, almost 24% of the world's population (World Bank 2021). This high population density, coupled with socio-economic vulnerability and rapid urbanization, exacerbates the impact of these disasters. South Asia has suffered 50% of global disaster-related deaths in the last two decades, with earthquakes and floods being the most devastating (UNDRR 2019). This vulnerability is exacerbated by rapid urbanization, population growth, and climate change, which together increase the frequency and intensity of these hazards.

South Asia's vulnerability to natural hazards is largely due to its geographical location and climatic conditions. For example, the Himalayan region is highly prone to earthquakes due to the collision of the Indian and Eurasian tectonic plates. The earthquake in Nepal in 2015 with a magnitude of 7.8 claimed almost 9,000 lives and caused economic losses of around 10 billion dollars, which corresponds to around 50% of Nepal's GDP (Government of Nepal 2015; Goda et al. 2015). The 2004 Indian Ocean tsunami also affected 14 countries and claimed over 230,000 lives, with the coastal regions of India, Sri Lanka, and the Maldives being severely affected (Mimura et al. 2011). Similarly, the extensive coastline of South Asia, particularly along the Bay of Bengal, is frequently affected by cyclones. Cyclone Amphanin in May 2020, caused around $13 billion worth of damage in India and Bangladesh and displaced millions of people (World Bank 2020). The region also faces significant flood risks as the annual monsoon rains cause widespread flooding, affecting millions of lives and livelihoods. Flooding continues to be a recurring problem, especially in Bangladesh, where nearly 80% of the country is floodplain, making it highly vulnerable to monsoon flooding.

Effective disaster and risk management in South Asia requires a multi-faceted approach that integrates technological, social, economic, and environmental strategies (Wen et al. 2023; Waheed et al. 2023; Orru et al. 2023). The use of remote sensing and geospatial technologies has been instrumental in improving early warning systems, hazard mapping, and disaster relief coordination (Krishnamurthy et al. 2014; Singh et al. 2017; Yang et al. 2023). For example, the integration of satellite data with ground-based observations has improved the accuracy of flood forecasting in the Ganges–Brahmaputra-Meghna basin, benefiting millions of people in Bangladesh and India (Hossain and Katiyar 2006). In addition, the application of artificial intelligence and machine learning in predictive modeling has shown promise in improving the accuracy of disaster risk assessments and early warning systems, thereby reducing vulnerabilities and increasing resilience (Rahman et al. 2023; Das et al. 2023; Talukdar et al. 2023; Akash et al. 2023; Islam et al. 2023; Rihan et al. 2023; Khan et al. 2024).

However, technological advances alone are not enough (Shaw and Kishore 2023). Effective disaster risk reduction (DRR) strategies must also include community-based approaches that involve the local population in preparedness and response activities (Lin and Lee 2023). In Nepal, post-earthquake reconstruction efforts have shown the importance of building local capacity and incorporating traditional knowledge into the reconstruction process. A comparative study of the post-earthquake reconstruction process of houses in Nepal found that local engagement and capacity building significantly improved resilience and shortened the reconstruction phase. Community-led initiatives in India, such as the development of disaster-resilient homes in

Odisha, have also proven effective in mitigating the effects of cyclones and floods (Ray-Bennett 2009).

The aim of this book is to explore and document the various strategies and innovations in disaster and risk management across South Asia to provide a comprehensive understanding of the challenges and solutions in the region. By compiling research findings and case studies from leading experts, this volume aims to bridge the knowledge gap and promote a deeper understanding of effective DRR practices. It aims to provide policymakers, researchers, and practitioners with insights into the integration of advanced technologies, community-based approaches, and policy frameworks that are essential for building resilience in disaster-prone areas. Through a multidisciplinary perspective, this book will emphasize the importance of collaborative efforts and knowledge sharing in improving disaster risk reduction and preparedness in South Asia.

1.2 Recent Trends of Disaster Management in South Asia

Disaster management in South Asia has evolved considerably in recent decades, reflecting the region's increasing vulnerability to disasters such as earthquakes, landslides, floods, cyclones, and droughts (Kafle 2017; Amarnath et al. 2017; Mohanty et al. 2022; Mukherjee et al. 2022; Bera et al. 2020; Mirdda et al. 2022). Despite considerable progress, disaster management in South Asia remains fragmented and less advanced than in regions such as Japan and European countries, where disaster preparedness and response mechanisms are more sophisticated and better-coordinated (Yamaguchi et al. 2013; Ngaire 2023; Ishjwatari 2021; Kanbara and Shaw 2021; Fokaefs and Sapountzaki 2021; Andreassen et al. 2020; Bouckaert et al. 2020).

One of the recent trends in disaster management in South Asia is the shift to an integrated and community-based approach (Gautam and Phaiju 2013; Islam et al. 2013, 2020; Wedawatta et al. 2016; Sanyal and Routray 2016; Tanwattana 2018; Seidler et al. 2018; Azad et al. 2019; Mall et al. 2019; Seddiky et al. 2020; Ali et al. 2021; Kodag et al. 2022). This trend is evident in countries such as Nepal and Bangladesh, where local communities are increasingly involved in disaster risk reduction (DRR) activities (Shaw et al. 2013; Smith et al. 2017). In Nepal, for example, following the devastating earthquake in 2015, there was an increase in community-led initiatives aimed at strengthening resilience and improving local capacity for DRR (Dixit and Shaw 2023). Similarly, the Cyclone Preparedness Program (CPP) in Bangladesh, which involves local volunteers in the dissemination of early warnings and evacuation processes, has helped to reduce the number of cyclone-related deaths (Ruszczyk et al. 2020). Another notable trend is the use of technology and innovation in disaster management. Geographic information systems (GIS), remote sensing, and early warning systems have become an integral part of disaster risk assessment and management (Garg and Eslamjan 2017; Gusain et al. 2019; Paul et al. 2020, 2020; Mohammad and Bishnoi 2021; Bera et al. 2021). India, for example, has developed a robust cyclone warning system that uses satellite data to provide timely warnings and facilitate evacuations. In addition, the use of mobile technologies and social media to disseminate information in real-time during disasters across the region has become increasingly important (Sukhwani and Shaw 2020; Yudarwati et al. 2022; Arfanuzzaman 2021; De and Bandyopadhyay 2020; Dahal et al. 2021; Shrestha et al. 2021).

The analysis of these trends is supported by a systematic review of 60 articles published between 2013 and 2024, focusing on keywords such as South Asia, disaster risk reduction, community-based approach, geoscience, and latest technology. In the word cloud in Fig. 1.1, terms such as "community'," "assessment'," "remote sensing'," "GIS'," "vulnerability" and "resilience' appear," reflecting the growing importance of community involvement and the use of advanced technological tools in disaster management. The frequent occurrence of terms

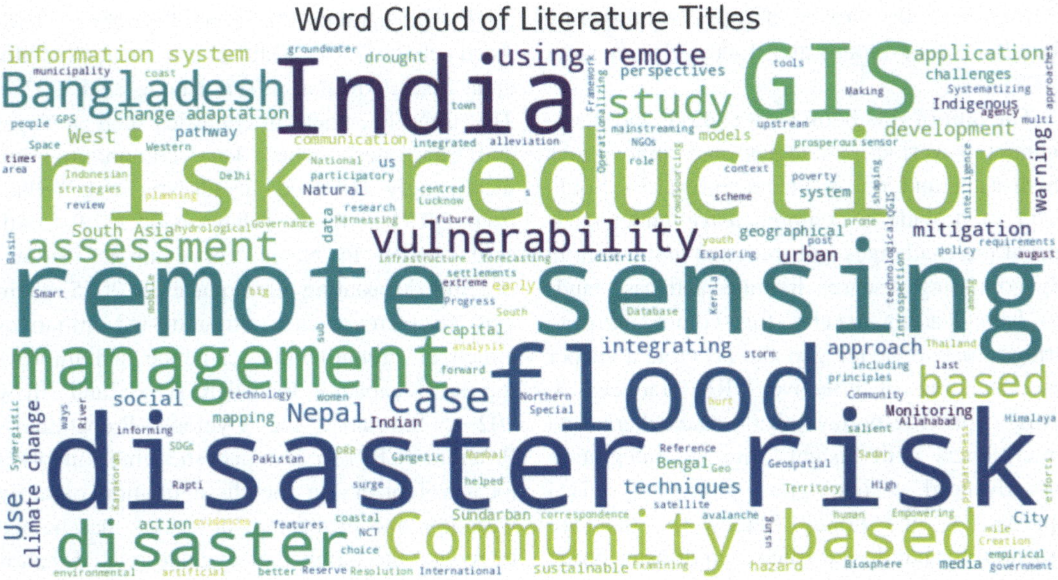

Fig. 1.1 Word Cloud of Literature Titles (2013–2024)—This word cloud illustrates the predominant themes and keywords in the reviewed literature on disaster management in South Asia

referring to specific countries such as “India’,” “Bangladesh” and “Nepal” highlights the regional focus of the literature, which emphasizes localized strategies tailored to the unique socio-economic and environmental circumstances of South Asia.

Furthermore, the temporal distribution of publications (Fig. 1.2) shows a notable increase in research output, peaking in 2021. This is likely due to the increased awareness of disaster risks and the urgent need for improved disaster management strategies amidst the COVID-19 pandemic. This increase indicates significant research interest and investment in disaster management technologies and community-based approaches during this period. However, the subsequent decline in 2022 suggests a potential research gap or shift in funding priorities and highlights the need for continued focus and support in this important area. This trend analysis highlights the need for sustained research efforts and the integration of innovative technologies and community engagement to improve resilience to disasters in South Asia.

Figure 1.3 shows the key trends in disaster management research in South Asia from 2005 to 2022, with a focus on community-based approaches and technology and innovation. The graph (Fig. 1.3a) shows a striking trend where the number of publications rises sharply from 2018 and peaks in 2021 with eight publications on community-based approaches and three on technology and innovation. This peak indicates increased research interest, likely triggered by recent significant disaster events and the increasing integration of technological innovation into disaster response practice. The decline in 2022, while notable, may reflect the cyclical nature of research funding and priority areas. The bar chart (Fig. 1.3b) showing the top 20 keywords in the literature titles also highlights the themes observed. Keywords such as ‘disaster’, ‘risk’, ‘remote sensing’, ‘mitigation’, and ‘India’ are most frequently found, indicating a strong emphasis on geospatial technologies in disaster risk management and a focused geographical interest in India. The frequency of these keywords underscores the critical areas of research and intervention in the region and highlights the ongoing efforts to utilize advanced technologies and methodologies to effectively mitigate disaster risks.

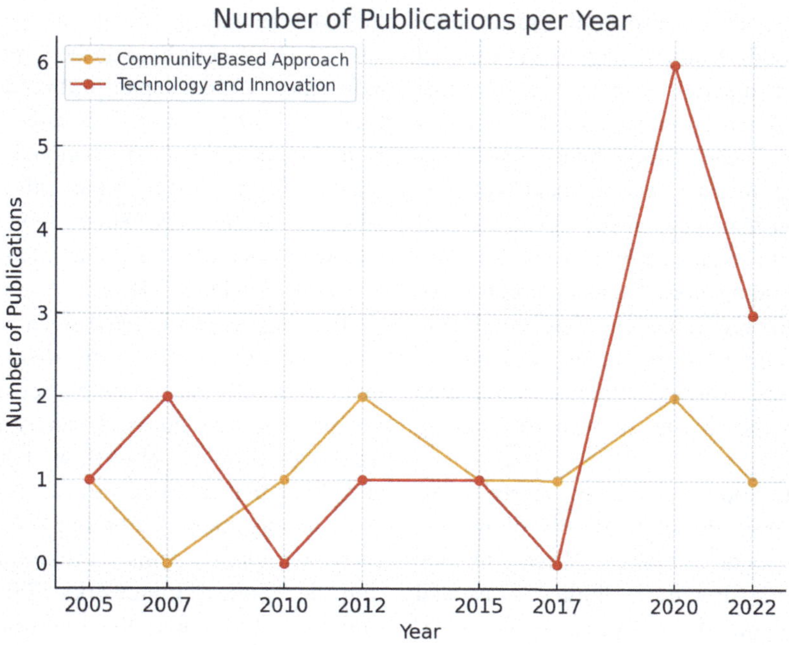

Fig. 1.2 Number of Publications per Year from 2005 to 2022—This graph shows the annual publication trend in disaster management research within South Asia

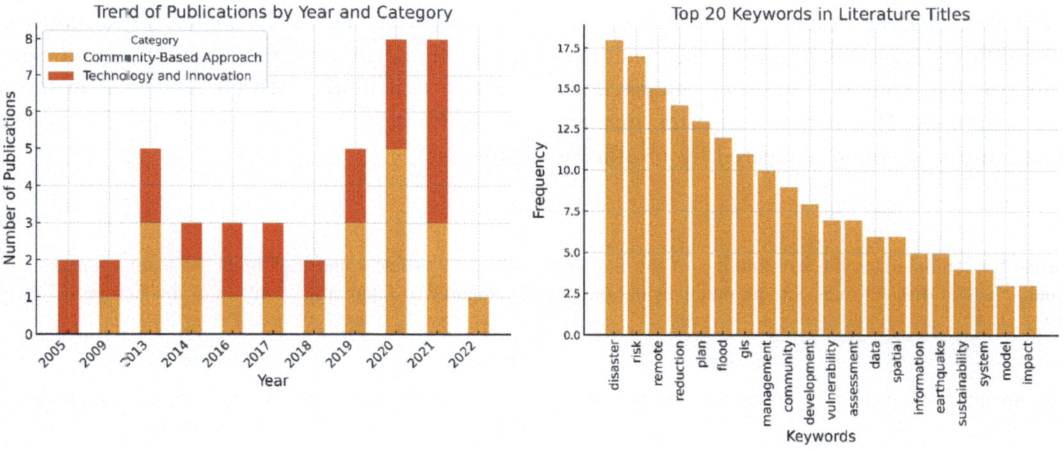

(a) (b)

Fig. 1.3 **a** Trend of Publications by Year and Category (2005–2022)—This graph shows the yearly distribution of disaster management research publications categorized into 'Community-Based Approach' and 'Technology and Innovation,' highlighting a significant peak in research activity in 2021 followed by a decline in 2022. **b** Top 20 Keywords in Literature Titles (2013–2024)—This bar graph displays the frequency of the top 20 keywords appearing in disaster management literature titles, emphasizing the dominant focus on disaster risks, technology use, and specific regional contexts such as India

Despite these advances, overall disaster management in South Asia still lags behind global standards. A comparative analysis shows that Japan among Asian countries and the countries of Europe have more comprehensive and better-coordinated disaster management systems. Japan's disaster management strategy is a global benchmark and is characterized by careful planning, strict building codes, and comprehensive public education programs. The European Union (EU) is also characterized by its civil protection mechanism, which facilitates coordinated responses and the sharing of resources among member states.

In contrast, South Asia faces several challenges that hinder the effectiveness of its disaster management efforts. These include inadequate infrastructure, limited financial resources, lagging in effective communication strategy, and social innovation. In addition, the region's high population density and rapid urbanization exacerbate its vulnerability to natural hazards. The floods in Pakistan in 2010, for example, which affected more than 20 million people, highlighted the shortcomings in the country's disaster preparedness and response. The region's disaster management efforts are often hampered by insufficient financial resources, limited technical capacity, and a lack of coordination between the various authorities. Unlike Japan in Asia and the countries of Europe, where disaster management is highly integrated into national development plans and supported by robust infrastructure and advanced technology, South Asia often struggles with fragmented approaches and inadequate resources (UNDRR 2019).

To underscore the need for a more focused and coherent approach to disaster management in South Asia, a comprehensive literature review was conducted. It analyzed 50 peer-reviewed articles published between 2010 and 2023 that address various aspects of disaster management, including risk assessment, early warning systems, community participation, and policy frameworks (Table 1.1). The literature shows that the importance of integrating scientific research and technological advances into disaster management practice is increasingly recognized. However, it also reveals significant gaps in implementation and coordination. However, policy and governance-related studies are underrepresented, suggesting that more research is needed in these areas to inform and improve policy-making processes.

Table 1.1 summarizes the key findings of the literature review (60 articles) and highlights the main strengths and weaknesses of disaster management practices in South Asia compared to Japan and countries of the European Union (EU)

Aspect	South Asia	Japan	Countries of Eu
Community participation	Increasing but fragmented	Well-integrated into national strategy	Strong local and regional involvement
Technology adoption	Growing use of GIS, remote sensing, mobile tech	Advanced early warning systems, robust infrastructure	Comprehensive use of technology, coordinated systems
Policy and communication	Weak implementation, lack of coordination	Strong policy frameworks, stringent building codes	EU Civil Protection Mechanism, effective resource sharing
Infrastructure	Inadequate, often outdated	State-of-the-art, resilient infrastructure	Advanced infrastructure, focus on sustainability
Financial resources	Limited, heavily reliant on international aid	Significant government investment	Well-funded through national and EU budgets
Public education	Growing awareness but inconsistent	Extensive public education programs	Strong emphasis on public awareness and education

This analysis emphasizes the need for South Asian countries to improve their disaster management by adopting best practices from Japan and Europe. Key recommendations include strengthening political and government structures, increasing investment in resilient infrastructure, and encouraging greater community participation. In addition, leveraging technology and innovation to develop more effective early warning systems and response mechanisms is critical. While South Asia has made progress in improving its disaster management, there is a clear need for a more coordinated and comprehensive approach. This edited book aims to contribute to this endeavor by providing insights into successful case studies, innovative practices, and interdisciplinary approaches that can improve the region's resilience to disasters. By fostering collaboration between researchers, policy makers, and practitioners, the book aims to create a more unified and effective framework for disaster management in South Asia.

Organization of the Book

This edited volume titled "Impact of Disaster and Risk Management in South Asia" is organized to provide a holistic understanding of disaster risk reduction (DRR) through four main themes: Impact of Disasters and Risk Management, Communication in Disaster Management, Science and Technology in Disaster Risk Management, and Innovation in Disaster Risk Management. Each theme is designed to explore different dimensions of DRR and contribute to the existing body of knowledge with specific case studies, empirical research, and theoretical insights.

The first theme, Disaster Impacts and Risk Management, looks at the practical aspects of disaster risk reduction and resilience in South Asia. Chapter 1, "Disaster Risk Reduction Capacity Development in the Built Environment: A Comparative Study of Post-Earthquake Housing Reconstruction Process in Nepal", examines post-earthquake reconstruction strategies and their effectiveness in rebuilding resilient infrastructure. This chapter provides important insights into how capacity development in the construction and housing sectors can significantly mitigate future risks. Chapter 2, "The Consequences of the Interaction of Flood and COVID-19 Pandemic for Surjapuri Municipality: A Case Study of Puria, Bihar" examines the combined impact of simultaneous disasters and emphasizes the need for integrated disaster management plans that address multiple hazard scenarios. Chapter 3, "Impacts of Climate Variability on Fishermen's Livelihoods: Evidence from the Western Coast of India"," focuses on the socioeconomic impacts of climate change on vulnerable communities and emphasizes the need for adaptation strategies to protect livelihoods. Finally, Chapter 4, "Assessment of factors influencing agricultural risks and resilience strategies of the farming community in Borda village of Kalahandi district in Odisha"," provides a comprehensive analysis of agricultural resilience and suggests strategies to improve the capacity of farmers to withstand climate-related risks. Together, these chapters underscore the importance of building resilient communities and infrastructure for better disaster risk reduction in South Asia.

The second theme, Communication in Disaster Management, addresses the critical role of effective communication in improving disaster mitigation and preparedness. Chapter 5, "Disaster Management in India: An Assessment Study for Disaster Communication" assesses the current communication strategies and identifies gaps that need to be addressed to improve response times and efficiency during disasters. Chapter 6, "Social Media: Potential Communication Tool in Disaster Response—Cyclone Ockhi and the Sri Lanka Floods using AI and ML" demonstrates the innovative use of social media platforms enhanced by artificial intelligence and machine learning to disseminate real-time information during disasters. This chapter illustrates the potential of using modern technology to improve communication strategies. Chapter 7, "Administrative Response to Cyclone Nilam: A Case Study of Tiruvallur District Administration" provides an in-depth analysis of the administrative response to the cyclone and highlights the importance of coordinated efforts and effective communication channels in managing the impact of disasters. These chapters highlight the importance of robust communication networks that enable timely and effective disaster management in South Asia.

The third theme, Science and Technology in Disaster Risk Management, focuses on the application of scientific research and technological advancement in disaster risk reduction. Chapter 8, "Assessing Dynamics of Exposed Elements to Natural and Anthropogenic-Induced Landslide Hazard Model for Land Use Policy," explores the use of advanced modeling techniques to understand and mitigate landslide hazards. Chapter 9, "Cyclone Induced Hazard Zonation of Bhadrak District, Odisha Using Analytical Hierarchy Process (AHP) and Geospatial Technology: A Tropical Cyclone Dynamics of East Coast of India Adjacent to Bay of Bengal," demonstrates the application of geospatial technology and AHP in cyclone hazard assessment. Chapter 10, "Spatiotemporal Flood Impact Assessment Using Google Earth Engine: Case of Darrang District, Assam," utilizes Google Earth Engine for comprehensive flood impact assessments. These chapters emphasize the critical role of science and technology in improving risk assessment, planning, and response to disasters.

The fourth theme, Innovation in Disaster Risk Management, presents innovative approaches and solutions to improve disaster risk reduction. Chapter 11, "Use of Unmanned Aerial Vehicles (UAVs) for Disaster Risk Reduction and Management in Nepal"," highlights the use of UAVs for effective disaster assessment and management. Chapter 12, "Collectivization and Local Leadership for Resilience to Disasters," discusses the importance of community-led initiatives and local leadership in building resilience to disasters. Chapter 13, "A Study of Biophilic Design and Resilience in the Malabar Region: An Innovation Towards Organizational Sustainability," explores the concept of biophilic design as a sustainable approach to disaster resilience. Chapter 14, "Co-Designing Public Spaces with Youth: Dual Pathways to Adolescent Engagement within Disaster Risk Mitigation and Adaptation" emphasizes the role of youth engagement in disaster risk reduction. These chapters present innovative strategies and practices that can significantly improve disaster resilience in South Asia.

By organizing the book around these themes, each section contributes unique insights and practical recommendations for improving disaster risk reduction in South Asia. The various case studies and research findings presented in this volume are a valuable resource for policymakers, researchers, and practitioners seeking to develop more effective and resilient disaster management strategies.

References

Akash SH, Sarkar SK, Bindajam AA, Kumari R, Talukdar S, Mallick J (2023) Assessment of coastal vulnerability using integrated fuzzy analytical hierarchy process and geospatial technology for effective coastal management. Environ Sci Pollutn Res, 1–18

Ali T, Paton D, Buergelt PT, Smith JA, Jehan N, Siddique A (2021) Integrating Indigenous perspectives and community-based disaster risk reduction: a pathway for sustainable Indigenous development in Northern Pakistan. Int J Disaster Risk Reduc 59:102263

Amarnath G, Alahacoon N, Smakhtin V, Aggarwal P (2017) Mapping multiple climate-related hazards in South Asia. International Water Management Institute (IWMI). 170

Andreassen N, Borch OJ, Sydnes AK (2020) Information sharing and emergency response coordination. Saf Sci 130:104895

Arfanuzzaman M (2021) Harnessing artificial intelligence and big data for SDGs and prosperous urban future in South Asia. Environ Sustain Indic 11:100127

Azad MAK, Uddin MS, Zaman S, Ashraf MA (2019) Community-based disaster management and its salient features: a policy approach to people-centred

risk reduction in Bangladesh. Asia-Pac J Rural Dev 29(2):135–160
Bera S, Guru B, Chatterjee R, Shaw R (2020) Geographic variation of resilience to landslide hazard: a household-based comparative studies in Kalimpong hilly region, India. Int J Disaster Risk Reduc 46:101456
Bera S, Melo R, Guru B (2021) Assessment of exposed elements XE "Exposed elements" in a changing built environment by using an integrated model of debris flow initiation and runout (Kalimpong region, Himalaya). Bull Eng Geol Env 80(9):7131–7152
Bouckaert G, Galli D, Kuhlmann S, Reiter R, Van Hecke S (2020) European coronationalism? A hot spot governing a pandemic crisis. Public Adm Rev 80(5):765
Dahal L, Idris MS, Bravo V (2021) "It helped us, and it hurt us" The role of social media in shaping agency and action among youth in post-disaster Nepal. J Contingencies Crisis Manag 29(2):217–225
Das T, Talukdar S, Shahfahad Baig MRI, Hang HT, Siddiqui AM, Rahman A (2023) Assessing vulnerability to cyclones in coastal Odisha using fuzzy logic integrated AHP: towards effective risk management. Spatial Inform Res, 1–19
De J, Bandyopadhyay S (2020) Smart governance in times of extreme natural disasters: an introspection. The 'after phase'of super cyclone amphan in india and bangladesh. South Asia Democratic Forum (SADF).
Dixit A, Shaw R (2023) Smart cities in Nepal: The concept, evolution and emerging patterns. Urban Governance 3(3):211–218
Fokaefs A, Sapountzaki K (2021) Crisis communication after earthquakes in Greece and Japan: effects on seismic disaster management. Sustainability 13(16):9257
Garg V, Eslamian S (2017) Monitoring, assessment, and forecasting of drought using remote sensing and the geographical information system. In: Handbook of Drought and Water Scarcity. CRC Press. pp 217–252
Gautam DK, Phaiju AG (2013) Community based approach to flood early warning in West Rapti River Basin of Nepal. IDRiM Journal 3(1):155–169
Goda K, Kiyota T, Pokhrel RM, Chiaro G, Katagiri T, Sharma K, Wilkinson S (2015) The 2015 Gorkha Nepal earthquake: insights from earthquake damage survey. Frontiers in Built Environment 1:8
Government of Nepal (2015) Post disaster needs assessment. Government of Nepal, Kathmandu
Gusain HS, Negi HS, Dhamija S, Mishra VD (2019) Development of avalanche information system using remote sensing and GIS technology in the Indian Karakoram Himalaya. Curr Sci 117(1):104–109
Hossain F, Katiyar N (2006) Improving flood forecasting in international river basins. EOS Trans Am Geophys Union 87(5):49–54
Ishiwatari M (2021) Institutional coordination of disaster management: engaging national and local governments in Japan. Nat Hazard Rev 22(1):04020059
Islam MN, Malak MA, Islam MN (2013) Community-based disaster risk and vulnerability models of a coastal municipality in Bangladesh. Nat Hazards 69:2083–2103
Islam S, Chu C, Smart JC (2020) Challenges in integrating disaster risk reduction and climate change adaptation: Exploring the Bangladesh case. Int J Disaster Risk Reduc 47:101540
Islam ARMT, Bappi MMR, Alqadhi S, Bindajam AA, Mallick J, Talukdar S (2023) Improvement of flood susceptibility mapping by introducing hybrid ensemble learning algorithms and high-resolution satellite imageries. Nat Hazards 119(1):1–37
Kafle SK (2017) Disaster risk management systems in South Asia: natural hazards, vulnerability, disaster risk and legislative and institutional frameworks. J Geogr Nat Disasters 7(207):2167–2587
Kanbara S, Shaw R (2021) Disaster risk reduction regime in Japan: an analysis in the perspective of open data, open governance. Sustainability 14(1):19
Khan NS, Roy SK, Talukdar S, Billah M, Iqbal A, Zzaman RU, Mallick J (2024) Empowering real-time flood impact assessment through the integration of machine learning and Google Earth Engine: a comprehensive approach. Environ Sci Pollut Res, 1–16
Kodag S, Mani SK, Balamurugan G, Bera S (2022) Earthquake and flood resilience through spatial planning in the complex urban system. Progress Disaster Sci 14:100219
Krishnamurthy RR, DasGupta R, Chatterjee R, Shaw R (2014) Managing the Indian coast in the face of disasters & climate change: a review and analysis of India's coastal zone management policies. J Coast Conserv 18:657–672
Lin BC, Lee CH (2023) Constructing an adaptability evaluation framework for community-based disaster management using an earthquake event. Int J Disaster Risk Reduc 93:103774
Mall RK, Srivastava RK, Banerjee T, Mishra OP, Bhatt D, Sonkar G (2019) Disaster risk reduction including climate change adaptation over south Asia: challenges and ways forward. Int J Disaster Risk Sci 10:14–27
Mimura N, Yasuhara K, Kawagoe S, Yokoki H, Kazama S (2011) Damage from the Great East Japan Earthquake and Tsunami-a quick report. Mitig Adapt Strat Glob Change 16:803–818
Mirdda HA, Bera S, Chatterjee R (2022) Vulnerability assessment of mountainous households to landslides: a multidimensional study in the rural Himalayas. Int J Disaster Risk Reduc 71:102809
Mohammad S, Bishnoi P (2021) Use of GIS and remot e sensing as risk reduction techniques in disasters with special reference of India. Int J Environ Climate Change, 119–124
Mohanty A, Dubey A, Singh RB (2022) Cyclonic disasters and resilience: An empirical study on South Asian coastal regions. Springer Nature

Mukherjee M, Wickramasinghe D, Chowdhooree I, Chimi C, Poudel S, Mishra B, Shaw R (2022) Nature-based resilience: experiences of five cities from South Asia. Int J Environ Res Public Health 19(19):11846

Ngaire M (2023) The role of non-governmental organizations in disaster management: case studies from Japan and New Zealand. Stud Soc Sci Human 2(11):20–29

Orru K, Klaos M, Nero K, Gabel F, Hansson S, Nævestad TO (2023) Imagining and assessing future risks: a dynamic scenario-based social vulnerability analysis framework for disaster planning and response. J Contingencies Crisis Manag 31(4):995–1008

Paul P, Aithal PS, Bhimali A, Kalishankar T, Saavedra MR, Aremu PSB (2020) Geo information systems and remote sensing: applications in environmental systems and management. Int J Manag Technol Soc Sci (IJMTS) 5(2):11–18

Rahman A, Roy SS, Talukdar S, Shahfahad. (2023) Remote Sensing and Artificial Intelligence for Urban Environmental Studies. Advancements in Urban Environmental Studies: Application of Geospatial Technology and Artificial Intelligence in Urban Studies. Springer International Publishing, Cham, pp 1–12

Ray-Bennett NS (2009) The influence of caste, class and gender in surviving multiple disasters: a case study from Orissa India. Environ Hazards 8(1):5–22

Rihan M, Bindajam AA, Talukdar S, Naikoo MW, Mallick J, Rahman A (2023) Forest fire susceptibility mapping with sensitivity and uncertainty analysis using machine learning and deep learning algorithms. Adv Space Res 72(2):426–443

Ruszczyk HA, Upadhyay BK, Kwong YMC, Khanal O, Bracken LJ, Pandit S, Bastola R (2020) Empowering women through participatory action research in community-based disaster risk reduction efforts. Int J Disaster Risk Reduc 51:101763

Sanyal S, Routray JK (2016) Social capital for disaster risk reduction and management with empirical evidences from Sundarbans of India. Int J Disaster Risk Reduc 19:101–111

Seddiky MA, Giggins H, Gajendran T (2020) International principles of disaster risk reduction informing NGOs strategies for community based DRR mainstreaming: the Bangladesh context. Int J Disaster Risk Reduc 48:101580

Seidler R, Dietrich K, Schweizer S, Bawa KS, Chopde S, Zaman F, Khaling S (2018) Progress on integrating climate change adaptation and disaster risk reduction for sustainable development pathways in South Asia: evidence from six research projects. Int J Disaster Risk Reduc 31:92–101

Shaw R, Kishore K (2023) Disaster risk reduction and G20: a major step forward. Prog Disaster Sci 17:100274

Shaw R, Mallick F, Islam A (eds) (2013) Disaster risk reduction approaches in Bangladesh (Vol. 103). Springer, New York, NY

Shrestha MS, Gurung MB, Khadgi VR, Wagle N, Banarjee S, Sherchan U, Mishra A (2021) The last mile: flood risk communication for better preparedness in Nepal. Int J Disaster Risk Reduc 56:102118

Singh C, Deshpande T, Basu R (2017) How do we assess vulnerability to climate change in India? A systematic review of literature. Reg Environ Change 17:527–538

Smith PJ, Brown S, Dugar S (2017) Community-based early warning systems for flood risk mitigation in Nepal. Nat Hazard 17(3):423–437

Sukhwani V, Shaw R (2020) Operationalizing crowdsourcing through mobile applications for disaster management in India. Prog Disaster Sci 5:100052

Talukdar S, Pal S, Naikoo MW, Rahman A (2023) Exploring the Flooding under Damming Condition in Punarbhaba River of India and Bangladesh. Floods in the Ganga–Brahmaputra–Meghna Delta. Springer International Publishing, Cham, pp 161–182

Tanwattana P (2018) Systematizing community-based disaster risk management (CBDRM): case of urban flood-prone community in Thailand upstream area. Int J Disaster Risk Reduc 28:798–812

UNDRR (2019) Global assessment report on disaster risk reduction. United Nations Office for Disaster Risk Reduction

Uniyal A, Shamiuddin PN, Priyanka LS, Pipil S, Rao S, Mohan R, Haldar AL () High resolution remote sensing, GPS and GIS based geospatial database creation for disaster risk reduction in lucknow city. Disaster Dev, 81

Waheed A, Fischer TB, Kousar S, Khan MI (2023) Disaster management and environmental policy integration in Pakistan—an evaluation with particular reference to the China-Pakistan Economic Corridor Plan. Environ Sci Pollut Res 30(48):105700–105731

Wedawatta G, Kulatunga U, Amaratunga D, Parvez A (2016) Disaster risk reduction infrastructure requirements for South-Western Bangladesh: perspectives of local communities. Built Environ Project Asset Manag 6(4):379–390

Wen J, Wan C, Ye Q, Yan J, Li W (2023) Disaster risk reduction, climate change adaptation and their linkages with sustainable development over the past 30 years: a review. Int J Disaster Risk Sci 14(1):1–13

World Bank (2020) Poverty and shared prosperity 2020. World Bank, Washington, D.C.

World Bank (2021) World development indicators. World Bank, Washington, D.C.

Yamaguchi N, Kotani T, Fukushima T, Yoshitomi N, Fouse D, Ear J (2013) Asia-Pacific Center For Security Studies Honolulu HI. Enhancing trilateral disaster preparedness and relief cooperation between Japan, US and Australia: Approaches from Various Civil-Military Perspectives

Yang Z, Barroca B, Laffréchine K, Weppe A, Bony-Dandrieux A, Daclin N (2023) A multi-criteria framework for critical infrastructure systems resilience. Int J Crit Infrastruct Prot 42:100616

Yudarwati GA, Putranto IA, Delmo KM (2022) Examining the Indonesian government's social media use for disaster risk communication. Asian J Commun 32(1):1–20

Disaster Risk Reduction Capacity Development in the Built Environment: A Comparative Study of Post-earthquake Recovery Process of Houses in Nepal

2

Md. Shahid Akhter, Shibu K. Mani, and Somnath Bera

Abstract

The present study attempts to explore and understand the impacts of 2015 Gorkha Earthquake Nepal through the lens of "Build Back Better" approach. The study has looked into the recovery of people's housing in Barpark and Laprak in Gorkha and Sitapaila and Dahachowk in Kathmandu district. The study adopted a qualitative and a multi-cited case study approach. The data were collected through Focus Group Discussions (FGDs), key informants' interviews and field observation. Semi-structured questionnaires were developed for a detailed field survey, and that was also conducted through FGDs. The study finds that despite the hardships the people had to face, tremendous indomitable spirit and community solidarity made them resilient in many ways. The distance between affected villages and district headquarters, seismic complexities, ineffective political leadership, and difficult transportation and communication system, poor public awareness and insufficient resources were found to be impacting the recovery process in the villages of Gorkha district. Contrarily, villages in Kathmandu district, particularly adjacent to Kathmandu city were found to be speedy in recovery process due to coordination for resource mobilization, efficient communication and transportation system. Along with this, they also have high literacy rate, opportunities for alternative occupations, high alert about government's aid scheme, and better risk consciousness. The study supports the approach that disasters should be taken as tools of opportunities to bring changes and to build back better to a society or community for long-term sustainable development.

Keywords

2015 Nepal earthquake · Built back better environment · Housing recovery · Sustainable development

M. Shahid Akhter (✉)
Department of Humanities and Social Sciences, BITS, Pilani, Hyderabad Campus, Hyderabad, Telengana 500078, India
e-mail: shahidpu10@gmail.com; p20180437@hyderabad.bits-pilani.ac.in

S. K. Mani
Department of Civil Engineering, School of Engineering and Technology, Christ University, Kengeri Campus, Kumbalgodu, Mysore Road, Bengaluru 560074, India
e-mail: shibuk.mani@christuniversity.in

S. Bera
Department of Geography, Central University of South Bihar, Gaya, Indiae-mail: somnath@cusb.ac.in

S. Talukdar et al. (eds.), *Communication, Science, Technology and Innovation in Disaster Risk Management*, GIScience and Geo-environmental Modelling, https://doi.org/10.1007/978-3-031-77189-7_2

2.1 Introduction

The concept of Disaster Risk Management have developed in beginning of the twentiethtwentieth century and it has gradually become institutionalized (Rollnick 2006). Asian (2002) has defined Disaster Risk Reduction as 'the conceptual framework developed with the possibilities of reducing vulnerabilities and kinds of disaster risk through preventing and mitigating the adverse impacts of hazards under the broad context of sustainable development". "Build Back Better" is an emerging approach of disaster risk reduction. It is a concept of building resilient communities throughout the reconstruction phase following a disaster. It is necessary to reduce the potential disaster risk and construct homes in lower-risk areas, and need to build resistant structures. A disaster can be an opportunity to experience and learn lessons such as land-use planning and the construction of disaster-resilient structures (Song 2022).

2.1.1 Post-Disaster Recovery

Historically, the recovery process has comprehensively been understood as setting a disaster-stricken community back to its pre-disaster state (Mileti 1999). It is a process of returning to "normalcy" after the disaster strikes. The recovery is the transforming process which passes through various stages with its stakeholders such as survivors, organizations, and communities. This process may show varied pace depending on the context, both spatial and temporal (Phillips 2009). But it has increasingly been accepted that recovery is not only about rebuilding what there was before a catastrophe but also, it's about creating safer and developing better communities which can be considered as a new normal. Recovery, one key facet of disaster management can open windows for shaping the future relationship between the local government, civil society and state government (Qadir 2015). The term "recovery" needs to be utilized interchangeably with rebuilding, reconstruction, restoration and rehabilitation in post-disaster redevelopment (Quarantelli 1998). The post-disaster period could be used for the preparedness towards future catastrophes. Mileti (1999) sees disaster recovery as a transform of the collaboration of choice-making among a variety of institutions including businesses, households and finally the community at large.

In a pragmatic sense, many specialists consider that disaster recovery time normally needs no reasonable endpoint. Others (Blaikie et al. 2014) have argued that in certain cases, especially the most vulnerable households and individuals would not recover fully from the impacts of disasters. However, the process of recovery can be categorized into four stages: development, relief, rehabilitation, and reconstruction. The process of recovery would vary from disaster to disaster and country to country, as well as between rural and urban areas. It depends on certain physical and human factors, such as the level of damage, accessibility, and availability of resources, existing cultural and infrastructural conditions as well as political settings (Frerks et al. 1995). Active support from state and federal emergency agencies with the guidance of local government entities introduce new hazard reduction measures and implement building codes and zoning regulations to develop all-around resilience for future disasters (Prater and Lindell 2000; Perry et al. 2001).

2.1.2 Post-Disaster Housing Recovery

The essential aspect that constitutes the resilience of the built environment is the involvement of efficient engineers, adequate design, and appropriate construction. It has been stated that constructing and designing a resilient built environment needs in-depth knowledge and expertise to mitigate the adverse consequences of disasters (Hamelin and Hauke 2005). The post-disaster housing recovery can be categorized into: (i) Emergency shelter; (ii) Temporary shelter; (iii) Temporary housing, and (iv) Permanent housing. There are significant demands for the involvement of varied stakeholders in each approach. But, the DRR measures can be implemented only in Permanent housing (Quarentelli 1982).

It has been concluded that chartered surveyors with efficient and appropriate training play key roles during all disaster phases, from the very first preparedness to immediate relief, temporary and traditional recovery to long-term reconstruction. According to Barenstein (2012), the post-disaster permanent housing recovery has five different approaches such as owner-driven approach, subsidiary approach, participatory approach, external contractor-drove approach and contractor-driven approach in situ. Further, Barenstein and Iyanger (2010) elucidate that post-disaster recovery programme is by the state government at the state level, national government at national level and civil society and community people at local level. Owner-driven approach is a recent phenomenon (Karunasena and Rameezdeen 2010) and it should be considered as the most important, empowering, natural and dignifying approach for encouraging people to build their houses (Barenstein and Iyanger 2010). The importance of owner-driven approach is increasingly getting attention among the donor communities.

Nevertheless, disaster generates windows of opportunities for some low-income housing communities by getting assistance from numerous government and non-government agencies. This privilege is extended to their housing resettlement in adopting mitigation measures. This could help them to create a resilient state in the future and physically build back better (Kreimer 1980). The significance of integrating and implementing DRR approach into reconstruction process has rightly been reinforced by Wu & Lindell (2004). Post-disaster construction is involved with a range of critical activities: immediate temporary shelter before and after the disaster; restoration and recovery of public services such as water supply, schools, hospitals, power, communication, environmental infrastructure and increasing secure income-earning opportunities for disaster victims in the affected areas (World Bank 2001). Sevin and Little (1998) suggest that highly technical modern and computerized building plans, building and infrastructural analysis program and damage assessments models can enhance rapid rescue and restoration as well as recovery of earthquake victims.

2.1.3 Nepal Earthquake 2015

Nepal suffered a massive loss of lives and property by the devastating earthquake with a magnitude of 7.8 on Saturday 25 April 2015. The quake caused avalanches on Mount Everest and in the Langtang Valley. Villages were destroyed, and people were made homeless across 31 districts, among them 14 districts were severely damaged. Transportation systems devastated throughout the earthquake zone due to an earthquake-induced landslide in various parts of the country. So many historic heritages like stupas, temples, and Missionaries were destroyed. More than 9,000 people died, and 22,300 people were severely injured. The Nepal government reports showed that at least 498,852 private houses and 256,697 government buildings were fully damaged. Another 3,622 government buildings and 256,697 private homes were partially damaged. Further, 19,000 school classrooms were fully impaired.

In spite of various drawbacks in technology and urbanization on human lives, the Sendai Framework for Disaster Risk Reduction 2015 (Pearson and Pelling 2015) in recovery approach is based on technocratic and bureaucratic approach. Resilient society comes from the action of individualization. Beck's analysis on individualization says that risks bearing people need to take much more responsibility themselves. The adaptation of DRR in recovery is depended mainly on the individual's economic condition, understanding, and awareness for probable risks and disasters (Zinn 2008). Focusing on the aforementioned aspects, the study tries to understand the recovery process and implementation of DDR in recovery as well as the interface between communities and the

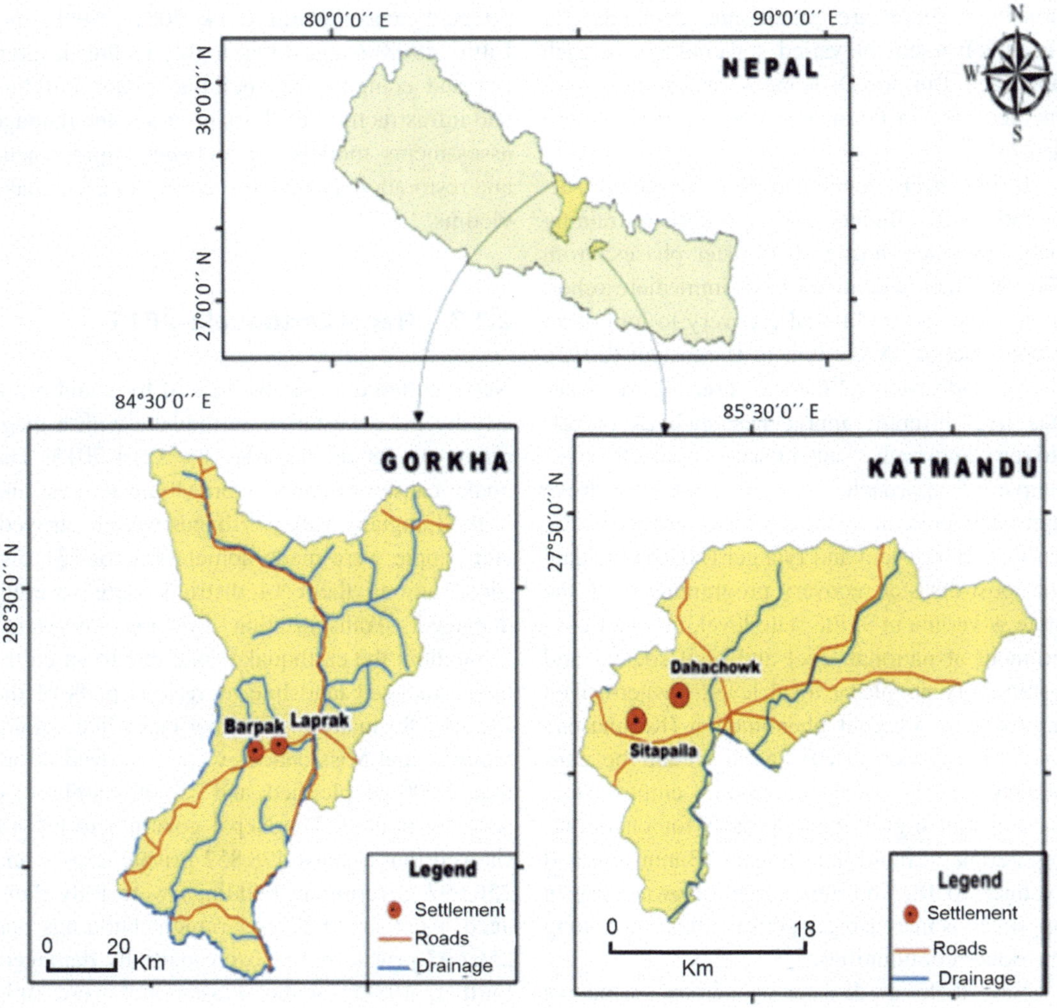

Fig. 2.1 Location of the study area

state mechanisms, and the challenges faced during recovery process. The study is carried out in the context of DRR Capacity Development in the Built Environment: A comparative study of the post-earthquake recovery process in Nepal.

2.2 Methodology

2.2.1 Site Selection

The Gorkha and Kathmandu districts were selected as the study area because the comparative study of both districts could show the status of the recovery process (Fig. 2.1). The important reason to have chosen Gorkha and Kathmandu districts as study area were its severity of impacts in 2015 earthquake. Seismically active, instability of topography, frequent weather changes, weak infrastructure, poor water supply, insignificant proportion of literacy, traditional agricultural practices and humdrum transportation logistic were significant factors for the selection of Laprak and Barpark VDC in Gorkha District. Less distance from Kathmandu city, economically enriched, high accessibility and feasibility of different supporting organizations, high proportion of educated people, social bonding, more awareness among people, strengthen infrastructure systems, less dependency members in a family, more advanced scientifically and technologically were the key reasons to have made the selection of Sitapaila and Dahachowk in Kathmandu district as a study area. The field visit was conducted for a period of three months (from June to August 2017) in the Kathmandu and Gorkha districts of Nepal.

Dahachowk and Sitapaila: Dahachowk is a village development committee in the western part of Kathmandu District, Nepal. The sharing borders are Ramkot VDC on the north, Mahadevstan VDC on the south, Balambu VDC on the east and Barbhanjyang VDC on the west. It has an area of 5018 square km. Dahachowk VDC is one of the most affected VDC in Kathmandu but the process of recovery was almost completed during the survey. Few rubble houses in the hilly areas are suffering from construction and repairs due to the financial crisis. It had a total population of 3860 as per the 2001 census. The literacy rate is 40% which is very less as compared to Kathmandu city. The total household in the VDC is 770 and an average family size of 5.01 members.

Sitapaila is a village development committee and is a part of Nagarjun Municipality in province No. 3 of central Nepal. The VDC is located between 27.72° North latitude and 85.28° East longitude. According to the 1991 Nepal census, the VDC had a population of 5, 156 and had 1, 008 households in it. The office of Nagarjun Municipality is located at Harisiddhi, Sitapaila.

The district, Gorkha is geographically located to the North-West of the Kathmandu City. The latitudinal and longitudinal position of the district is 28.47° N and 84.68° E, respectively. The average elevation of the district is 4,322 m. The total area covered by the Gorkha district is 361,470 ha. The population of Gorkha district mainly depends on cultivation as a principle source of income and livelihood. The total cultivated agricultural land is 41,482 ha. Total population according to 1991 census is 252,524. The entire part of the Gorkha is seismically active, and the geomorphology is hence very fragile. Such complex geology made thedistrict very less density populated district. The socio-cultural development of any region depends on the enrichment of economic condition. Higher the source of income is higher the social status of human being and vice versa. Poor financial condition affected their education system severely, and only 43.6% literacy is seen in this district.

Barpark: Barpark is a village situated in the northern part of the Gorkha district of Nepal, in the Barpark village development committee. The latitude and longitude of Barpark VDC are 28.21° N and 84.74° E respectively. It is inhabited by Gurungs, Ghales, Sunars, and others. Being close to the epicentre and lodged on a steep hillside, the village was extremely hard hit by the 2015 Nepal earthquake. Reports say that less than 10 of 1,200 homes remain standing after the earthquake. Barpark is situated on the hilltop about 6,200 ft above sea level of Gorkha and approximately 45 km away from Gorkha Bazar.

Laprak: Laprak is a village development committee in Gorkha district in the Gandaki Zone of northern-central Nepal. The latitude and longitude of Laprak VDC are 28.22° N and 84.79° E respectively. At the time of 1991 Nepal census, it had a population of 2,165 and had 462 houses in the village. Laprak is near to Barpak VDC and also got heavily affected by 2015 earthquake. It is so difficult to reach by vehicle. People use foot trails to reach the village. Laprak has no machinery and limited electricity to help rebuild their devastated houses.

2.2.2 Research Design

- The study is based on qualitative approach. The proposed research is carried out in the context of "DRR Capacity Development in the Built Environment: A comparative study of the post-earthquake recovery process in Nepal." The broader goal of this study is to understand the implementation of DRR in the recovery process in the context of post-disaster. The qualitative approach has been taken as a research method because it has given the flexibility and ability to the understand people's behaviours, beliefs and their characters ith respect to disaster. Qualitative method explicitly provides a holistic understanding of event, problem, and situation about nature and people by using inductive logic (Flick 2009).

- This study has also used a multi-cited case study approach based on information of primary and secondary data. The secondary data was collected from the website of National Reconstruction Authority (NRA), the study of Post-Disaster Needs Assessments (PDNA) reports, relevant districts, municipalities and various agencies.
- The process in which primary data and information were collected: (i) Key Informants Interviews with earthquake victims, municipal authorities, and other relevant government authorities. (ii) FGDs with a stratified group of people in both areas. Direct field observation of selected earthquake affected settlements.

2.2.3 Sampling and Respondents

In case of study methods, the unit of analysis is considered as the basis of the case. In addition to this, in order to gain an overall understanding of participation in the post-earthquake recovery process, the study covered informants from villages like the community leaders and its members along with the beneficiary of the government scheme. The study had involved the views of members from a government department, external agencies, district administration and village level involved in post-disaster recovery programme.

The table below gives the details of participants of the study involved at the time of data collection (Table 2.1).

2.2.4 Data Collection Methods

The study used both the primary and secondary sources of data. The primary data were collected through FGDs, key informants' interviews; field observation, in-depth case studies, and analysis of photographs. The secondary data and information were collected from literature, analogue maps, satellite images, some departments and relevant stakeholders such as the Department of Urban Development and Building Construction, Nepal Reconstruction Authority and municipalities.

2.3 Results and Discussions

2.3.1 Building Damages

Housing damages were considerably high in Gorkha district compared to other districts. Equally, the death toll of Ballpark and Laprak were also very high. Almost every household has encountered death or injuries of their family members. Many Unreinforced Masonry Buildings in Barpark and Laprak Village Development Committee (VDC) as well as Sitapaila and Dahachowk periphery of Kathmandu, were badly affected by the earthquake (Fig. 2.2).

(i) The houses built with stone and mud mortar in the study area were either completely collapsed or were severely damaged and reached beyond inhabitation. Lack of timely maintenance of the houses was also a reason for such devastation.

(ii) Houses were collapsed in certain cases due to the excessive weight of the mud-made roof.

(iii) The collapse of side walls (partially or fully) was noticed in certain houses. Some of the houses were built during 1200 century which had no load-bearing walls on any side.

(iv) Weak masonry bonding at the corner walls and lack of corner ties have led to the development of vertical cracks on the walls in many traditional houses in the study area.

(v) Diagonal cracks across the masonry walls have developed in many old houses due to lacking strength of the structures to resist

Table 2.1 Respondents for the study

District	Study village	Mode of data collection	Number of respondents / participants	Details
Gorkha	Barpark	Focus group discussion (One)	8	Respondents were selected based on age, sex, highly affected, aid beneficiaries
		Key informants interviews	7	A primary teacher, A Civil Engineer, Ex-Indian Army, two community leaders
	Laprak	Focus group discussion (One)	10	Respondents were selected based on age, sex, highly affected, Govt. aid beneficiaries
		Key informants interviews	5	Ex-Indian Army, NGO worker, community leader, NRA officer
Kathmandu	Sitapaila	Focus group discussion (One)	10	Respondents were selected based on age, sex, highly affected, Govt. aid beneficiaries
		Key informants interviews	9	A widow woman, a local journalist, building contractor, NRA officer, local political leader, a victim
	Dahachowk	Focus group discussion (One)	12	Respondents were selected based on age, sex, highly affected, Govt. aid beneficiaries
		Key informants interviews	6	VDC leader, an old quake victim, two farmers

(a) Cracks in the building due to the earthquake in Dahachowk, Kathmandu district

(b) Damaged houses in Sitapaila VDC, Kathmandu district.

Fig. 2.2 Partially and fully damaged houses

the tensile stressed due to horizontal and vertical loads during an earthquake.

(vi) Many old houses were of wood with mud mortar were collapsed because of poor in horizontal stiffness.

2.3.2 Temporary Shelter

Nepal National Planning Commission (NPC) has shown in Fig. 2.3 that recovery of temporary shelters from earthquake impact is very slow in the most affected districts. Only 13% people have returned their own houses from temporary shelters in 14 months of duration. 52% of people were living in temporary shelter till February 2016. In September 2016, after seven months from the first survey, around 45% people were living in temporary shelter while 39% people were living in temporary shelters till April 2017. It is being estimated that complete return to their house will take more than ten years.

Economically backward and physically vulnerable people are more likely to remain in temporary shelters than economically developed and physically and socially robust people. In the Barpark and Laprak VDC in Gorkha district,

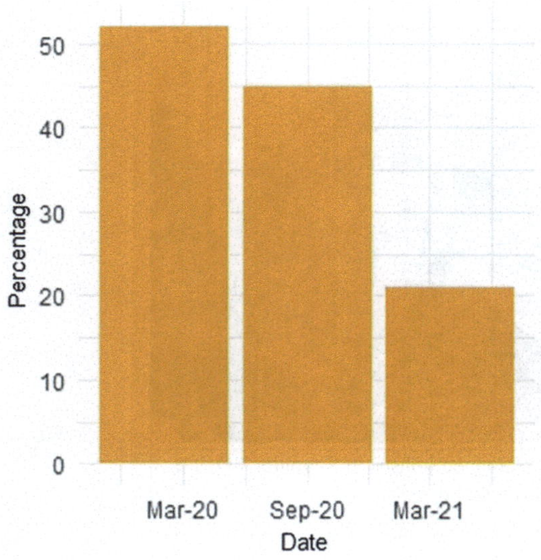

Fig. 2.3 The proportion of people living in temporary shelter-over time (Source: Nepal National Planning Commission 2017)

the majority people are in temporary shelters because of the difficulties in getting construction materials to rebuild houses. This was primarily due to the constraints with respect to transportation and communication infrastructures. The resilience of the built environment depends upon the capacity of each facility and infrastructure system to maintain acceptable levels of functionality during and after a disruptive event and to recover full functionality within a specified period of time. The political unawareness and lack of effective leadership made hurdle in the recovery process. It is also being seen that people those are living far away from the main road and district headquarters are more in number living in temporary shelters. Nearly forty five% people are still in temporary shelters those are living 26 km away from district headquarter, followed by 37% in 18 km and only 6% in 10 km. The basic premise is that critical facilities and infrastructure systems such as power, emergency response, hospitals, and transportation play pivotal role during and after a hazard event to support other aspects of community resilience.

As per the report of NPC (2017), the percentage of people in temporary shelters is very high in Gorkha district (44%) compared to Kathmandu district (3%). According to the Impact Recovery Monitoring survey, many people in temporary shelters are consistently struggling due to heat, coldness or wetness in the respective period of summer, winter and rainy seasons. A large number of temporary shelters in a small confinement region make lives difficult especially in cooking smokes and in extreme weather condition.

2.3.3 Governmental Housing Reconstruction Program

As per Earthquake Affected Infrastructure Reconstruction Act 2015, the National Reconstruction Authority (NRA) has played remarkable role in creating integrated housing and relocation of vulnerability settlement. Apart from this, there are also various international NGOs, local community organizations, and donor agencies that played

(a) (b)

Fig. 2.4 **a** An American tent given by UNICEF just after the earthquake but after two years of earthquake, people are still using it as shelter. **b** A widow woman got help from Oxfam to build her temporary shelter

important role in the reconstruction of destructed Nepal (Fig. 2.4).

People living in the Laprak and Barpark VDC are isolated and marginalized in respect to education, economic and socio-cultural dimension. Further, the remote location is the biggest hurdle to access information about allocation of resources and assistance-related information from the government and other donors. Their dependency on critical transportation systems, utilities, and other infrastructure systems created different types of hazard and vulnerabilities in the communities. The seismic investigation team of NRA identified the location of Laprak VDC as a landslide-vulnerable zone. Observing these vulnerable factors, NRA attempted to shift the entire Laprak VDC to seismically safe and less landslide-prone zone (Fig. 2.5).

The National Reconstruction Authority is also supported by Non-Residential Nepalis (NRNA) in reconstructing a model village of Laprak in Gorkha district. This master initiative of National Reconstruction Authority fostered the development of new settlement, where all basic public services such as supply of power, drinking water, best sanitation facility and all others luxurious facilities will be available. Total 573 new houses are under process of reconstruction that will cover the houses of all the earthquake victims of Laprak village. Standards hazard design and earthquake resistant code are followed in the construction of houses. Although, the construction of new settlement for Laprak VDC is quite praiseworthy but other areas for example Barpark, Sitapaila and Dahachowk are remained untouched from such initiatives.

2.3.4 Typology of Reconstructing Buildings

The technical survey conducted by Nepal government has tried to categorize reconstructing buildings as (i) Engineered design based seismic resistant building; (ii) Low strength masonry building (Figs. 2.6 and 2.7). By observing the buildings, it can be estimated that about 45% of newly constructed buildings in urban settlement of Kathmandu is engineered based and almost 55% of reconstructed buildings are based on Mass Rapid Transit (MRT) in Kathmandu urban region. However, Low strength masonry buildings are not found in settlements of Sitapaila and Dahachowk of Kathmandu district. Further, the majority of people in Gorkha district have

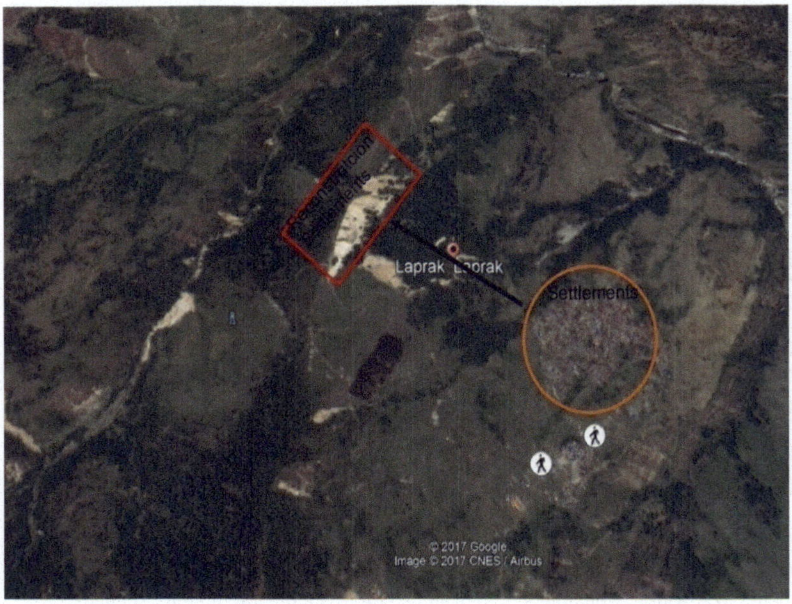

Fig. 2.5 Shifting of Laprak VDC from landslide-prone zone to seismic and landslide-free zone

(a) The construction of the seismic resistant building in new site of Laprak, Gorkha

(b) Using procures stone and mud to construct a new settlement in Laprak, Gorkha district

Fig. 2.6 Rehabilitation and Resettlement of Laprak VDC by NRA and NRNA

reconstructed their houses based on MRT, which was limited to masonry buildings. It is estimated that almost 90% newly constructed houses in urban settlements of Gorkha district are MRT buildings. The poor economic condition made Gorkha people to reconstruct low strength masonry buildings (Figs. 2.6 and 2.7).

Although there are provisions to build earthquake resistant house but people faced hurdle to achieve it so. As a result, huge illegal construction and non-safety norms houses come up which are highly risk prone and demanded to monitor. The factors affecting to follow the resilient houses are: (i) lack of financial resources; (ii) legal issue of land; (iii) lack of technical skills or guidance; (iv) waiting for government policy/ decision; (v) lack of building permit; and (iv) risk of dismantling neighbours building.

According to Nepal National Planning Commission 2015, 100% the newly constructed houses in Kathmandu city are based on Reinforced Concrete (RCC) frame whereas

(a) (b)

Fig. 2.7 **a** Seismic Resilience Building structure in Dahachowk, Kathmandu. **b** Lack of Seismic robustness structure in Laprak VDC, Gorkha district

nearly 80% houses in Urban Gorkha are based on RCC frame. Table 2.2 shows building level category of reconstructing building in settlements of Kathmandu and Gorkha district (percent).

Just after the earthquake, a large-scale survey was conducted by Ministry of Home Affair (MoHA) and it is found that a total of 498,852 houses fully damaged which is beyond the repair and 256,697 houses were partially damaged. The effects of the catastrophic event on the built environment of Nepal were because of the high seismic vulnerability of unreinforced masonry buildings which can be seen throughout the country. There is lack of relevant building codes and awareness of seismic risk in communities. Most of the buildings that fully and partially collapsed were old constructions in mud mortar, without seismic resilient detailing (deficient construction practices and non-engineered buildings).

Table 2.2 Building typologies and damage

Typology of buildings	Fully collapsed or beyond repairs (%)	Partially damaged (can be repaired/retrofitted) (%)
Low strength masonry	95	67.7
Cement based masonry	3.7	25.6
Reinforced concrete frame	1.7	6.7

Source Nepal National Planning Commission (2015)

Additionally, extending the house both vertically and horizontally is a culture in Nepal without considering the structural strength of the original buildings. The National Society for Earthquake Technology has classified the fully and partially damaged building into three types such as Low strength masonry, Cement based masonry, Reinforced concrete frame. Lack of integrity between structural members and use of weak materials in low strength masonry was the main cause of failures. 95% of low strength masonry building in study regions are completely collapsed or reached beyond repairable condition whereas only 3.7% and 1.7% Cement based Masonry and Reinforced Concrete Frame got damaged respectively.

2.3.5 Impacts on Unreinforced Masonry Construction

The main causes of huge damaged of unreinforced structural houses in the Barpark and Laprak VDCs are the shallow foundation, poor bonding of mud mortar, the absence of damp proof course, lack of tie at corner walls, and poor quality of building materials. There are numerous reasons for collapsing of RCC structures in peri-urban of Sitapaila and Dahachowk in Kathmandu district that are lack of skilled labour, non-engineering construction with poor quality material and changing of building use.

2.3.6 Monitoring of Reconstructing Houses

Financial facility for concessional home loan scheme has announced by Nepal Government to support earthquake victims through financial institutions and banks. The earthquake victims who are eligible to get this loan from Bureau of Financial Institutions (BFIs) need to pay back with an interest rate of only 2%.

Moreover, yet Barpark and Laprak VDCs in Gorkha district are considered the most affected ones but people in these VDCs struggle the most to get financial aid from the Government and NGOs. However, the financially supported people also face challenges to reconstruct their houses as the allocation of grant do not reach at once. The allocated grant amount is Rs 3,00,000 but this amount would be given in three instalments. In the first instalment, the receive amount is Rs 50,000 that is given to construct the foundation of the building, and second and third instalments are Rs 1,50,000 and Rs 1,00,000 respectively that are subsequently given to construct the rest which takes the maximum times and most of the times, victims do not receive the second and third instalments and that causes the serious problem.

Once financial support for the beneficiaries is completed, the Department of Urban Development and Building Construction (DUDBC) takes led to provide technical support over the newly constructed buildings. In addition to this, community coalitions that address infrastructure resilience, land-use planning, and adoption and enforcement of building codes and standards; a risk management strategy that includes complementary structural and non-structural risk reduction measures are required to follow subsequently.

2.3.7 Dependency on Local Resources

Due to the terrain characteristics of the region and long distance from the district headquarter and markets, the price of construction materials are very high for both Barpark and Laprak VDCs. As a result, they use local resources such as stone and mud mortar for the housing construction. But people in the Sitapaila and Dahachowk of Kathmandu district use brick with cement mortar to build their houses (Fig. 2.8). The interviewers and participants of FGD shared their concern for earthquake for being constructed their houses using stone and mud mortar (Fig. 2.8). A large number of houses in Gorkha district have built by stone with mud mortar. The numbers of such houses were specifically more in Laprak and Barpark VDCs.

2.3.8 Information Dissemination and Capacity Development

Residents of Barpark and Laprak VDCs were totally unknown about the importance of seismic resistant buildings until earthquake hit them hard in 2015. The remoteness and backwardness of all communities could be attributed to less or no access to the knowledge and techniques for safer construction. The use of appropriate seismic resistant building is considered very essential for hazard prone region. Focusing this issue, the utmost importance is to distribute brochures, posters, pamphlets, books on seismic resistant design among the residents to raise the awareness. For the dissemination of such knowledge and techniques, NRA and Department of Urban Development and Building Construction (DUDBC) take a leading role. However, the main causes of failing to reach the Build Back Better (BBB) position is to lack cooperation and coordination among various disaster

Using brick and cement to construct an earthquake resilient house in Dahachowk,

House without RCC pillar but using brick and cement to construct in Sitapaila,

Fig. 2.8 Types of housing construction in Sitapaila and Dahachowk VDC

management-related agencies; adequate funds and resources; a lack of roads, transportation and communication facilities; a lack of modern technology; the absence of early-warning systems; and inadequate rehabilitation programmes.

2.3.9 Relocation and Rehabilitation Planning

Each VDC/ Municipality is responsible for making planning for relocation and rehabilitation. Technical support to each VDC/ Municipality (e.g. the hazard maps) is provided by the district level. The displaced communities are required to resettle on their land. The investigation of relocating sites regarding landslides, the source of livelihoods (arable lands) and basic services is needed to conduct as frequent as possible. The process of identification of hazardous areas is to investigating damage condition, landslide hazard map, field survey, and analysis, etc. It is observed that affected communities are still residing and cultivating in the landslide-prone region. Before relocating the earthquake victims, it is essential to consider several other countermeasures such as development of early-warning system, preventing landslides, and the identification of evacuation sites and routes. Figures 2.9 and 2.10 shows the reconstruction and resettlement of houses in Barpark VDC, Gorkha district, and Sitapaila and Dahachowk Kathmandu district on landslide-prone areas.

2.4 Challenges in Recovery Process

The reasons for delaying reconstruction process have emerged through the Focus Group Discussions as: (i) lack of financial resources; (ii) legal issues pertaining to land; (iii) lack of technical skills or guidance; (iv) delay in government policy/ decisions; (v) decision to stop giving building permits; (vi) risk of vulnerable neighbours building. The studies conducted in the past have pointed out that complete recovery of housing sector from the devastating impacts of a disaster can only happen if participation of community members, local government, and INGO/ NGO play an active role (Paul 2011). Slow pace rebuilding has been observed where voices of the community are soft, inactive and where community involvement is either absent or at a low level (Kamel and Loukaitou-Sideris 2004). It is observed, through the Housing recovery initiatives that have been shaping the post-2015 earthquake recovery of the communities in the area of this study, that pre-existing vulnerabilities and social relationships have played an important role in the post-disaster recovery process.

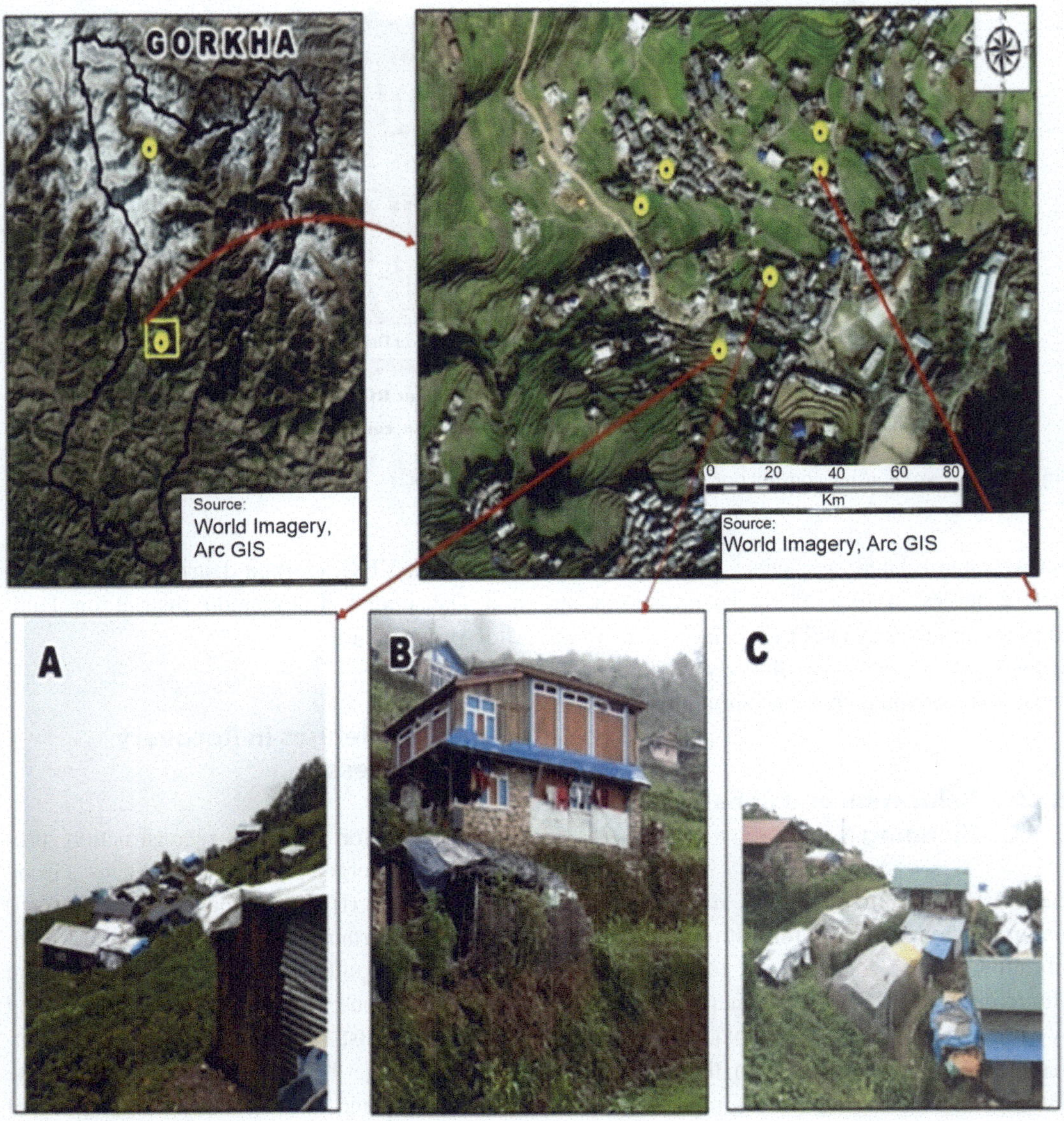

Fig. 2.9 Rehabilitation and resettlement of houses in Barpark VDC on landslide-prone areas

The study shows that the economic condition of peri-urban Kathmandu is mostly depends on secondary and tertiary activities as well as they are enrich from social, economical and cultural background compared to the Gorkha people. Therefore, the construction of Dahachowk and Sitapaila VDCs in Kathmandu District are based on RCC seismic resistant whereas the people from Gorkha suffers lots of difficulties such as high angel of slopes, variable climatic conditions, very high peaks, rugged and fragile geophysical structure, active tectonic processes and unplanned settlement which makes constraints in constructing seismic resistant houses. The previous study says that the density of population also matters in the housing recovery. Areas with high densities may recover more slowly due to difficulties of temporary and permanent housing (Tobin and Montz 2017). In this study, it is not applicable because the density did not hinder in the reconstruction of Kathmandu city. As Wright et al. (1979) stressed that the

Fig. 2.10 Reconstruction and resettlement of houses in Sitapaila and Dahachowk on landslide-prone areas

reconstruction of an affected region is controlled by the economic condition of a neighbourhood and the education, employment, and job skills. Similarly, the recovery process of Gorkha district is slow especially Laprak and Barpark VDC due to the financial crisis and heavy transportation cost than Sitapaila, and Dahachowk of Kathmandu District. Most of the reconstruction in Barpark and Laprak is made by stone with mud mortar, and brick with cement in Kathmandu's study area.

Bera et al. (2020) depicted the resilience of household varies with geographic distances. In this study, Barpark and Laprak VDCs are 59 km and 65 km, and Dahachowk and Sitapaila VDCs are 15 km and 7 km away from the district headquarter respectively. The variation of distance from the district headquarter and main market play significant role in recovery process. It is found that the affected region closer to market or headquarter have a speedy recovery than region located far. Although the damages of earthquake in Dahachowk and Sitapaila VDCs were more and less equal to Barpark and Laprak VDCs but due to relatively easy accesses to Kathmandu city helped them to come back in normal condition. Not only construction materials but people those are living in long distance deprives to get information such as Government Schemes/loans and aids from NGOs, etc.

2.5 Conclusion

The study examines the housing recovery process in post-earthquake 2015 in Nepal. Housing was analyzed to ascertain proceedings of the recovery process. It is observed, through the housing recovery initiatives that it has been shaping the post-2015 earthquake recovery of the communities in the area of this study, that pre-existing vulnerabilities and social relationships have played an important role in the post-disaster recovery process. The recovery and resettlement program and policies have rigid and very little room for incorporating local needs. Disasters have been utilized as opportunities by various nations to expand the country's social welfare schemes and integrate the peripheral and distant areas, but this assimilation is yet to be observed in Nepal even in villages that lost almost all standing buildings. Despite a severe loss of housing and other infrastructure in the villages under study, it is observed that the villages have not been in a position to demand political focus onto their plight.

The recovery process of study area depends on the coordination between different stakeholders includes forest, IT, experts, professionals, transport; local government functionaries such as Panchayat, BDO, SDO, DM, AE, JE and most importantly beneficiaries (community). The successful recovery of build environment of any earthquake affected community lies on the activeness of local leader, awareness of community people and occupation of people with the efficient allocation of resources. However, most earthquake victims were unable to complete the house on time due to physical, social and economic constraints they encountered. Constraints were occurred because of their geographical location especially for the VDC of Barpark and Laprak which is 59 km and 65 km away from the main market respectively.

Finally, this research has pointed out the major findings after the Gorkha earthquake, with an emphasis on recovery and rehabilitation. Cooperation among different national and international organization has analyzed and also shown efficient management in the post-disaster scenario. The role of National Reconstruction Authority and others technical skills of the different organization in recovery process are the main pillars for successful post-disaster management. The people owned initiative is considered the best option for the accomplishment of recovery.

References

Asian DRC (2002) Living with risk. A global review of disaster reduction initiatives. Preliminary version

Barenstein JD, Iyengar S (2010) India: From a culture of housing to a philosophy of reconstruction. Building Back Better, 163

Barenstein JD (2012) Communities' perspectives on housing reconstruction in Gujarat following the earthquake of 2001. Post-Disaster Reconstruction and Change Communities' Perspectives, 71–100

Bera S, Guru B, Chatterjee R, Shaw R (2020) Geographic variation of resilience to landslide hazard: a household-based comparative studies in Kalimpong hilly region, India. Int J Disaster Risk Reduc 46:101456

Blaikie P, Cannon T, Davis I, Wisner B (2014) At risk: natural hazards, people's vulnerability and disasters. Routledge

Flick U (2009) Qualitative methoden in der evaluationsforschung. Z Qual Forsch 10(1):9–18
Frerks GE, Kliest TJ, Kirkby SJ, Emmel ND, O'Keefe P, Convery I (1995) A'disaster'continuum? Disasters 19(4):362–366
Hamelin JP, Hauke B (2005) Focus areas: quality of life: towards a sustainable built environment. European Construction Technology Platform, Paris
Kamel NM, Loukaitou-Sideris A (2004) Residential assistance and recovery following the Northridge earthquake Urban Studies 41(3):533–562
Karunasena G, Rameezdeen R (2010) Post-disaster housing reconstruction: comparative study of donor vs owner-driven approaches. Int J Disaster Resilience Built Environ 1(2):173–191
Kreimer A (1980) The role of the mass media in disaster reporting: a search for relevant issues. In: Committee on disasters and the mass media workshop. National Academy of Sciences. 9–31
Mileti D (1999) Disasters by design: a reassessment of natural hazards in the United States. Joseph Henry Press
Montz BE, Tobin GA, Hagelman RR (2017) Natural hazards: explanation and integration. Guilford Publications
NPC (2017) National Review of Sustainable Development Goals. Government of Nepal. https://www.undp.org/sites/g/files/zskgke326/files/migration/np/National_Review-SDGs.pdf
Paul BK (2011) Environmental hazards and disasters: contexts, perspectives and management. John Wiley & Sons
Pearson L, Pelling M (2015) The UN sendai framework for disaster risk reduction 2015–2030: Negotiation process and prospects for science and practice. J Extreme Events 2(01):1571001
Perry RW, Lindell MK, Tierney KJ (2001) Facing the unexpected: Disaster preparedness and response in the United States (eds). Joseph Henry Press
Phillips BD (2009) Disaster recovery. Auerbach Publications
Prater CS, Lindell MK (2000) Politics of hazard mitigation. Nat Hazard Rev 1(2):73–82
Qadir U (2015) UN development programme (UNDP). Human development report 2015-work for human development. Pak Dev Rev 54(3):277–278
Quarantelli EL (1998) What is a disaster?: perspectives on the question (ed). Psychology Press
Quarantelli EL (1982) Sheltering and housing after major community disasters: case studies and general observations
Rollnick R (2006) The aftermath of natural disasters and conflict. Habitat Debate 12(4):4–5
Sevin E, Little RG (1998) Mitigating terrorist hazards: the engineering community has a critrical role to play in finding and promoting rational, balanced solutions to terrorist violettee. Bridge-Washington 28:3–8
Song J, Pandey R, Dong G, Sharifi A, Subedi BP (2022) Urban-rural disparity in community resilience: a multilevel analysis of the relief progress after the 2015 Nepal Earthquake. Sustain Cities Soc 79:103698
World Bank (2001) World Bank and Asian Development Bank complete preliminary Gujarat Earthquake damage assessment and recovery plan, http://www.worldbank.org/gujarat
Wright JD (1979) After the clean-up: long range effects of natural disasters. (No Title)
Wu JY, Lindell MK (2004) Housing reconstruction after two major earthquakes: the 1994 Northridge earthquake in the United States and the 1999 Chi-Chi earthquake in Taiwan. Disasters 28(1):63–81
Zinn JO (2008) Risk society and reflexive modernization. Social theories of risk and uncertainty: an introduction, 18–51

Consequence of Coupling Flood and Covid Pandemic on Surjapuri Community: A Case Study of Puria, Bihar

3

Sohail Akhtar and Smitha S. Nair

Abstract

The Surjapuri community has endured substantial floods over the past six decades. This research aims to explore the impacts of both the pandemic and floods on the community. Employing a descriptive qualitative research approach, the data was gathered through interviews and focused group discussions. The study elucidates the intertwined challenges posed by floods and the pandemic in Seemanchal, shedding light on the multifaceted repercussions on people's lives and the obstacles faced in providing assistance. These insights can aid stakeholders in better addressing such issues in the future.

Keywords

Flood resilience · COVID-19 impact · Surjapuri community · Disaster management · Bihar flooding

S. Akhtar (✉) · S. S. Nair
School of Social Work, Centre for Health and Mental Health, Tata Institute of Social Sciences, Mumbai, India
e-mail: sohailakhtartiss@gmail.com

3.1 Introduction

Bihar, a northern state in India, grapples with an acute and persistent challenge: the annual devastation caused by flooding. Despite accounting for a mere three percent of India's geographical area and an eight percent share of its population, Bihar bears a disproportionately heavy burden in terms of flood-related disasters (Choudhary 2020). Approximately 17% of the state's landmass is flood-prone, and a staggering 36% of the country's flood-affected population resides in Bihar. Bihar shares around 23% normal yearly flood harm in the country. The recurring problem of floods in North Bihar has been a persistent and deeply impactful challenge for the residents of the region. Over the past few decades, these devastating flood events have left a trail of destruction and disruption in their wake, affecting the lives of millions of individuals (Singh et al. 2022). The scale of these flood disasters is both remarkable and concerning, reflecting the vulnerability of the region to this natural calamity (Senapati 2022).

To provide a historical perspective, during the year 2017, a staggering 17.1 million individuals were adversely affected by flooding. In 2015, the floodwaters disrupted the lives of 3 million people, while in 2013, a total of 20 districts were affected, impacting the livelihoods of 7.23 million residents. In 2010, the floodwaters took a toll on 0.718 million individuals, and

S. Talukdar et al. (eds.), *Communication, Science, Technology and Innovation in Disaster Risk Management*, GIScience and Geo-environmental Modelling, https://doi.org/10.1007/978-3-031-77189-7_3

in 2008, an alarming 3.3 million people were affected by the floods. The year 2007 witnessed a devastating flood event that impacted a staggering 25 million individuals, while in 2004, flood-related destruction affected the lives of 21 million people. Going further back, in 1987, a calamitous flood event influenced 28.2 million residents, and in 1974, the lives of 16.39 million individuals were disrupted by floods.[1] These recurring flood disasters have posed a formidable challenge to the region's resilience and infrastructure (Acharya et al. 2023). The sheer magnitude of the affected population over the years underscores the urgent need for comprehensive flood mitigation strategies, infrastructure development, and disaster preparedness in North Bihar. The historical data serves as a compelling reminder of the critical importance of proactive measures to address and mitigate the impact of floods on the lives and livelihoods of the people in this flood-prone region. In 2020, during the monsoon season, Bihar grappled with floods and the COVID-19 pandemic. The flooded areas faced dual challenges: the pandemic affected daily wage earners, many of whom were stranded in cities due to lockdowns, while the flood posed life-threatening risks to village residents. The nationwide lockdown made it difficult for people to return home in time (Pal and Yadav 2020; Laxminarayan et al. 2020). Consequently, thousands lost their jobs, exacerbating the economic impact (Haleem et al. 2020; Tandon 2020). Additionally, the Mahananda flood damaged infrastructure, roads, and housing, compounding the devastation in the region (Alim 2017). The Surjapuri community, a distinct ethnic group, primarily resides in specific districts within the Indian state of Bihar, including Kishanganj, Katihar, Purnia, and Araria. They are also found in certain areas of West Bengal, namely North Dinajpur, Dakshin Dinajpur, and some segments of Siliguri. The Surjapuri community's existence in these regions is characterized by a host of formidable economic challenges, primarily stemming from the unique socio-geographical conditions prevalent in the broader Seemanchal region. The Seemanchal region, which encompasses the aforementioned districts, is characterized by an unfortunate annual phenomenon—severe flooding. This recurring natural calamity is predominantly attributed to the swelling waters of the Mahananda and Kosi rivers, which inundate the region with striking regularity, once every year. The consequence of these annual flood events is a harsh cycle of economic deprivation, backwardness, and social exclusion, which particularly impacts the Surjapuri community. The economic challenges faced by the Surjapuri people in their native regions are exacerbated by the relentless flooding, which renders agriculture and livelihoods precarious. The disruption caused by these floods perpetuates a state of economic instability, making it challenging for the community to break free from the shackles of underdevelopment and social marginalization. The past studies revealed the flood impact, mitigation challenges and inequality of social vulnerability (Cuny 1991; Mishra 1997, 1999, 2001, 2008; Sinha 2008; Rorabacher 2008; Pandit 2009; Shrestha et al. 2010). However, less studies explored the impact of dual disaster like COVID-19 and flood on local community such as the Surjapuri.

The objective of this study is to assess the impact of floods on the Surjapuri community living in proximity to the Mahananda River in Bihar. The research endeavors to establish an association between the flood-related challenges and the additional burden imposed by the COVID-19 pandemic, which collectively have had a compounding effect on the well-being, infrastructure, and local businesses of the Surjapuri people within the chosen study area. Beyond the evident physical damages caused by floods, the study also aims to shed light on their profound influence on both the physical and mental health of the community, as well as their access to essential services. Furthermore, this research will provide insights into the

[1] The damage data of flood are derived from Bihar State Disaster Management Authority. See https://state.bihar.gov.in/disastermgmt/SectionInformation.html?editForm&rowId=441.

repercussions of the flood and the COVID-19 pandemic on the lives of the Surjapuri people in the Seemanchal region, while also exploring their resilience in the face of such adversities.

3.2 Methodology

The study adopts a descriptive research design to identify the consequences of the flood, and it is carried out qualitatively. The research conder Puria village as a case study area.

3.2.1 Area of Selection

Puria village, Bhawanipore panchayat, situated in Barsoi block within the Katihar district of Bihar, occupies a total land area of 94 hectares. As per the 2011 census, the village is home to 360 families, totaling a population of 1710 individuals, with 891 males and 819 females. Among them, 317 are children aged 0–6 years, comprising 18.545% of the total population. The village sustains 470 individuals engaged in various work activities, with 77 as cultivators (owners or co-owners) and 385 as agricultural laborers. Agriculture constitutes the primary occupation, while self-employment ventures like general stores, vegetable vending, daily wage labor, and fishing are also prevalent.

These villages were chosen for their proximity to the Mahananda River, a tributary of the Balason and Kankai rivers, located around 100 m away. The villages face annual flooding, motivating the selection for study due to the recurrent flood risks they encounter (Fig. 3.1).

3.2.2 Data Collection

The research utilized a combination of primary and secondary data. Primary data collection methods included semi-structured interviews, oral histories, and focus group discussions. Five interviews were conducted with individuals residing near the Mahananda River to assess the conditions of the affected area. Oral histories were obtained from senior citizens aged 60–65, who possessed memories of past floods. Two

Fig. 3.1 Landscape of puria village (Source: Google Earth Image, 4/9/2024)

such oral histories were recorded. Additionally, four focus group discussions were conducted in Ward No. 10, one each with men and women, and two in Ward No. 11, following the same gender division. Each focus group consisted of 8–9 participants, comprising homemakers, agricultural laborers, and daily wage laborers, including those working in cities. In total, 26 respondents were interviewed. Purposive and convenience sampling techniques were employed for data collection. Secondary data relevant to the research were gathered from various governmental and non-governmental sources.

3.3 Result and Discussion

The annual Seemanchal floods inevitably impact the residents each year. Many lack permanent housing, with some residing in structures built on state government land. Despite their long-term occupancy, these residents are often overlooked by the government. The indicators suggest that their vulnerability stems not solely from poverty but also from marginalization. Poverty compels individuals to inhabit risky areas, where the absence of proper infrastructure leaves them susceptible to significant property and life losses.

3.3.1 Migration

Construction ranks as the second largest employment sector in India, drawing individuals from underdeveloped states like Bihar to major metropolitan hubs such as Mumbai and Delhi for employment opportunities. In Katihar, residents primarily engage in labor-intensive work in cities, such as head loading and construction, earning monthly incomes ranging from 8000 to 14,000 rupees, contingent upon their skill levels and experiences. Payment structures vary, with some laborers receiving wages daily, weekly, or monthly, often without negotiation with contractors, who dictate terms of compensation. Illiteracy among workers often leads them to accept terms without question, fearing job loss otherwise. These construction workers, comprising a marginalized segment of society, endure inadequate access to essentials like drinking water, sanitation, and canteen facilities. Additionally, they are denied maternity benefits, social security, healthcare, and prenatal/postnatal care. Their migrant status often precludes them from government registration and basic entitlements.

The pandemic has profoundly disrupted their livelihoods. In March 2020, the initial wave prompted a mass exodus of workers back to their villages due to job scarcity and fear of contracting COVID-19. With the nation under strict lockdown, returning home became the sole option. Many undertook arduous journeys, covering distances of up to 1500 km on foot, taking up to seven days to reach destinations like Katihar from cities such as Delhi. Approximately 80% of village men work in urban areas as construction laborers or daily wage earners at eateries and gas stations. During their journeys home, they exhausted their savings paying exorbitant rents to vehicle operators. By April 14, 2020, thousands of migrant laborers had amassed at Bandra Station in Mumbai, desperate to return home. Md Munna, a villager of puria from the same village says that:

> *We come to work and send money to our parents at home but this time the opposite was happening. We asked for money from my parents so that we can lead our life in Mumbai in the lockdown till it gets over. We were cashless as our salary was stopped. We heard many NGOs are working and providing meals but we didn't get any from anyone here at Mumbai central where we stay.*

Residents of Puria village primarily engage in daily wage labor or migrate to cities like Delhi, Mumbai, or Ahmedabad for work opportunities. Following the flood and the onset of the first wave, traveling to metropolitan areas became a necessity. Despite the lockdowns in cities, individuals sought to return to their workplaces after the initial wave to support their families and improve their livelihoods. Typically, they spend four seasons working in cities and two seasons at home for agricultural activities. Securing work through the Mahatma Gandhi

National Rural Employment Guarantee Act (MGNREGA) is not guaranteed for everyone; it relies on the decisions of local authorities such as ward members or village heads. Employment opportunities under MGNREGA are contingent upon the approval of projects like road, bridge, or toilet construction. However, such projects often require bribes for approval, and there is no certainty of work until they are sanctioned by the local administration. Consequently, many villagers do not rely on MGNREGA for employment and instead seek work in cities. During their absence, family members receive support from relatives and neighbors.

3.3.2 Impact on Health

Health is a crucial component of human capital formation, greatly influenced by environmental factors such as floods. Post-flooding, water contamination emerges as a significant threat to public health. Floodwaters carry various chemicals and minerals, contaminating drinking water sources like hand pumps and wells, leading to the spread of waterborne diseases such as diarrhea, jaundice, and skin conditions. Consequently, the community faces additional financial burdens due to healthcare expenses. The situation exacerbates when the primary breadwinner falls ill. Around three-quarters of the village population experienced health issues during and after the floods, compounded by a lack of COVID-19 testing awareness.

Local healthcare services suffered damage during the flood, while those situated farther away became inaccessible due to road and bridge damage. Disrupted access to healthcare services results in an increased incidence of diseases due to shortages of essential medicines. Common ailments reported during the floods included diarrhea, cough, malaria/fever, and skin rashes, particularly affecting households near the river, which rely on it for drinking water, making them susceptible to disease outbreaks due to heightened pollution. Furthermore, challenges like clothes not drying during the monsoon contribute to skin rashes, while overflowing pit latrines lead to open defecation, exacerbating the risk of diarrheal diseases and community exposure to fecal contamination.

The flood-induced health issues compound the already challenging livelihood responsibilities of the community. Various challenges were encountered during the flood period, exacerbating the difficulties faced by the population. In 2017, as recounted by Begam, an elderly resident of the village, "*My daughter in law was pregnant, and we were fully wet the whole day. We tried to relocate our important things. We kept her on the rooftop in one of our villager's house. The roof was full of people. Some were cooking, some were talking about the flood, and some were trying to help others. In the meantime, she had pain in her stomach, after that we tried to help her for the delivery. Luckily, she gave birth to a child.*"

3.3.3 Impact on Crops

Rice, jute, and wheat are the primary crops cultivated by farmers in Seemanchal, Bihar. In March–April, farmers typically sow kharif crops like rice, reaching full cultivation by July and August when the plants grow to a height of 3 feet. However, the onset of floods during this crucial period resulted in significant losses for villagers as the inundation destroyed the rice crops, exacerbated by excessive waterlogging during the 2020 flood event, which persisted for four months, weakening the plants. Some farmers also planted jute during the monsoon season, but these crops also suffered damage from waterlogging. This year, floods washed away jute plants that had been harvested and stored on the land. Bulo said, "*Due to the flood we didn't get a profit from the Jute plantation as it was destroyed. Some people cultivated rice this year as Kharif crops. Every year we got 5 quintals (500 kg) of paddy but this year we got only 50–60kgs of Paddy (rice) which is a huge loss to them.*"

Ansari said, "*Earlier we sold paddy and then used the money for other expenses. As this year we don't have rice to eat for the year, we can't sell even, rather we have to buy rice for our meal. I have four children, my mother-in-law, my husband and my brother-in -law. Running a family for months with 50 kg rice is not easy work.*"

Most villagers possess 3–4 bighas of land, yielding around 4–5 quintals of crops after cultivation, which sustain them throughout the year. They typically sell rice and reserve a portion for personal consumption. However, due to low rice yields this year, they suffered significant losses, as selling agricultural produce constitutes their primary source of income. None of the respondents managed to sell rice this year. This income is crucial for their livelihood, enabling them to purchase basic necessities. For those without land to cultivate, employment opportunities in cities or as laborers on larger farms within the village become necessary for survival.

3.3.4 Resilience and Adaptation Strategies

Over time, the villagers have developed psychological resilience to cope with floods, adapting to their recurring impacts. They have established norms for dealing with flood situations, such as constructing temporary shelters using tarpaulin sheets on roads. As soon as waterlogging begins in their homes during floods, people attempt to evacuate to safer areas like roads, schools, or rooftops. While some individuals manage to procure boats to transport household materials, most do not. A few respondents have crafted "Bheda," handmade boats made from banana stems also called Boita/ Kolar Vela, which they use to navigate through flooded areas and reach markets. Some even transport motorcycles on these makeshift boats to traverse the water and reach roads.

To safeguard their grains, villagers store them in the homes of neighbors with permanent structures, as only one-fourth of the village has such dwellings. During emergencies, people seek refuge on rooftops if other spaces in relatives' or neighbors' homes are already occupied. They also erect tents to shield themselves from rain and heat. In dire situations, many are compelled to borrow money from neighbors. Additionally, the floods result in the loss of essential documents such as bank books, voter cards, BPL cards, and Aadhaar cards for numerous individuals. Raziuddin, a school teacher said, "*the whole village is dependent on BPL cards and rations. Only a few families might have APL cards*".

In Puria, residents construct elevated platforms called machans to allow water to flow beneath them during floods. Given the inevitability of annual floods, they prioritize preemptive measures. Historical accounts suggest that the people of Puria have long endured floods, dating back to the post-independence era, spanning over 70 years. The 2017 flood in Bihar stands out as particularly devastating, resulting in loss of life and extensive damage to kutcha houses. Domestic animals perished, homes were destroyed, and essential belongings such as motorcycles, vans, bicycles, chairs, and tables were swept away. With little time to prepare, people found themselves stranded in various locations, with food and shelter becoming critical concerns during that period. Sakera Begam, a 65 years lady says that, "*People were very hungry in the village, so we told some of the strong men to travel to the market area which was 1 mile away from our village. They took a boat to bring some rice, wheat, oil, potatoes and other eating materials for the villagers. It was seen that 8–9 people went on a small boat to bring a meal for us.*"

Numerous tales recount the challenges and barriers of the past. Over time, the residents have acclimated to life amidst floods, considering it a regular occurrence. Their approach has consistently been one of resilience, adapting to ensure safety and security.

3.4 Conclusion and Lessons Learning

The study has illuminated the profound and multifaceted impact of the recurrent floods and the added challenge posed by the COVID-19 pandemic on the Surjapuri community residing near the Mahananda River in Bihar. Through an exploration of various aspects, including migration, health, and the agricultural sector, the research has unveiled the intricate dynamics of the hardships faced by this marginalized community and their remarkable resilience in the face of adversity.

Migration stands out as a central theme in the lives of the Surjapuri people. The construction sector serves as a critical source of employment, drawing laborers from underdeveloped states like Bihar to major metropolitan cities. The people from Katihar, for instance, predominantly work as laborers, earning modest incomes that vary based on their experiences. This migration, however, comes with its set of challenges, as workers often face difficult working conditions, a lack of basic amenities, and social security benefits. The COVID-19 pandemic further exacerbated the challenges faced by these migrants, compelling them to return home due to job losses and health concerns. This mass return created a precarious situation as individuals struggled to secure their basic needs, and the absence of government support compounded the difficulties faced by this community.

The impact on health within the Surjapuri community is significant. The floodwaters carry contaminants, which compromise drinking water sources and result in waterborne diseases like diarrhea and jaundice. The lack of access to health services, coupled with the absence of testing for COVID-19, poses a serious threat to the community's health and well-being. With the flooding of their homes, the residents are forced to adapt to unhygienic conditions, exacerbating the risk of disease transmission. These issues are further complicated by the lack of healthcare facilities and the inaccessibility of services due to road and bridge damage caused by the floods.

Agriculture, a vital sector for the livelihoods of the Surjapuri community, has also been severely impacted by the floods. The primary crops cultivated, such as rice, jute, and wheat, have suffered significant losses due to the inundation of floodwaters. This loss of agricultural yield has far-reaching consequences, affecting not only the residents' ability to feed their families but also their economic stability. The absence of a proper system for crop insurance further compounds the financial hardships.

Despite these challenges, the Surjapuri community has demonstrated remarkable resilience and adaptation strategies. Over the years, they have developed coping mechanisms to deal with the annual floods, including constructing temporary shelters on roads and utilizing handmade boats, such as the "Bheda," to navigate through floodwaters. These strategies help them secure their basic needs during emergencies. Additionally, villagers rely on each other for support and often store grains in the homes of those with sturdier, pucca houses. The community is well-acquainted with the floods and has integrated flood preparedness into their way of life.

In conclusion, the Surjapuri community in Bihar faces significant challenges due to annual floods and the added crisis of the COVID-19 pandemic. These challenges span from economic hardship and health concerns to agricultural losses. Nevertheless, the community's resilience and adaptability are noteworthy, as they have devised strategies to cope with the recurrent disasters. This research underscores the need for proactive measures to address the multifaceted issues faced by the Surjapuri people, including bolstering healthcare facilities, promoting agricultural resilience, and providing economic support to migrants during crises. Additionally, it emphasizes the importance of recognizing the community's own strategies for dealing with adversity and integrating them into comprehensive disaster management efforts.

References

Acharya SK, Parida J, Singh BKN (2023) Exploring community disaster resilience after kosi floods 2008 in Bihar India. SAGE Open 13(3):21582440231192820

Alim MA, Ahmed F, Islam MS (2017) Seepage analysis of Mahananda earthen embankment at Chapai Nawabganj in Bangladesh. J Am J Eng Technol Manag 2(1):1–6

Choudhary S (2020) Factors affecting flood management in Bihar, India. Int J Environ Sci 11(1):72–76

Cuny FC (1991) Living with floods: alternatives for riverine flood mitigation. Land Use Policy 8(4):331–342

Haleem A, Javaid M, Vaishya R (2020) Effects of COVID-19 pandemic in daily life. Curr Med Res Pract 10(2):78

Laxminarayan R, Jameel S, Sarkar S (2020) India's battle against COVID-19: progress and challenges. Am J Trop Med Hyg 103(4):1343

Mishra DK (1997) The Bihar flood story. Econ Polit Weekly, 2206–2217

Mishra DK (1999) Flood protection that never was: case of Mahananda Basin of North Bihar. Econ Polit Weekly, 2013–2018

Mishra DK (2001) Living with floods: people's perspective. Econ Polit Weekly, 2756–2761

Mishra DK (2008) Bihar floods: the inevitable has happened. Econ Polit Weekly, 8–12

Pal R, Yadav U (2020) COVID-19 pandemic in India: present scenario and a steep climb ahead. J Prim Care Commun Health 11:2150132720939402

Pandit C (2009) Some common fallacies about floods and flood management. Curr Sci 97(7):991–993

Rorabacher JA (2008) Gerrymandering, poverty and flooding: a perennial story of Bihar. Econ Polit Weekly, 45–53

Senapati S (2022) Vulnerability and risk in the context of flood-related disasters: a district-level study of Bihar, India. Int J Disaster Risk Reduct 82:103368

Shrestha RK, Ahlers R, Bakker M, Gupta J (2010) Institutional dysfunction and challenges in flood control: a case study of the Kosi flood 2008. Econ Polit weekly, 45–53

Singh VK, Kumari G, Dhiman A, Kumari A (2022) Multi-stakeholder planning and coordination for flood risk reduction in Bihar. In: 5th World congress on disaster management, Taylor & Francis, Vol II, p 60

Sinha R (2008) Kosi: rising waters, dynamic channels and human disasters. Econ Polit Weekly, 42–46

Tandon PN (2020) COVID-19: impact on health of people & wealth of nations. Indian J Med Res 151(2–3):121

Impact of Climate Variability on Fishermen Livelihood: Evidence from the Western Coast of India

4

Shailendra Rai and Somnath Bera

Abstract

Fishermen heavily rely on natural resources for their livelihood, a reliance known to be affected by climate dynamics. However, most current research primarily concentrates on the effects of climate-driven changes in ocean productivity on fish distribution and production, neglecting to address the impact on fishing activities and the adaptive measures adopted in response. This study aimed to explore the impact of climate risks on fishing activities in Uran Koliwada, Raigad District of Maharashtra. The primary objectives are to assess changes in fishing efforts and production and investigate the impacts of weather patterns and climate variability on fish yields, fishing duration, and livelihoods. To achieve these objectives, a combination of surveys, focus group discussions, and in-depth interviews was conducted to gather primary and secondary data from various sources. The study revealed that catch per unit efforts serve as a basic measure of abundance, and weather and climate variability significantly affect fishermen, leading to risks such as loss of life, boat submergence, and adverse health impacts.

Keywords

Climate risk · Fishermen livelihoods · Extreme events · Uran Koliwada

4.1 Introduction

The global sea level experienced an average rise of 1–2 mm per year between 1901 and 2018 (IPCC 2019), and it is anticipated to continue rising in the near future. This rise in sea level poses various threats such as beach erosion and altered flooding, impacting economic sectors, the environment, and the infrastructure of adjacent coastal communities (FitzGerald et al. 2008) (Nicholls 2011).

Climate change has already caused changes in ocean conditions, such as alterations in water temperature, biological dynamics, and physical characteristics. These changes are anticipated to affect fisheries production, distribution, and species composition (Sumaila et al. 2011). As

S. Rai (✉)
Natural Resources Institute Finland (LUKE), Paavo Havaksen Tie 3, 90570 Oulu, Finland
e-mail: rajanshailendrarai@gmail.com; ext.shailendra.rai@luke.fi

University of Jyväskylä, Kokkola University Consortium Chydenius, P.O. Box 567, 67701 Kokkola, Finland

S. Bera
Department of Geography, Central University of South Bihar, Gaya, India

S. Talukdar et al. (eds.), *Communication, Science, Technology and Innovation in Disaster Risk Management*, GIScience and Geo-environmental Modelling, https://doi.org/10.1007/978-3-031-77189-7_4

fishermen venture further out to sea, they face increasing risks, including social distancing from their families and intensified competition for fishing grounds (Poggie 1980). Such risks disproportionately affect smaller, less efficient vessels. Conflicts among different gear users and other socio-economic risks further compound the challenges faced by fishermen.

Fisheries provide significant employment opportunities and boost local economies (Sathiadhas 2009). Small-scale fisheries play a crucial role in food security (Maire et al. 2021), livelihoods (Isaacs 2012), and global economies (Teh and Sumaila 2011), particularly in low-income countries, where vulnerability to climate change is the highest (Bindoff et al. 2019; Rogers et al. 2019; Ojea et al. 2020). Small-scale fishing communities, despite their significance, face a shortage of assistance in dealing with the effects of climate variability and change on their ways of life (Salgueiro-Otero and Ojea 2020) (Cheung et al. 2009, 2010; Lam et al. 2014, 2016).

In India, annual fisheries production has surged to 3 million tons, with a production-level fish catch valued at US$2.8 billion and an additional US$1.8 billion earned through fish and fishery product exports (Sathianandan 2010). The fisheries sector directly employs 1 million people and indirectly supports another 0.6 million individuals. The west coast of India contributes 67% of the average fish yield, with the east coast contributing the remaining 33%. Marine fish landing revenue at the point of first sales amounted to approximately Rs. 19,753 crores during 2009–10 (CMFRI 2010). However, the challenge lies in safeguarding coastal communities, a significant portion of which lives below the poverty line and relies on traditional fishing methods. The majority of fishing operations are conducted by small-scale fishermen in undeveloped coastal areas, where erosion and habitat destruction endanger both landing sites and fishing communities. Vulnerable fishing communities residing in low-lying areas face the imminent threat of submersion due to rising sea levels (Mahon 2002). Moreover, worsening storms and changing weather patterns poses additional risks, potentially disrupting fishing operations. Extreme events disrupting other sectors, such as agriculture and manufacturing, could lead to conflicts over labor opportunities and heightened fishing pressure. The livelihoods of small-scale fishermen are already vulnerable to various non-climatic risks, including loss of resources and access, market fluctuations, changes in markets, strife, and inadequate governance (Allison et al. 2009). These insecurities hinder investments in sustainable fisheries long-term strategies and are likely to be to worsen by climate change impacts.

India holds the position of the world's second-largest fish producer, contributing approximately 5.43% to global fish production and ranking second in aquaculture sector. In the country, this sector supports the livelihoods of approximately 14 million people (Das et al. 2014). While some studies have assessed the vulnerability of inland (Das et al. 2014) and coastal fisheries (Patnaik and Narayanan 2005; and Shyam et al. 2014), most research has centered on documenting fish population trends (Vivekanandan 2009; Salagrama 2012). Changes in weather patterns, including wind, storms, temperature, and rainfall, are likely to affect fish production, distribution, and species composition. While there has been extensive research on how climate change affects aquatic ecosystems and fish populations (Sarojini et al. 2022; Muruganandam et al. 2023; Çelekli et al. 2023; Pande et al. 2023; Jonsson 2024), there is a lack of knowledge about small-scale fishing communities' responses to climate variability. This study aimed to examine the impacts of climate risks on fishing activities by considering a case study from coastal India.

4.2 Study Area

Raigad District, situated in the Konkan region of Maharashtra, had a population of 2,635,394 in 2011, which increased from 2,207,929 in the preceding decade. Raigad encompasses the significant harbor of Pen-Mandwa, forming a contiguous landmass with Mumbai harbor. Uran, located east of Mumbai across the Dharamtar Creek, lies within the Raigad District. It is primarily a fishing village contributing 80% of Mumbai's fish production. According to the 2011 India census, Uran had a population of 30,439, with males constituting 53% of the population. Notably, JNPT stands as India's largest container terminal, alongside other container terminals like APM Terminals and DP World.

This study was conducted in Uran Koliwada of Raigad District (Fig. 4.1), Maharashtra, commonly referred to as Koli communities, in India. The common reasons behind the selection of this village are as follows:

- Majority of the villagers form a part of the fishing community and a high number of people involved in actual fishing.
- Highly reliance on fishing for their livelihood
- A good mix of mechanized and non-mechanized boats

The area is near to JNPT port and Karanja. In Karanja majority of the fishermen own trawlers. The selection of Uran Koliwada is also based on the shift of small-scale fishermen toward fishing labor in trawlers.

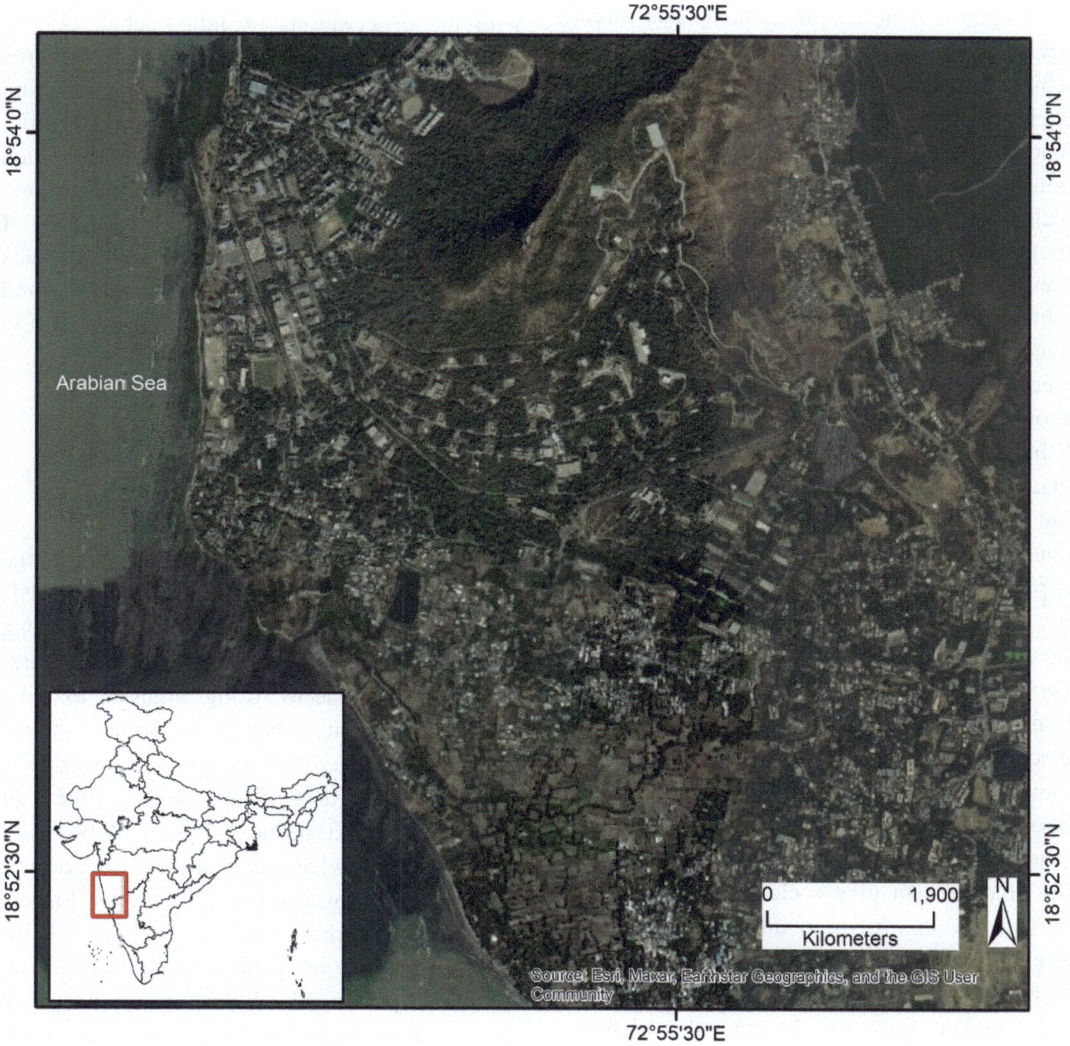

Fig. 4.1 A synoptic view of study area

4.3 Methodology

The target population was fisherfolk in Uran Koliwada, Raigad District. There are 44 fisherfolk households in Uran Koliwada with a population of 194 populations. The survey was conducted of all the fishermen households. Data was collected from fieldwork through surveys, focus group discussions, and conducting in-depth interviews. The primary data collection began in the month of November 2015, when a pilot visit had been undertaken. Before the data collection began, preliminary data from government offices such as the Maharashtra fisheries department has already been collected. The pilot visit lasted for 7 days which helped to identify the key factors influencing the livelihood. The second visit was made in the month of December 2015 to collect household data. For primary data, three survey techniques were used i.e., household surveys, focus group discussions, and in-depth interviews. A structured questionnaire was prepared which had both closed-ended and open-ended questions. In which, the close-ended questions were used for information regarding socio-economic characteristics, profit sharing pattern, fishing duration, etc. To gather opinions on climate variability's impact on fish catch and fishing duration, the survey utilized closed-ended questions. Prior to finalizing the questionnaire, a pilot survey was conducted to ensure clarity, comprehension, and relevance of the questions in obtaining pertinent information. To verify the quantitative data collected through household's survey questionnaire; focus group discussions were conducted. In-depth interviews were carried out with a few elderly members of the village to understand the scenario/situations, elderly members being evidence of the changes can give a clear picture for comparison. Semi-structured interviews were conducted with people of Maharashtra Fisheries Department to understand the administrative/policymakers view on climate risk and fishermen livelihoods.

4.4 Result and Discussion

4.4.1 Climate Variability Impact on Fishermen

The results presented in Fig. 4.1 indicate that all respondents (100%) observed adverse effects of strong winds on fishermen.

This is supported by the fact that about 81.81% experienced the loss of five individuals from boat submergence, 13.63% reported fishermen becoming ill from physical injuries, and 4.56% of respondents observed the loss of fishing equipment in the water. These findings align with the observations of Jallow et al. (1999), which highlighted that changes in wind patterns and the frequency of severe storms resulting from climate variability can lead to the destruction of valuable assets and infrastructure such as boats, gear, and landing sites. Furthermore, climate risks impact fishermen, with 57.57% of respondents indicating that high rainfall leads to loss of life due to boat submergence, and 21.21% reporting the loss of fishing gear in water as a consequence of climate risks.

4.4.2 Climate Variability Impact on the Fish Duration

Based on the findings presented in Fig. 4.2, the majority of respondents, accounting for 89.4%, noted that no fishing occurs during strong winds, while 7.57% mentioned a shorter duration of fishing due to strong winds or extreme weather conditions. Changes in rainfall patterns are anticipated to affect fishing durations differently; during periods of high rainfall, 78.8% of respondents observed no fishing activity at all, and 15.15% reported a decrease in fishing duration due to changes in rainfall patterns. Fishers often adapt to these changes without significant economic consequences in the short term. However, prolonged climate variability may result in more enduring alterations in the availability of certain fish stocks over time.

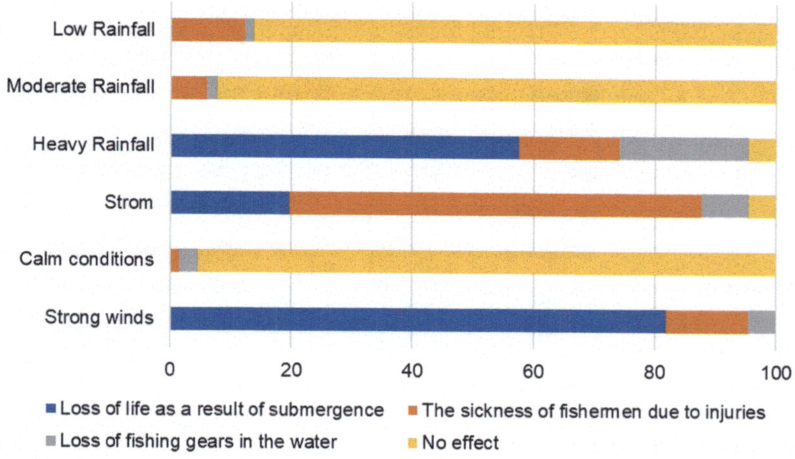

Fig. 4.2 Climate variability impact on fishermen

4.4.3 Effect of Climate Variability on the Fish Catch

The influence of climate variability on the fish catch is depicted in Fig. 4.3. According to the data presented earlier, 62.11% of respondents noted that strong winds lead to a decrease in fish catch. This corresponds to the findings indicating reduced fishing duration and reports of no fishing activity at all during adverse weather conditions. The storms had an even more adverse impacts on fish catch where 96.96% reported a reduction in fish catch. Around 48.48% of the fishermen observed low catch during high temperatures (Fig. 4.4).

4.4.4 Impact of Climate Variability on Fishermen's Livelihoods

Other effects of weather and climate variability mentioned encompass security concerns, boat losses, and diminished income. A significant proportion, approximately 96%, of fishermen now supplement their fishing activities with additional endeavors such as boat repair, fish

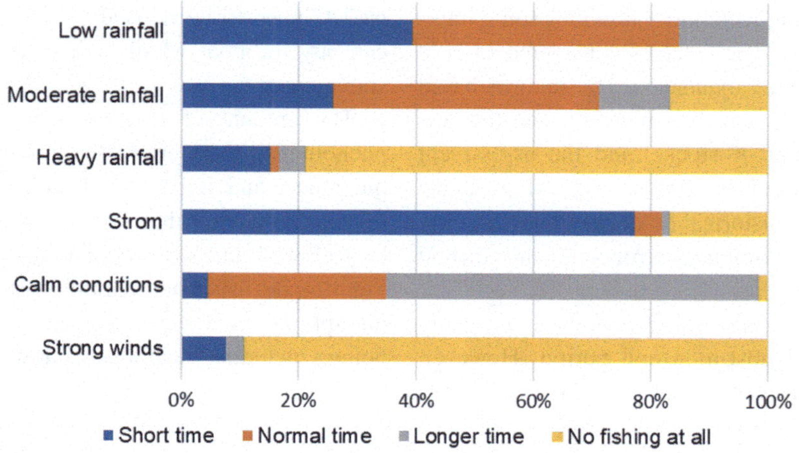

Fig. 4.3 Climate variability impact on the fish duration

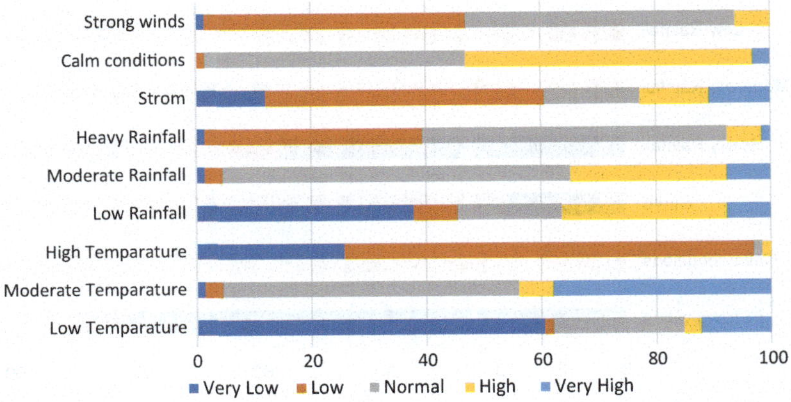

Fig. 4.4 Climate variability impact on the fish catch

marketing, and small businesses, which differs from past practices. According to FAO (2009), communities reliant on fisheries are susceptible to heightened vulnerability, experiencing safety hazards and less secure livelihoods due to fishing activities in adverse weather conditions.

Climate change is forecasted to have diverse effects on the fisheries and their dependent communities. While there is ample understanding of the biophysical repercussions of climate change on aquatic ecosystems, there remains a lack of knowledge concerning the socio-economic aspects of the fisheries impacts. The effects of the climate risk ranges from loss of lives and livelihoods to the destruction of fishing equipment, change in fishing efforts and long duration of fishing or fish catch. A comparison of the production, the effort and the catch per hour across different fishing techniques provides some interesting insights. The catch is considered to be a crude measure of abundance. The analysis based only on CPUE data does not represent the correct assessment of stocks, and the assessment should be based on considering all available data, and also historical data should be carefully examined to eliminate erroneous and invalid assumptions.

The bulk of the fishing revenue is typically earned in the post-monsoon period. However, due to fishermen's lack of savings habits and often insufficient income to meet daily expenses and loan repayments, many are compelled to take out additional loans during lean seasons. Fishermen deal in highly perishable commodities, and the absence of storage facilities, markets, and cooperative societies leads them to sell their catch at discounted rates to local traders. The fishermen are employed only for a few months because of seasonality and there are less alternate employment opportunities available. The fishing sector is highly unorganized, the fishing labors who work on the boats are temporary, and the wages paid are also minimal. Due to low catch, certain fishermen have been compelled to pursue alternative means of earning a living. But they are limited and some of these are not sustainable due to the size of the investment. The loan conditions of NCDC need to be revised so that fishing labor can buy the boat which will help in earning a stable income.

The amount of risk the fishermen take is phenomenal; there are losses of lives, boat submergence, and impact on health. The occupational safety and other safety standards need to be followed and strict regulations are required. Small-scale fishermen endure poverty and face disruptions in various aspects of their well-being, including health, education, and safety (Table 4.1).

Table 4.1 The influence of climate variability on the livelihoods of fishermen

Impact climate variability	Frequency	Percentage
Income loss from fishing	63	96
Switch to alternative livelihoods	54	82
Conflict with other fishermen over fishing areas	52	80

Source Author field survey (2015)

4.5 Conclusion and Policy Implementation

During the monsoon season, fishing days are often lost due to inclement weather. Brackish water aquaculture development emerges as a fitting scheme for coastal regions. However, the low fish production during the monsoon diminishes household income and results in inadequate fish supply for consumers at reasonable prices, creating an opportunity for middlemen exploitation. This situation can be rectified by establishing sufficient storage and processing facilities at major landing centers. A public agency could be instituted to purchase fish at minimum support prices and distribute them after necessary processing during the monsoon. The agency must also ensure an efficient distribution system to transport and market fish in inland areas. Furthermore, extensive waterlogged regions along the coastline offer ideal conditions for prawn aquaculture. With sufficient technical and financial backing from governmental bodies, these areas can be effectively utilized for prawn farming, thus generating employment opportunities for marine fishermen.

It is vital to establish strong institutions to govern resource utilization. While centralized or inflexible methods may appear attractive at first, they may not possess the adaptability required to maintain resilient fisheries systems and communities amidst climate change. Resilient fisheries require adaptive co-management strategies, which seek to tackle uncertainty and harness the insights and commitment of resource users across different levels. It also requires finding a balancing support for vulnerable populations with efforts to promote effective adaptation is crucial.

Appendix

See Table 4.2

Table 4.2 Details of respondents

Characteristic	No. of respondents	Percentage
Gender	46	70
Male	20	30
Female		
Age	0	0
Below 21 years	7	11
21–30 years	9	14
31–40 years	18	12
41–50 years	32	29
51 + years		
Education level	16	24
None	35	53
Primary	6	9
Secondary	7	11
Above Secondary	2	3
Graduation		
Number of people	49	74
living in a household	16	24
Below 5	1	2
5–10		
10+		
Income per month	13	20
Below 10,000	20	30
10,000–20,000	33	50
20,000+		

References

Allison EH, Beveridge MCM, van Brakel M (2009) Climate change, small-scale fisheries and smallholder aquaculture. In: Culberg M (ed) Fish, Trade and Development. Stockholm, Royal Swedish Academy of Agriculture and Forestry (in press), pp 73–87

Bindoff NL, Cheung WW, kairo JG, Arístegui J, Guinder VA, Hallberg R et al (2019) Chaning ocean, marine ecosystems, and dependent communties. In: IPCC special report on the ocean and cryosphere in a changing climate Intergoverrnmental Panel on Climate Change, 477–587

Çelekli A, Zariç ÖE (2023) Hydrobiology and ecology in the context of climate change: the future of aquatic ecosystems

Cheung WWL, Lam VWY, Sarmiento JL, Kearney K, Watson R, Pauly D (2009) Projecting global marine biodiversity impacts under climate change scenarios. Fish Fish 10(3):235–251. https://doi.org/10.1111/j.1467-2979.2008.00315.x

Cheung WWL, Lam VWY, Sarmiento JL, Kearney K, Watson R, Zeller D et al (2010) Large-Scale redistribution of maximum fisheries catch potential in the global ocean under climate change. Global Change Biol 16:24–35. https://doi.org/10.1111/j.1365-2486.2009.01995.x

CMFRI (2010) Maharashtra Marine Fisheries Census 2010. Cochin: CMFRI

Das MK, Srivastava PK, Rej A, Mandal ML, Sharma AP (2014) A framework for assessing vulnerability of inland fisheries to impacts of climate variability in India. Mitig Adapt Strat Glob Change. https://doi.org/10.1007/7s11027-014-9599-7

FAO (2009) Climate change implications for fisheries and aquaculture: overview of current scientific knowledge. FAO Fisheries and Aquaculture Technical Paper. No. 530. Rome

FitzGerald D, Fenster M, Argow B, Buynevich I (2008) Coastal impacts due to sea-level rise. Annu Rev Earth Planet Sci 36:601–647

Isaacs M (2012) Recent progress in understanding small-scale fisheries in southern Africa. Curr Opin Environ Sustain 4(3):338–343. https://doi.org/10.1016/j.cosust.2012.06.002

IPCC (2019) Climate change and land: an IPCC special report on climate change, desertification, land degradation, sustainable land management, food security, and greenhouse gas fluxes in terrestrial ecosystems

Jallow BP, Toure S, Barrow MM, Mathieu AA (1999) Coastal zone of The Gambia and the Abidjan region in Côte d\'Ivoire: sea level rise vulnerability, response strategies, and adaptation options. Climate Res 12:129–136

Jonsson B (2024) The effect of climate change on salmonid fishes in rivers. Fishes 9(1):29

Lam VWY, Cheung WWL, Sumaila UR (2014) Marine capture fisheries in the Arctic: winners or losers under climate change and ocean acidification? Fish Fish 17(2):335–357. https://doi.org/10.1111/faf.12106

Lam VWY, Cheung WWL, Reygondeau G, Sumaila UR (2016) Projected change in global fisheries revenues under climate change. Nat Pub Group 6:32607. https://doi.org/10.1038/srep32607

Mahon R (2002) Adaptation of fisheries and fishing communities to the impacts of climate change in the CARICOM region

Maire E, Graham NAJ, Macneil MA, Lam VWY, Robinson JP, Cheung WWL et al (2021) Micronutrient supply from global marine fisheries under climate change and overfishing. Curr Biol 31(18):1–7. https://doi.org/10.1016/j.cub.2021.06.067

Muruganandam M, Rajamanickam S, Sivarethinamohan S, Reddy MK, Velusamy P, Gomathi R, Munisamy SK (2023) Impact of climate change and anthropogenic activities on aquatic ecosystem–a review. Environ Res, 117233

Nicholls R (2011) Planning for the impacts of sea level rise. Oceanography 24(2):144–157

Ojea E, Lester SE, Salgueiro-Otero D (2020) Adaptation of fishing communities to climate-driven shifts in target species. One Earth 2(6):544–556. https://doi.org/10.1016/j.oneear.2020.05.012

Pande A, Pandey PK (2023) The perspective of climate change on the aquatic environment and fish production. In: Outlook of climate change and fish nutrition. Singapore: Springer Nature Singapore. pp 3–22

Patnaik U, Naryanan K (2005) Vulnerability and climate change 2005. An analysis of the eastern coastal districts of India. Paper presented at the International workshop on human security and climate change, Oslo, 21–23

Poggie JJ (1980) Maritime anthropology: socio-cultural analysis of small-scale fishermen's cooperatives: introduction. Anthropol Quart 53(1):1–3

Rogers LA, Griffin R, Young T, Fuller E, Martin K, Pinsky ML (2019) Shifting habitats expose fishing communities to risk under climate change. Nat Climate Change 9:512–518. https://doi.org/10.1038/s41558-019-0503-z

Salagrama V (2012) Climate change and fishery: perspectives from small-scale fishing communities in India on measures to protect life and Livelihood, SAMUDRA Monograph International Collective in Support of Fishworkers, Chennai, India

Salgueiro-Otero D, Ojea E (2020) A better understanding of social-ecological systems is needed for adapting fisheries to climate change. Mar Policy 122:9. https://doi.org/10.1016/j.marpol.2020.104123

Sarojini S, Umesh M, Jagannath N, Sivakumar B, Kaloor A, Jayaram S (2022) Ramifications of climate change induced by anthropogenic factors on global fish population scenario. Aquac Sci Eng, 1–33

Sathiadhas R, Sangeetha KP (2009) Employment scenario and labour. Migr Mar Fish Asian Fish Sci. 22:713–727

Sathianandan T (2010) Marine fisheries sector in indiaresource endowments, infrastructure, intensities, and stakeholder analysis. Cochin: Central Marine Fisheries Research Institute

Shyam SS, Kripa V, Zacharia PU, Mohan A, Ambrose TV, Rani M (2014) Vulnerability assessment of coastal fisher household in Kerala: a climate change perspective. Indian J Fish 61(4):98–103

Sumaila UR, Cheung WWL, Lam VWY, Pauly D, Herrick S (2011) Climate change impacts on the biophysics and economics of world fishery. Nat Clim Change 1301:1–8

Teh LCL, Sumaila UR (2011) Contribution of marine fisheries to worldwide employment. Fish Fish 14(1):1–12. https://doi.org/10.1111/j.1467-2979.2011.00450.x

Vivekanandan E (2009) Climate Change, Marine Ecosystem, and Fisheries. Sagar Sangamam. Indian Fishermen and Fisheries Workers Co-Operative Society Ltd., Thiruvananthapuram, pp 70–82

5 Assessment of Factors Affecting Agricultural Risks and Resilience Strategies of Farmer Community in Borda Village, Kalahandi District of Odisha

Ramakanta Naik

Abstract

Agriculture sector is susceptible to multiple sets of risks such as disaster, climate, pest, disease, institutional, market and policy risk. Developing resilience strategies in a risk-prone area is a difficult task. A multiple risk management approach requires to develop resilience strategies Assessment of vulnerabilities, hazards, and risks is crucial for risk management, and it also contributes to resilience building. The intensity and severity of the impact of a disaster depend on the vulnerability and resilience of a place, community, or individual. Farmers are particularly vulnerable to severe weather conditions like drought and flooding. To some extent, though, the community is able to adjust and cope with disaster-like circumstances. The agricultural community is equally affected by the same dangers, despite having varying levels of vulnerability and resilience. Community-based disaster risk reduction (CBDRR) is a tool that studies, understand, and assesses the vulnerability, risk and resilience factors incorporated with community participation. This article is based on a case study of Borda village in the Kalahandi district. The study has adopted a quantitative and exploratory case study. A descriptive statistical analysis of 201 households discussed various socioeconomic, land, assets, credit, and insurance parameters. The paper also discusses regional vulnerabilities, risk and resilience factors from farmers' and community perspectives. It is found that droughts occur almost every alternative year in the village. Nearly ninety percent of households are indebted and have taken loans from financial institutions. It is also found that the insurance claims are not compensated for crop loss due to disaster even after several complaints were filed. The study recommends implementing and utilizing various risk management tools concerning community participation.

Keywords

Agricultural risk · Disaster · Vulnerability · Resilience strategy · CBDRR · Community

5.1 Introduction

Agriculture, a highly dynamic and vulnerable sector, is subjected to various risks and uncertainties that significantly impact the livelihoods of farming communities (Harvey et al. 2014).

R. Naik (✉)
Jamsetji Tata School of Disaster Studies, Tata Institute of Social Sciences, Mumbai, India
e-mail: ramakantnaik.8@gmail.com

S. Talukdar et al. (eds.), *Communication, Science, Technology and Innovation in Disaster Risk Management*, GIScience and Geo-environmental Modelling, https://doi.org/10.1007/978-3-031-77189-7_5

The existing study emphasizes the factors influencing agricultural risks and resilience strategies adopted by farmers. By examining the existing literature, this assessment seeks to explore the challenges faced by farmers and the existing strategies to enhance their resilience. Economic development and advancement of technology bring more risk to human lives. It is not wrong to say that high technology creates high risk and high profit creates high risk in real life. Disasters and climate change factors are becoming more complex with the rise of humans' cognitive ability (Jian and Rehman 2016). It is difficult to find a unique concept of risk as the perception of risk is different by persons globally. Farmers face multiple risks such as drought, floods, heat waves and cold waves, hailstorms and cyclones ; however, there is a limited number of study conducted on investigating the impact of multiple risks on agricultural productivity (Birthal 2022). Farmers face multiple agriculture risks and cope with them to manage them. However, the existing literature and research have limited scope for the overall risk management strategy. Komarek et al. (2020) conducted an analysis of over 3283 peer-reviewed articles pertaining to five distinct categories of risk: production, market, institutional, personal, and financial risk. Still, the analysis discovered that only eighteen studies had addressed all five categories of risk. Rather than evaluating how these hazards impact the farmer's yield and revenue, these studies focus more on conceptual issues. The multiple risk management strategy in agriculture can be a significant scope to cope with and manage risk (Komarek et al. 2020). Many South Asian countries prioritize building resilience in communities against climate-related disasters (Naik and Murari 2023; Haque et al. 2022). Community-based initiatives are being started since South Asian nations have likewise committed to achieving sustainable development goals by 2030. Community-based disaster risk reduction is a useful strategy that aids in community attempts to become more resilient to disasters and better able to bear the consequences of them (BDRCS 2022). The paper discusses regional vulnerabilities, risk and resilience factors from farmers' and community perspectives which may help develop suitable resilience strategies and policy framework. Kalahandi district has been historically known for the deprivation and food crisis but the agricultural development has now changed the district's outlook. The district has made significant progress in agricultural development in recent years. The cropping intensity has increased from 120% in 2000–01 to 156% in 2021–22. The productivity of major crops has also increased. For example, the productivity of paddy has increased from 2.2 tons per hectare in 2000–01 to 3.2 tons per hectare in 2021–22 (DAFPO 2020).

Historically, the major debate in the field of risk is the degree to which it is objective and measurable versus the degree to which it is subjective and sometimes immeasurable (Etkin 2016). According to Laura (2012), there are six main categories of risks associated with agriculture: financial, political, human, production, and economic risks. Risks in agriculture come from a variety of sources. However, all the risks are related to each other. Dohlman (2020) describes five broad forms of risk: financial, institutional, pricing and market, production, and human or personal risk. He claims that the unpredictable natural growing process of crops and livestock is the source of production risk. Weather, diseases, pests, and other factors can affect both the quantity and quality of production.

In her study, Laura (2012) discussed various types of risks in agriculture. The identification of risk and its evaluation is necessary to get a positive result on the farm, for which scientific literature has suggested many methods. Even though many methods are being used to evaluate the risks in agriculture, it is necessary to use the most appropriate method which discusses integrated risk evaluation in a logical framework. The study by Komarek et al. (2020) presents the complexity of definitions, interpretations and glossary convincingly related to agriculture risk management. They define five types of agriculture risks as:

i. Production risk caused due to the impact of climatic factors and pests and diseases uncertainly impact the natural growth of the crop.
ii. The main areas of market risk uncertainty are price, cost, and market access. However, risks like uncertain weather, price increases, or limited market access factor influence farmers' decision-making.
iii. Unpredictable changes in agricultural policies, norms, and regulations are referred to as institutional risk, which is another significant risk in agriculture.
iv. A person's health or a relationship that has an impact on the farm or farm household constitutes personal risk. The use of machinery, chemicals, and technology in agriculture are significant sources of personal danger that have an immediate effect on health.
v. Financial risk relates to the credit, cash flow, and finances of the farm management. Over 70% of Indian farmers are small and marginal farmers who encounter difficulties as a result of rising agricultural input costs. They run the possibility of experiencing a severe financial crisis due to production loss. Numerous research studies on risk management have been carried out, and various methods and instruments are employed to minimize risk. The scientific research currently in being published, however, does not address integrated and multiple risk management methods and techniques in great detail.

The effects of climate change and disasters will only get worse in the lack of a robust resilience strategy. A community can be equipped to deal with the vulnerability caused by climate change and recover from a shock caused by disaster if it has the right approach and resilience. Resilience building within a community is crucial to improving its sustainable way of life (BDRCS 2022). The Latin verb "resilire," which meaning to rebound or recoil, gave the meaning of the word resilience in the seventeenth century (Mbae 2014). The usage of the term found that "why some types of wood were able to accommodate sudden and severe loads without breaking." The term resilience had been used in different disciplines like material science, environmental studies, and ecology. Resilience has since been applied to many other fields, including ecology, engineering, psychology, and disaster studies, where it has been utilized since the 1970s. Resilience, which mostly refers to a group or community's capacity to withstand external dangers brought on by social, political, and environmental change, has gained attention in the development field in recent years. It is also argued that vulnerability is a risk and resilience concerns coping (Moret 2014).

Another definition of resilience is the capacity to foresee, control, endure, and recover from calamities. It consists of the available resources and strength. Resilience is seeing, preparing for, and reducing the risk of disasters in order to sustainably protect people, communities and their livelihoods, health, cultural heritage, socio-economic assets, and ecosystems (Das and Paul 2021). In addition, resilience draws attention to the possibility of non-linearity and transformation processes, as well as the gradual forces of change that impact systems. It emphasizes capacities at the institutional, community, and individual levels, with an emphasis on learning, creativity, and resilience in adaptation. (Birkmann 2007). Taking a risk is equivalent to putting oneself at danger of injury or loss. Risk usually means an aversion to some of the potential impacts and is not without benefit (Hardarker et al. 2004).

5.1.1 The Relationship Between Risk and Resilience

In the context of farmers' studies, the relationship between risk and resilience is a subject of continuous discussion and investigation. While the exact nature of this relationship is still being explored, there are some insights from the available literature:

i. Discussion and integration: There is disagreement on how risk and resilience are related to one another and how they need to be related (Aven 2021). Nonetheless, risk and resilience are becoming more widely acknowledged as being intrinsically linked and need to be taken into account simultaneously (Logan et al. 2022). The integrated approach to resilience is emphasized by risk science, and the current discourse on risk can offer an integrated framework for comprehending resilience (Logan et al. 2022).

ii. Complementary notions: Risk and resilience are related concepts that can be used to comprehend and overcome the risks and challenges that farmers face. Resilience is the ability to recover from disasters or changes in circumstances, while risk is the probability and impact of disaster. In a world full of risks, resilience is crucial, and developing the capacity to avoid damage and recover from disasters in the future is necessary. (Integrate 2022).

iii. Risk as a resource for resilience: Although they may go unnoticed, risk concepts and related literature can offer insightful resources for resilience research. The current understanding of risk science provides a framework for comprehensive investigation into resilience's multiple components (Logan et al. 2022). Understanding and managing risks can contribute to building resilience in agricultural systems.

iv. Individual factors: Resilience factors shield people from the damaging effects of stress, but personal risk factors make people more vulnerable to everyday stress (Diehl et al. 2021). These personal factors can influence farmers' ability to cope with risks and build resilience in their farming practices.

Research on the relationship between risk and resilience is a diverse and emerging field. Although there is still disagreement, it is acknowledged that risk and resilience are related and need to be taken into account simultaneously. In a world full of risk, building capacity for resilience is essential. Risk concepts can offer useful tools for resilience analysis. The perception and resilience of risk among farmers are also influenced by personal characteristics (Aven 2021; Logan et al. 2022); Integrate 2022; Diehl et al. 2021; Vanderbilt and Shaw 2008; Marks 2021).

5.1.2 Assessment of Factors Affecting Risk and Resilience Studies

Multiple factors are considered while assessing the variables influencing agricultural risks and resilience. Because of the unpredictable nature of the weather, farmers are exposed to several risks. It is also true that crop productivity is impacted by climate change, and the likelihood of extreme weather events is increased. Droughts and floods are seriously harming crops (Kumar et al. 2021). Again, diseases and pests have a big impact on agricultural output. The resilience of farming systems can be impacted by their complexity. The ability to maintain a system's functionality in the face of increasing and complicated problems is known as resilience, and it may be examined via a framework designed to evaluate the resilience of farming systems (Srinidhi et al. 2023). For agricultural systems to be resilient, they must be sustainable. A few examples of the variables that affect agricultural systems' resilience and sustainability are soil quality, water availability, and biodiversity (Lee et al. 2022). Economic factors that can impact an agricultural system's resilience include input costs, market pricing, and loan availability. Agricultural systems' resilience can be impacted by policies and structures that facilitate risk mitigation and adaptation (Kumar et al. 2021). Assessing agricultural risks and resilience requires a systematic literature review (Talukder et al. 2020).

India has a number of case studies on risk and resilience assessment in agriculture. The Agriculture Resilience Assessment of Central India under climate change is one such study (Kumar et al. 2021). This study evaluates how climate change is affecting Central Indian agriculture and suggests ways to strengthen farming systems' resilience. An additional study evaluates climate resilience in the past using two case studies from Maharashtra, India (Srinidhi et al. 2023). The goal of the study is to make the farming systems in the area more resilient to climate change. A number of potential elements could impact India's agricultural systems' resilience and sustainability: weather fluctuations, natural disasters, pests and diseases, farming systems' complexity, sustainability, economic considerations, and institutional and regulatory concerns (Lee et al. 2022; Talukder et al. 2020; OECD 2023). Crop production is particularly susceptible to the consequences of climate change and water scarcity, argued by Talukder et al. (2020).

5.2 Methodology

This study employs a quantitative research approach characterized by an exploratory case study design. The theoretical framework for community-based disaster risk reduction (CBDRR) as a tool for vulnerability and resilience assessment was developed using secondary literature and information. The study area selection was informed by a pilot survey conducted in collaboration with the Assistant Agriculture Officer of the Bhawanipatna block, utilizing secondary data on disaster occurrences and agriculture production. The village under study comprises 882 households, with farmers constituting 40% of the population, as per the 2011 Census. From this population, a sample of 201 farmer households was selected for detailed examination. Borda village, situated within the Bhawanipatna block, emerged as the most disaster-prone region due to its geographical characteristics, notably its dry hilly terrain and lack of irrigation infrastructure. Consequently, the village experiences recurrent droughts, adversely impacting crop yields on an annual basis. Data collection took place during the period of April to June 2022, focusing on household information pertaining to the financial year 2021. Primary data collection involved the administration of a household survey questionnaire tool. Initial data gathering involved the identification of self-help groups and other community organizations. Household surveys were conducted among members of these community groups, encompassing various aspects such as personal information (e.g., name, age, gender, education status, occupation, family details), socio-economic conditions (including income sources, insurance, and credit), agricultural practices (crop cultivation), disaster occurrences, and awareness of government agricultural schemes. Prior to interviews, ethical consent forms were completed and signed by the respondents. Data analysis was carried out using JASP software, employing simple descriptive statistics and cross-tabulation methods to analyze the collected data (Fig. 5.1).

5.3 Results and Discussion

In this section of the article, findings of the study have been discussed in detailed. The study is an attempt to explore the existing risk and resilience factors that are in cognitive mode. Risk and resilience coexist, when risk reduces resilience increases and resilience decreases if risk increases. The following discussion on finding emphasis on the potential of risk and resilience factors in Borda village. The data were collected from people's age ranging from 20 to 65 years old. The average age of the respondents is 38 and the maximum respondents are female among them. Maximum people are living in nuclear family in Borda, which is 53 and 47% are living in joint family.

Nearly 89% of people are married, 8% are single who never married and 3% are either widows or divorced. Table 5.1 shows the educational qualifications of respondents in Borda village. It is reported that 54.23% of the respondents are illiterate in the village. People who have completed high school is 15.92% and nearly 7% have completed intermediates. Only 5.47% of the people having a college degree are listed. More than 165 of people have completed primary education in the village. The table shows that the educational attainment of this group of people is relatively low. The illiteracy rate is high; only a small percentage have a college degree. It is reported that people face challenges in the workforce, as they do not have the necessary skills and knowledge to compete for jobs in the region.

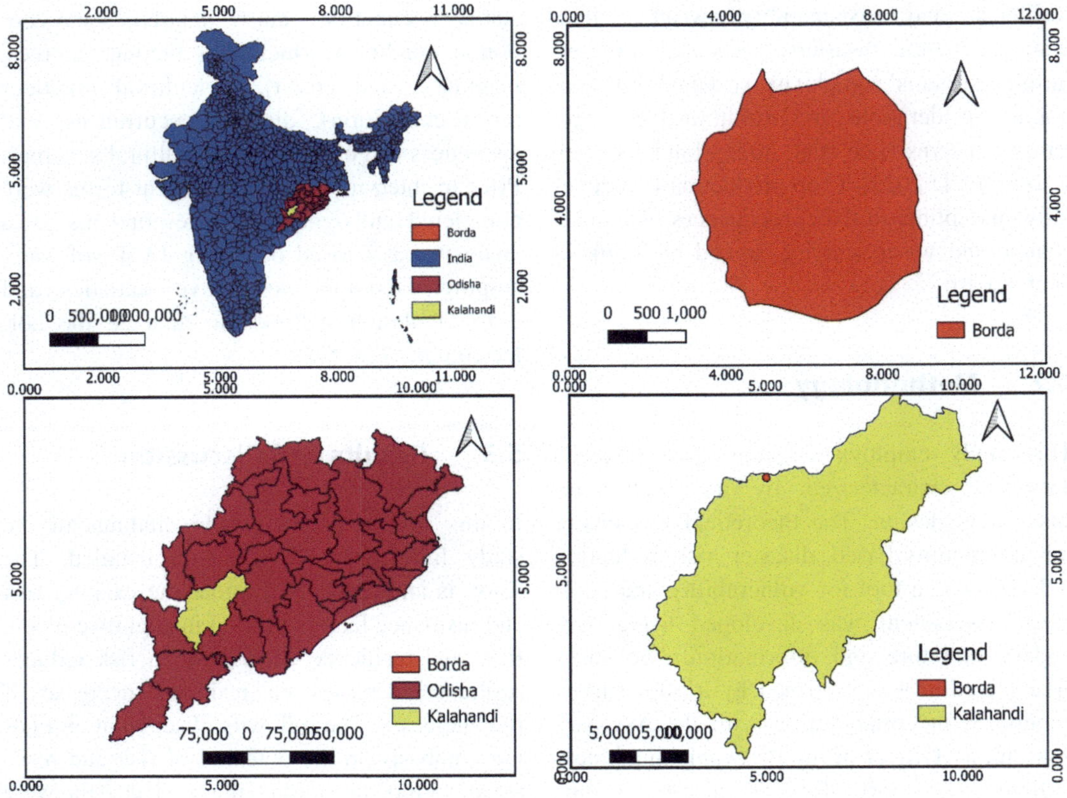

Fig. 5.1 Study area map of Borda village

Table 5.2 shows the frequency distribution of occupations for a sample of 201 people. The first column shows the occupation, the second column shows the frequency, and the third column shows the percent. The most common occupation is daily wage labor, with 86 people (42.78%) in this category. The second most common occupation is farmer, with 55 people (27.36%) in this category. The third most common occupation is semi-professional, including clerk, manager, and assistant, with 10 people (4.97%) in this category. The least common occupations are defense, such as police, army, navy, air force, paramilitary, teacher, lecturer, agriculture labor, and business, with 1 or 2 people in this category. Table 5.2 also shows that 14 people (6.96%) are unemployed. Overall, it shows that the most common occupations are daily wage labor and farmer, and that there is a significant number of unemployed people. It has been reported that a large number of people are in the informal sector, as evidenced by the high number of daily wage laborers. The table suggests a gender imbalance in the workforce, as men dominate most occupations.

Table 5.3 represents mean, standard deviation, minimum, and maximum values of the numbers of household members. The number of adult men, people living in the household, adult women, children below the age of 11, men and

Table 5.1 Educational qualification of the respondents

Educational qualification	Frequency	Percent
Graduation	11	5.47
High school	32	15.92
Illiterate	109	54.23
Intermediates	16	7.96
Primary school	33	16.42
Total	201	100

($N = 201$)

Table 5.2 Occupation of the respondents

Occupation	Frequency	Percent
Unemployed	14	6.96
Semi-professional (clerk/manager/assistant)	10	4.97
Defense (police/army/navy/airforce/para military)	1	0.49
Teacher/lecturer	2	0.99
Agriculture labor	2	0.99
Business	2	0.99
Daily wage labor	86	42.78
Farmer	55	27.36
Self employed	6	2.98
Skilled labor	23	11.44
Total	201	100

($N = 201$)

Table 5.3 Numbers of household members

Household members	Mean	Std. deviation	Minimum	Maximum
Total in the household	4.53	1.60	1	10
Adult men	2.02	1.00	0	5
Adult women	1.83	0.81	0	5
Children below 11	0.67	0.96	0	4

($N = 201$)

women earning an income have also been listed. The mean value for the number of people living in the household is 4.53 indicates that the average household has 4.53 people living in it. The standard deviation of 1.60 means a moderately wide range of values for the number of people living in households, with some households having as few as 1 people and others having as many as 10 people. The mean value for the number of adult men is 2.02, indicating that the average household has 2.02 adult men. The standard deviation of 1.00 means a relatively wide range of values for the number of adult men in households, with some households having 0 adult men and others having as many as 5 adult men. The mean value for the number of adult women is 1.83 and indicates that the average household has 1.83 adult women living in it. The standard deviation of 0.81 means a relatively wide range of values for the number of adult women in households, with some households having 0 adult women and others having as many as 5 adult women. The mean value for the number of children below the age of 11 years is 0.67 and indicates that the average household has 0.67 children below the age of 11 living in it. The standard deviation of 0.96 means a moderately wide range of values for the number of children below the age of 11 in households, with some households having as few as 0 children and others having as many as 4 children.

Table 5.4 shows that the mean value for the number of men earning an income is 1.78, indicating that the average household has 1.78 men earning an income. The standard deviation of 0.83 means a relatively wide range of values for the number of men earning an income in households, with some households having 0 men and others having as many as 5 men. The mean value for the number of women earning an income is 1.51 indicates that the average household has 1.51 women earning an income. The standard deviation of 1 means a moderately wide range of values for the number of women earning an income in households, with some households having 0 women and others having as many as 5 women. Overall, Table 5.4 shows that the average household in Borda village has 1.87 adult men, 4.15 people living in it, 1.75 adult women, 0.49 children below the age of 11, 1.78 men who are earning an income, and 1.51 women who are earning an income. However, each variable has a wide range of values, with some households having significantly more or fewer people than the average.

Table 5.5 represents the mean, standard deviation, minimum and maximum land holding of households in Borda village. The mean amount of land owned by households is 0.92 acres. The standard deviation is 1.58 acres, meaning most households own 0 and 3 acres of land. The minimum amount of land owned is 0 acres, and the

Table 5.4 Earning members in the household

Earning members in the household	Mean	Std. deviation	Minimum	Maximum
No. of men are earning	1.54	0.67	0	3
No. of women are earning	1.12	0.55	0	3
Total annual income (INR)	95,049.75	89,288.84	30,000	800,000

($N = 201$)

maximum is 10 acres. The mean amount of land rented in by households is 0.01 acres. The standard deviation is 0.14 acres, meaning most households rented-in very little land. The minimum amount of land rented-in is 0 acres, and the maximum is 1.5 acres. The mean amount of land rented out by households is 0.06 acres. The standard deviation is 0.43 acres, meaning most households rent very little land. The minimum amount of land rented out is 0 acres, and the maximum is 5 acres. The mean amount of land cultivated last year by households was 0.86 acres. The standard deviation is 1.53 acres, meaning most households cultivated between 0 and 3 acres of land last year.

Table 5.5 Landholding of the respondents

Landholding of the households	Mean	Std. deviation	Minimum	Maximum
Total land own (in acre)	0.92	1.58	0	10
Land rented-in (in acre)	0.01	0.14	0	1.5
Land rented-out (in acre)	0.06	0.43	0	5
Land cultivated last year (in acre)	0.86	1.53	0	10

($N = 201$)

The minimum land cultivated last year is 0 acres, and the maximum is 10 acres. Table 5.5 shows that the amount of land owned, rented in, rented out, and cultivated last year by households is relatively low. The mean values for all four categories are below 1 acre, and the standard deviations are all relatively high, indicating a large variation in the amount of land households hold. The table also shows that some households own, rent in, rent out, or cultivate much land. For example, the maximum amount of land households own is 10 acres, and the maximum amount cultivated last year is 10 acres. However, these households are the exception, not the rule.

Table 5.6 Irrigation and road access to cropland status in Borda village

Variables	Mean	Std. deviation	Minimum	Maximum
Irrigated land (%)	1.74	11.60	0	100
Land has road access for mechanization (%)	41.64	48.50	0	100

($N = 201$)

Table 5.6 represents the mean, standard deviation, minimum, and maximum values for the percentage of land that is considered irrigated and the percentage of land that has road access for mechanization in Borda village. The mean percentage of land that is considered irrigated is 1.74%. The standard deviation is 11.60%, which means that a large part of the land is not irrigated. The minimum percentage of land that is considered irrigated is 0%, and the maximum is 100%. Some farmers supply water to the vegetable crop and paddy crops through lift irrigation and groundwater through a pump set. The mean percentage of land with road access for mechanization is 41.64%. The standard deviation is 48.50%, meaning there is a large variation in the amount of land with road access for mechanization. The minimum percentage of land with road access for mechanization is 0%, and the maximum is 100%. Table 5.6 shows that the percentage of land considered irrigated and land with road access for mechanization is relatively low. The mean values for both variables are below 50%, and the standard deviations are relatively high, indicating a large variation in the amount of land that is irrigated and the amount of land that has road access for mechanization.

Table 5.7 shows that more than 75% of households face deficit to meet family requirements by the family's annual income. It obvious that the labor class and farmer households are the maximum in the village who struggle to meet their family need. Only 24% of households sufficiently meet their family requirements.

Table 5.7 Household requirements and sustainability

Adequacy of annual income to meet family requirements	**f**	%
Deficit	151	75.12
Sufficient	50	24.88
Total	201	100
Adequacy of farm production to meet family food	**f**	%
0–3 months	83	41.29
3–6 months	17	8.46
6–9 months	48	23.88
9–12 months	35	17.41
More than 12 months	4	1.99
Missing	14	6.97
Total	201	100
Management strategies for deficit months	**f**	%
Spending saving	21	10.45
Spending saving, takes loan	67	33.33
Spending saving, takes loan, borrowing food, minimize consumption	3	1.49
Spending saving, take loan, minimize consumption	47	23.38
Spending saving, borrowing food	1	0.50
Take loan	4	1.99
Take loan, borrowing food	3	1.49
Take loan, minimize consumption	8	3.98
Not applicable	47	23.38
Total	201	100

($N = 201$)

More than 41% of respondents say that their farm production lasts only 3 months to meet the family food requirements. Only 8% of them can manage for maximum 6 months and 24% manage not more than 9 months. It shows that food crisis is a primary concern of the village. People manage to fulfill their requirements by supplying rice by the gram panchayat authority. Maximum 33% of the households spend from savings or take monetary loans to manage the family requirement for the deficit months. Again, 23% of households says that they spend from saving, take a monetary loan, and decrease consumption to meet the family requirement for deficit months.

Table 5.8 shows that 91.54% of people have ongoing loans in Borda village indicates a relatively high percentage, and it suggests that many people in this village struggle to make ends meet. The vast majority of people who have ongoing loans got them from self-help groups which indicate that self-help groups are a popular source of loans for people who cannot get loans from traditional financial institutions. Only 1.00% of people got their ongoing loans from financial institutions. People are afraid of lengthy processes and documentation for loans in financial institutions and can't travel to Bhawanipatna or Karlapada bank for the loans as it is located nearly 40 km and 20 km, respectively. Self-help group authority comes to the community and provides them loans and the process is also easier for the people, mostly farmers and homemakers. Self-help groups provide loans at very low interest, such as 12% interest, to the group members, which attracts them to take loans and fulfill their needs. Although the loans were taken for farming or business purposes, they spend them on their daily expenses because they have no proper idea to start a business, nor the authority provides them with training and resources. People in Borda are taking continuous loans because they have limited sources of income, the rising cost of daily commodities, and household needs. It is observed that there is a need for more financial education and resources for people who are struggling financially. The self-help groups play

Table 5.8 Ongoing loans and the sources of the loans

Status of ongoing loans	*f*	%		Source of the loan	*f*	%
No	17	8.46		Financial institution	2	1.00
Yes	184	91.54		Self-help groups	184	91.54
Total	201	100		Total	201	100

($N = 201$)

an important role by providing financial support to people who are in need and traditional financial institutions are not accessible to them. The low percentage of people who got their ongoing loans from financial institutions suggests that they need to be more accessible, easy to process and willing to lend money to people already struggling financially.

The primary data shows that the average loan taken by people in the sample is 10069.65 INR. The maximum loan taken is 60000 INR, which is significantly higher than the average and some people did not take out any loans. The data also shows that the standard deviation of the loan taken is 5181.228 INR which says there is a large variation in the amount of loans people took out. The data says that some people in the village took out very large loans for farming and business purposes. It is reported that the purpose of the loans is maximum for farming which is nearly 44%, followed by business at 42%. Loans are also taken for medical purposes, indicating the risk of health issues and financial crises in the village. All the loans were taken from either financial institutions or self-help groups. The maximum number of loans taken from financial institutions is 91.54%.

Table 5.9 represents that the vast majority of respondents (99.5%) have had some savings, while only 0.5% have not, implying saving money is a common practice among people in Borda village. It is reported that most people save in a group (96.02%), followed by a financial institution (1.00%). A few people have saved their money in both a financial institution and a group (2.49%). Saving monthly amount ranges from 100 to 30,000 INR in Borda village, according to the data provided by the respondents. More than 92% of people save 100 INR monthly in self-help groups, followed by 500 and 1000 INR in financial institutions. The average amount of money saved in Borda is 438 INR monthly, which is relatively low from the highest amount of monthly savings, which is 30000 INR. The standard deviation is 2337.68, indicating a large variation of money saved by the people in Borda village. People are encouraged to save their money in groups because they get good returns and credit facilities from the group, along with a sense of community and support, which motivates saving money. The data also suggests that a few people do not have any savings because of factors such as low income, high expenses, not being involved in self-help groups, or a lack of financial knowledge. These individuals may need additional support to help them save money and develop more effective saving strategies.

Table 5.9 Savings of the households

Status of savings	**Frequency**	**Percent**
No	1	0.50
Yes	200	99.50
Total	201	100
Sources of savings	**Frequency**	**Percent**
Financial institution	2	1.00
Financial institution and Group	5	2.49
Group	194	96.51
Total	201	100

($N = 201$)

Numerous agricultural programs are being carried out in the area, including the Pradhan Mantri Kisan Samman Nidhi Yojana (PMKSNY), a federal government initiative that gives marginal farmers and landless laborers 6000 INR a year to cover their input costs. Pradhan Mantri Krishi Sinchai Yojana (PMKSY) is a central government scheme that provides irrigation facilities

Table 5.10 Agricultural schemes' awareness, benefits, risk management, and satisfaction

Agricultural schemes	Awareness		Avail benefits		Help in risk management		Satisfied	
	Yes (%)	No (%)	Yes (%)	No (%)	Yes (%)	No (%)	Yes (%)	No (%)
PMKSNY	99.01	1.00	1.00	99.01	3.48	96.52	1.00	99.01
KALIA	100	0	28.86	71.14	39.80	60.20	28.86	71.14
PMKSY	12.44	87.56	5.47	94.53	0.50	99.50	5.47	94.53
PMFBY	10.95	89.06	5.47	94.53	5.47	94.53	5.97	94.03
AIF	11.44	88.56	1.99	98.01	0	100	1.99	98.01

($N = 201$)

such as pump sets and lifts irrigation equipment and subsidies to farmers. Krushak Assistance for Livelihood and income Augmentation (KALIA) is a state government program in Odisha that gives 5000 INR to small, marginal farmers, and landless agricultural laborers. The agricultural infrastructure fund offers financial support and subsidies to farmers belonging to scheduled castes and tribes so they can develop their agricultural infrastructure. The central government also runs the Pradhan Mantri Fasal Bima Yojana (PMFBY), which is a crop insurance scheme that covers farmers. Table 5.10 displays the current state of knowledge, benefits available, assistance with risk management, and satisfaction with the agricultural programs being carried out in the area.

The study found that awareness of agricultural schemes in Borda village was maximum in the case of PMKSNY and KALIA Yojana. However, in Borda village, awareness of other schemes such as PMKSY, PMFBY, and agriculture infrastructure fund (AIF) Yojana is significantly lower, indicating the implementation and utilization of agricultural schemes are back-dropped in this region. PMFBY is a crucial risk management tool for markers. It was designed to provide insurance coverage to the farmers to get financial assistance in case of crop loss due to disasters or pest attacks. However, it is reported that farmers who purchased the crop insurance schemes didn't get crop loss compensation from PMFBY even after several complaints were filed. MPKSY is a very helpful scheme to the farmers that provides lift irrigation and pump sets, but the awareness and utilization are much less in this village, indicating the lower resilience capacity of the farmers in Borda village. Awareness of PMKSNY and KALIA is almost 100% in the village but only 29% are availing the benefits of KALIA and just 1% are availing the benefit of PM Kisan. It can also be seen that the amount of people availing benefits and awareness status is relatively low for other agricultural schemes. KALIA Yojana is significantly helping the farmers in risk management, followed by PMFBY and PMKSNY. It is also reported that only 29% of people are satisfied with the KALIA Yojana and 6% of people with PMKSNY and PMFBY.

5.4 Conclusion

In Borda village, almost thirty percent of the population directly makes their living from agriculture. The community is particularly vulnerable to drought in the Bhawanipatna block and has historically experienced drought for a very long time. Farmers rely on monsoon rains to cultivate their crops because they live in unirrigated areas; if unpredictable rainfall occurs, they may experience drought conditions. The rough terrain and high land make it difficult to connect irrigation facilities. The terrain is primarily composed of red, sandy soil that has a low water-holding capacity. The ongoing agricultural loss brought on by the drought and the scarcity of resources for managing it. The village is next to a river, but crop cultivation is not possible with river water. Thus, the severe water shortage in the area is the cause of the agricultural drought. Within their community, the vast majority of people depend on MGNREGA for daily wage labor and their means of subsistence. But the scarcity of employment puts them in a precarious position. The employment of agricultural machinery also eliminates the agricultural laborer's source of income. The region has also witnessed forced migration as a result of inadequate sources of income. Ten homes with male members who travel to different states for work, including Gujarat, Tamil Nadu, Andhra Pradesh, Kerala, and Maharashtra, were identified by the study.

The family's food needs are met by more than 23% of the household's farm output for six to nine months. The annual salary of almost 75% of people is insufficient to cover their families' needs. In order to manage the family's needs during deficit months, households borrow food grains, spend from savings, take out loans, and reduce consumption. 95% of the population in Borda village makes less than INR 5000 a month for five to six months out of the year. The maximum number of persons in Borda earn between INR 50,000 and INR 100,000 annually on average. Over 91% of the population has taken out bank or group loans. The loans are intended for use in business, agriculture, and health care.

Using agricultural machinery comes with a number of concerns, including harm to human health, deterioration of soil fertility, and death of farm animals. Modern technology and farm mechanization are helpful to farmers when it comes to labor shortages, time constraints, and simpler farming methods. But it also results in grain loss when harvesting. Grain loss is said to be at its highest when harvesting with a combined harvester equipment. A lot of paddy plants may be swiftly chopped by the combined harvester machine, but if the cropland is wet, has undulating terrain, or is low land upland and the paddy plants are laying on the ground, the procedure may result in grain loss. Using contemporary variations could be a long-term approach to risk control. High-yield seed types were introduced during India's green revolution and have since increased productivity and production. The goal of modified seed and variety diversification is to boost yield while demonstrating resilience against a range of hazards, including pest infestation, drought, and flooding. However, the local market's embrace of contemporary verities is not living up to the expectations that were set forth. It has also been noted that the contemporary varieties found in the local market are neither resistant to pests and illnesses, nor are they resistant to drought or flooding. The use of pesticides and fertilizers is becoming increasingly important in paddy production techniques. It is said that paddy crops, particularly the more recent types, do not grow well without fertilizers. It is inevitable for farmers to employ chemical fertilizers in order to increase yield. The cost of production goes up when the volume of inputs goes up. Youth migration is rising in the area since there are few options for a living and low pay in the agricultural sector when compared to other industries. For economic reasons, young people and laborers are moving to states like Maharashtra, Gujarat, Kerala, Tamil Nadu, and Karnataka. Farmers are finding it more and more difficult to finance their crop cultivation as the cost of agriculture production rises daily. Farmers are knowledgeable about several ways to adjust to calamitous circumstances. The district's community deals with a drought scenario almost annually. The majority of populations are extremely vulnerable because of disasters, illiteracy, disadvantaged castes, and protracted poverty. They are capable of overcoming such obstacles and surviving. The region that needs improvement is when risk management instruments including input facilities, seeds, crop insurance, irrigation systems, and disaster loss compensation are not functioning as intended.

The district's community deals with a drought almost every other year. Due to protracted poverty, disadvantaged castes, ignorance, and villagers' natural disasters, most communities are extremely susceptible. Farmers rely on monsoon rains to cultivate their crops because they live in unirrigated areas; if unpredictable rainfall occurs, they may experience drought conditions. In the areas where community-based planning and strategy adoption can improve the effectiveness of risk management measures, such as crop insurance, irrigation facilities, disaster loss compensation, seeds, and input facilities, they are not functioning as intended. Community organizations are eager to work hard, take calculated chances, and create resiliency plans. They still lack the knowledge, instruction, and support structure needed to understand and act.

The village is mainly lacking in agriculture program awareness and utilization. To increase the resilience of the communities, training facilities and awareness-raising are required. To create a resilient community, a community-based catastrophe risk reduction plan is required. Community organizations should be given resources and training on small business, livelihood generation, and contemporary farming techniques and technology. The village community organizations should be given a unique subsidy so they can create their own strategies for disaster risk reduction, which could increase the community's resilience.

Acknowledgements This project is funded by the Indian Council of Social Science Research (ICSSR) doctoral fellowship.

References

Aven T (2021) On some foundational issues concerning the relationship between risk and resilience. Special Ser Syst Risks Risk Anal 42(9). https://doi.org/10.1111/risa.13848

BDRCS, Bangladesh Red Crescent Society (2022) Baseline study report on integrated flood resilience program: phase two. Disaster Risk Management Department, Bangladesh Red Crescent Society (BDRCS), 684–686, Red Crescent Sarak, Bara Moghbazar, Dhaka-1217, Bangladesh

Birkmann J (2007) Risk and vulnerability indicators at different scales: applicability, usefulness and policy implications. Environ Hazards 7(1):20–31. https://doi.org/10.1016/j.envhaz.2007.04.002

Birthal PS (2022) Climate change and risk management in Indian agriculture. NABARD Research and Policy Series No. 4/2022. Department of Economic Analysis and Research. National Bank for Agriculture and Rural Development, Mumbai. https://www.nabard.org/auth/writereaddata/tender/2007223845Paper-4-Climate-and-Risk-Management-Dr.-Birthal.pdf

DAFPO (2020) 5 decades of Odisha agriculture statistics—2020. Directorate of agriculture and food production, Odisha. Government of Odisha. https://agri.odisha.gov.in/sites/default/files/2021-06/Five%20Decades%20of%20Odisha%20Agriculture%20Statisticss%20%282%29.pdf

Das S, Paul S (2021) An assessment of cultivators' perceptions about climate change and it's adaptation practice in agriculture of cooch henar sardar subdivision, West Bengal India. Appl Ecol Environ Sci 9(2):271–279. http://pubs.sciepub.com/aees/9/2/19

Diehl M, Hay EL, Chui H (2021) Personal risk and resilience factors in the context of daily stress. Annu Rev Gerontol Geriatr 32(1):251–274. https://doi.org/10.1891/0198-8794.32.251, https://www.ncbi.nlm.nih.gov/pmc/articles/PMC3462024/

Dohlman E (2020) Economic research service. Retrieved 9 Mar 2020, from https://www.ers.usda.gov/topics/farm-practices-management/risk-management/risk-in-agriculture

Etkin D (2016) Disaster theory: an interdisciplinary approach to concepts and causes. Elsevier's

Haque AKE, Mukhopadhyay P, Nepal M, Shammin Md R (2022) Climate change and community resilience insights from South Asia. ISBN 978-981-16-0680-9 (eBook). Springer. https://doi.org/10.1007/978-981-16-0680-9

Hardarker JB, Huirne RBM, nderson AR, Lien G (2004) Coping with risk in agriculture. Way W. Way Wallinggford, Oxfordshire, OX10 8DE, UK. CABI Publishing. Wallingford, U.K.

Harvey CA, Rakotobe ZL, Rao NS, Dave R, Razafimahatratra H, Rabarijohn RH, Rajaofara H, MacKinnon JL (2014) Extreme vulnerability of smallholder farmers to agricultural risks and climate change in Madagascar. Phil Trans R Soc B 369:20130089. https://doi.org/10.1098/rstb.2013.0089

Integrate (2022) What do we mean by risk and resilience? National science foundation (NSF) collaboration between the directorates for education and human resources (EHR) and geosciences (GEO) under grant DUE-1125331. https://serc.carleton.edu/integrate/workshops/risk_resilience/what_is_rr.html

Komarek Adam M, De Pinto A, Smith VH (2020) A review of types of risks in agriculture: what we know and what we need to know. Agric Syst 178:102738

Kumar N, Goyal MK, Gupta AK, Acharya P (2021) Agriculture resilience assessment over central India under climate change. Research case study under the CAP-RES (project funded by DST-GOI under the NAPCC/NKMCC) jointly with IIT Indore.

National Institute of Disaster Management (NIDM), Ministry of Home Affairs, Government of India. https://doi.org/10.13140/RG.2.2.15790.23360. CAP-RES/2020-21/01

Laura G (2012) Risks in agriculture and opportunities of their integrated evaluation. Procedia Soc Behav Sci 62:783–790

Logan TM, Aven T, Guikema SD et al (2022) Risk science offers an integrated approach to resilience. Nat Sustain 5:741–748. https://doi.org/10.1038/s41893-022-00893-w. https://www.nature.com/articles/s41893-022-00893-w

Marks N (2021) The concept of resilience: a new buzzword. CMSWIRE. https://www.cmswire.com/information-management/the-concept-of-resilience-a-new-buzzword/

Mbae JK (2014) Assessing resilient agriculture-based livelihoods: a case of conservation agriculture in Kanthonzweni thesis submitted in partial fulfilment of the requirement for the award of master of arts degree in planning. Department of Urban and Regional Planning

Moret W (2014) Vulnerability assessment methodologies: a review of the literature. United States Agency for International Development (USAID) Cooperative Agreement No. AID-OAA-LA-13-00001

Naik R, Murari KK (2023) Agriculture risk management and resilience building through community-based disaster risk reduction: a case study of Talmala village in Kalahandi district. In: Al Khaddar R, Singh SK, Kaushika ND, Tomar RK, Jain SK (eds) Recent developments in energy and environmental engineering, vol 333. Springer Nature, Singapore, pp 145–157. https://doi.org/10.1007/978-981-99-1388-6_11

OECD (2023) Taking a holistic approach to agricultural risk management. Risk management and resilience. Organisation for Economic Co-operation and Development. https://www.oecd.org/agriculture/topics/risk-management-and-resilience/

Srinidhi A, Werners SE, Dadas D, D'Souza M, Ludwig F, Meuwissen MPM (2023) Retrospective climate resilience assessment of semi-arid farming systems in India. Int J Water Resour Dev. https://doi.org/10.1080/07900627.2023.2207680

Talukder B, Blay-Palmer A, vanLoon GW, Hipel KW (2020) Towards complexity of agricultural sustainability assessment: main issues and concerns. Environ Sustain Indicators 6:100038. ISSN 2665-9727. https://doi.org/10.1016/j.indic.2020.100038. https://www.sciencedirect.com/science/article/pii/S2665972720300209

Vanderbilt-Adriance E, Shaw DS (2008) Conceptualizing and re-evaluating resilience across levels of risk, time, and domains of competence. Clin Child Family Psychol Rev 11(1–2):30–58. https://doi.org/10.1007/s10567-008-0031-2, https://www.ncbi.nlm.nih.gov/pmc/articles/PMC2683037/

van der Lee J, Kangogo D, Gülzari ŞÖ et al (2022) Theoretical positions and approaches to resilience assessment in farming systems. A review. Agron Sustain Dev 42:27. https://doi.org/10.1007/s13593-022-00755-x

Wang J, Abdur R (2016) Risk management in agriculture: theories and methods. Science PG. Science Publishing Group. 548 Fashion Avenue New York, NY 10018, U.S.A. ISBN: 978-1-940366-69-2. http://www.sciencepublishinggroup.com

Part II
Communication in Disaster Management

Disaster Management in India: An Assessment Study for Disaster Communication

6

Alisha Pathak

Abstract

The relevance of communication in our day-to-day lives has long been established. There is hardly any aspect of our day that's devoid of communication in one or the other form. In a country like India that is highly prone to disasters both natural and man-made, disaster management becomes an important area of interest both for academicians and policymakers. This study presents a perspective on how communication can aid in enhancing the efficiency of disaster management across the country during disaster preparedness, mitigation, resilience and recovery. It diverges from the dominant pattern of analyzing the role of social media as evident in available literature and goes on to conduct a contextual analysis of the terms 'communication' and 'media' in the documents issued by two major bodies involved in disaster management namely National Disaster Response Force and National Institute of Disaster Management. The objectives of this study are to explore the role of communication and media in policies related to disaster management in India and to suggest measures to improve communication in disaster management. The researchers adopt an interpretive paradigm and use triangulation method to collect and analyze data. The reports and policy documents available on the websites of NDRF and NIDM were analyzed using content analysis and the findings were discussed using thematic analysis in the subsequent section. The findings of this study suggest that communication is used in disaster management but mostly in infrastructural context. It is regarded as an infrastructural element more than a practice. Also, the role of media in disaster management is limited post infrastructural damages that cut off the community from authorities. This further suggests that involvement of communication strategies can be of vital importance in improving our existing disaster management performance and help us mitigate losses at different levels that disasters cause every year.

Keywords

Communication · Community · Disaster management · Media · NDRF · NIDM

A. Pathak (✉)
Department of Mass Communication and Media, Central University of South Bihar, Gaya, India
e-mail: alishapathak@cusb.ac.in

S. Talukdar et al. (eds.), *Communication, Science, Technology and Innovation in Disaster Risk Management*, GIScience and Geo-environmental Modelling, https://doi.org/10.1007/978-3-031-77189-7_6

6.1 Introduction

India constitutes 18.02% of world's population making itself the most populous country in the world. As per UN estimates, the population of India in 2024 is approximately 1.44 billion. The land of diversity shows variation in different forms. Due to its geographical features, topography and population, India is one of the ten most disaster-prone nations across the globe. 27 out of 29 states and nine union territories in India are affected by disasters such as earthquakes, floods, cyclones, landslides and droughts (UNICEF). In World Risk Report 2023 that ranks 193 countries based of the risk they possess for disaster published by Bündnis Entwicklung Hilft and the Institute for International Law of Peace and Armed Conflict (IFHV) at Ruhr University Bochum, India ranks third with only the Philippines and Indonesia ranking higher as first and second, respectively (Auer Frege et al. 2023).

Being at such severe risk of disaster possesses a threat to the overall development of the country thereby making disaster management one of the most important issues that require immediate attention by government as well as other stakeholders. The disaster management system in India has had a recent history. Unlike America that had its agency for Federal Emergency Management Agency in 1978, India got its National Institute of Disaster Management in 1995 and National Disaster Management Act in 2005 (Gilbert 2014).

Despite having a separate division for communication, there have been instances of communication failure on and off the record by individuals involved in disaster management. A better use of media and communication as a tool can aid in improving the disaster management system of the India. A firm disaster communication ecology if developed can play a vital role in mitigating the hazards of disasters in the country. It consists of communication resources at multiple levels. At the micro-level, interpersonal connections such as one's family, friends and acquaintances provide tangible and intangible resources majorly in the form of social support (Arnberg et al. 2012; Kaniasty 2019) to help one manage uncertainty and adversity emerging alongside a disaster. Meanwhile, news media and community-based organizations are two meso-level components in one's disaster communication ecology. News media, particularly local media, are vital in providing relevant information about the community (Waldman 2011). A proper network of professionals, experts and community is missing in India and that often aggravates the consequences of disasters resulting in hefty losses despite enormous efforts put in by the authorities. This paper tries to evaluate the status of disaster communication in disaster management endeavors of India and suggest measures that can be of use in reducing the impact of disasters.

6.2 Research Methodology

The researcher adopted an interpretative paradigm and triangulation approach to deduce the findings from the collected data. The method of content analysis was adopted where reports and policy documents from NDRF and NIDM were collected and analyzed. The frequency of terms 'communication' and 'media' were taken as the unit of analysis. The findings are transcribed and reported by the researcher using thematic analysis.

6.3 Result and Discussion

6.3.1 Communication as a Tool for Disaster Management

The findings of this study can be discussed under two major themes as deduced from collected data as shown in Table 6.1.

In Disaster Management Act 2005, the term communication is used eight times. In the aforesaid document undertaken for analysis, the term is used mostly to refer to communication system that denotes an infrastructural context of communication. Along with this, the document bestows the duties of information dissemination to the public and emergency communication but

Table 6.1 Frequency table for the term 'communication'

Name of the document	Issuing organization	Year	Total number of pages	Frequency of the term 'communication'
Disaster management Act 2005 (The Gazette of India 2005)	Ministry of Law and Justice (GoI)	2005	29	08
National policy on disaster management, 2009 (National Disaster Management Authority 2009)	NDMA	2009	44	16
Women's role in disaster risk management (Kumar and Soni 2023)	NIDM	2023	124	04
Documentation on dam leakage, breaching and disaster management at Karam Dam, Madhya Pradesh (Parkash et al. 2023)	NIDM	2023	72	27
Kashmir floods 2014: recovery to resilience (Khan 2023)	NIDM	2023	160	15
NDRF brochure	NDRF	N/A	20	00

doesn't outline any specific method or tool to be used for these purposes by authorities involved in disaster management.

National policy on disaster management 2009 uses the term communication sixteen times in the document. This policy mainly discusses the use of Information and Communications Technology (ICT) for disaster preparedness and disseminating information among the affected community. It states that:

> Communication and sharing of up-to-date information using state-of-the-art IT infrastructure remain at the heart of effective implementation of the disaster management strategy.

It suggests establishment of a National Emergency Communication Network and Emergency Operations Centre at national, state, metro and district levels that will be equipped with contemporary technologies and ensure last-mile connectivity for disaster-hit areas. However, it doesn't specifically emphasize the interventions that can be applied in order to improve communication before, during and after disaster.

In another document titled Women's role in disaster risk management published by NIDM, the word 'communication' is used thrice. It discusses the role of women in hilly areas and how their awareness about disaster management is crucial for the community living around. It talks about Malekudiya tribe and states that:

> Among the Malekudiya tribe the responsibility for taking care of the family is on women. However, most of the time, women are unaware of disaster management methods, as their interaction and communication outside their community is limited.

Any measure to cater to the information and communication needs of these women in the tribe is nowhere reported in the document hinting that there is a lack of proper channel in these areas to serve specific needs of the community (Kumar & Soni 2023).

Documentation on Dam Leakage, Breaching and Disaster Management at Karam Dam, Madhya Pradesh is another case study issued by NIDM in 2023 (Parkash et al. 2023). It highlights the course of and background of the disaster that took place in 2022. This document discusses two types of notifications for disasters related to dam breakage: "*The first one should include communication of an alert situation indicating that although failure or flooding is not imminent, a more serious situation could occur unless conditions improve. The second type of notification should include an alert situation followed by a warning situation. A warning situation should indicate that flooding is imminent due to expected dam failure. It should normally include an order for the evacuation of delineated inundation areas.*"

The grave consequence of failure of communication in such disaster-hit areas is evident in the statement that states, "*The communication network was very poor on the dam site. The mobile phone of SDM, Kukshi was the only point of contact between the headquarter and the dam top. He used to run between the spillway and the cutting channel to pass the information.*"

Kashmir floods 2014: Recovery to Resilience report published by NIDM mentions communication fifteen times (Khan 2023). It presents a case study of one of the worst floods in a century in India. This document discusses the infrastructural aspect of communication. It highlights the role of army personnel and other groups involved in restoring the broken communication channels by clearing roads and building bridges. This document also talks about the restoration of telecommunication services in the area and suggests a plan that can help the country in preventing such massive losses by floods in other parts of the country. The case study compiles both primary and secondary data to investigate and report how Kashmir went from recovery to resilience from 2014 to 2019. While highlighting disaster preparedness, it states that:

> Early warning system usually has four key components: Risk knowledge, monitoring and warning services, dissemination and communication and response capability.

However, it doesn't give a detailed outlook on how risk knowledge and information dissemination as well as communication could be carried out for community in disaster-prone areas.

Lastly, the brochure issued by National Disaster Response Force was analyzed to see frequency of the word 'communication' in its content but it was found that the brochure doesn't mention 'communication' anywhere. Although NDRF is entitled for disaster response and comprises best paramilitary forces of the country, yet, the importance of communication in its functioning seemed negligible from this document. However, from the NDRF website, the researcher was able to find the initiatives that NDRF has taken across the country to train various stakeholders and communicate with them.

6.3.2 Role of Media in Disaster Management

Disaster Management Act 2005 mentions media only twice in the document in its miscellaneous provisions as shown in Table 6.2. Instructing use of media in disaster management, Section 67 of Chap. 11 in this act states that:

> The National Authority, the State Authority, or a District Authority may recommend to the Government to give direction to any authority or person in control of any audio or audio-visual media or such other means of communication as may be available to carry any warning or advisers regarding any threatening disaster situation or disaster, and the said means of communication and media as designated shall comply with such direction.

National policy on disaster management 2009 uses media ten times in its document. One of the objectives of this policy as enumerated is to promote productive and proactive partnership with media for disaster management. It suggests to use both print and electronic media optimally in order to fulfill this objective. "*Effective partnership with the media will be worked out in the field of community awareness, early warning and dissemination, and education regarding various disasters.*"

This policy also talks about Incident Command System, which is a management system involving professionals to respond appropriately to various needs of disaster management. The text quotes:

> It will provide for specialist incident management teams with an incident commander and officers trained in different aspects of incident management, such as logistics, operations, planning, safety, media management, etc.

In another report released by NIDM highlighting the role of women in disaster management, the

Table 6.2 Frequency table for the term 'media'

Name of the document	Issuing organization	Year	Total number of pages	Frequency of the term 'media'
Disaster management act, 2005 (The Gazette of India 2005)	Ministry of Law and Justice (GoI)	2005	29	02
National policy on disaster management, 2009 (National Disaster Management Authority 2009)	NDMA	2009	44	10
Women's role in disaster risk management (Kumar and Soni 2023)	NIDM	2023	124	00
Documentation on dam leakage, breaching and disaster management at Karam Dam, Madhya Pradesh (Parkash et al. 2023)	NIDM	2023	72	14
Kashmir floods 2014: recovery to resilience (Khan 2023)	NIDM	2023	160	04
NDRF brochure (https://ndrf.gov.in)	NDRF	N/A	20	01

term 'media' was nowhere mentioned denying any sort of relationship between the role women play for disaster management and media itself. This suggests a significant gap in the application of media for disaster management.

Documentation on Dam Leakage, Breaching and Disaster Management at Karam Dam, Madhya Pradesh gives a detailed account of role of media and discusses media management practices in disaster management (Parkash et al. 2023). It highlights the successful contribution by media in disseminating correct information to the mass. It also mentions the common practice of misinformation circulation in disaster-hit areas and how social media misuse was tackled by establishing a direct virtual communication channel between the chief minister of state and public.

Kashmir Floods 2014: Recovery to Resilience mentions how Emergency Operation Centers (EOCs) were established in Kashmir to provide alerts using different types of media including print, electronic, social media, etc. It also suggests that a separate social media team was created by Department of Health Services Kashmir in order to update and directly communicate with public about the ongoing developments in Automatics Central Lab.

Communication is not only about a solid infrastructure that ensures smooth telecommunication services and uses of ICTs. It is rather an art that's acquired with experience, studying and observation as well as a science that helps predict outcomes at times (Pavel Slutskiy 2021). The way policies are framed in India for disaster management suggests a loophole in realizing the potential of communication as a tool. The training modules and programs being used currently are devoid of any specific communication specialists' role in disaster communication (Chaturvedi 2022; Parkash et al. 2023). Theoretically designing tools and content for training are bound to be of minimal use unless individuals learn to practice communication effectively in disaster preparedness, mitigation, recovery and resilience. By not giving ample consideration to disaster communication, we are hindering its significance in building up a better disaster management system.

In the documents analyzed above, use of print, electronic and social media was evident but when all the channels of communication are damaged, it is a herculean task to disseminate information among public (Carey 2003). This is where the role of communication specialists must come into picture. An individual who is well trained, equipped and assigned the duty of only communicating with the community is bound to be more productive and effective at their work than a team that is involved in more than one course of action in disaster management. Traditional media as well as interpersonal channels can be optimally used to convey information provided it is given significant attention. This is the high time that we begin accepting the relevance of communication in disaster management and stop seeing various tools of media only as a tool to increase panic. A systematically designed mechanism that specifically caters to disaster communication will not only aid in disaster management but also ensure that the issues of misinformation are tackled appropriately in the process. Therefore, the study proposes the framework acknowledging their roles and incorporating communication specialists that is shown in Fig. 6.1.

Due to temporal limitations, the study had to confine itself with limited set of documents found relevant to objectives but an extensive study of training modules that NIDM and NDRF use to prepare individuals for different stages of disaster management operations shall result in more validating gaps that will not only establish the lack of communication element in the process but also shed light on how integration of communication in the modules can improve their performances.

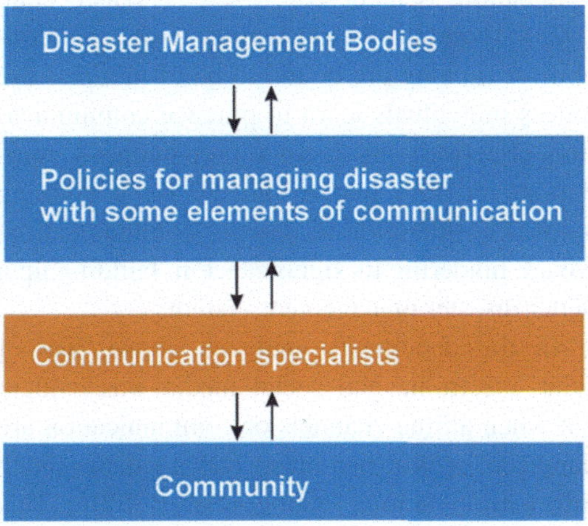

Fig. 6.1 Proposed framework for communicating disaster in India

6.4 Conclusion

From the above findings and discussion, it can be concluded that policies for disaster management focus on communication infrastructure before, during and after disaster. They train different stakeholders related to disasters across the country. Organizations like NIDM and NDRF aid in creating a strong and impactful disaster coping mechanism that has proven to be helpful in mitigating the effects of disaster in India but they fail to acknowledge the need of communication as a disaster management tool and utilise it optimally. The discussions of Information and Communications technology and telecommunication were evident in almost all the documents undertaken for analysis but the context in which 'communication' is used in all these documents is itself a limitation of its application thereby a significant consideration and exploration to exploit communication in disaster management is needed for fostering better co-ordination amongst the affected communities and its various stakeholders while mitigating the impact of disasters in India.

References

Arnberg FK, Hultman CM, Michel PO, Lundin T (2012) Social support moderates posttraumatic stress and general distress after disaster. J Trauma Stress 25(6):721–727. https://doi.org/10.1002/jts.21758

https://nidm.gov.in/PDF/Modules/NIDMSDMP_22.pdf

Auer Frege I, Bündnis Entwicklung Hilft, Radtke K, Kienzl P, Weller D, Schneider S, Bliecke V, Bradshaw S, Dijkzeul D, Funke C, Küsters K, Masuch L, Nasreen M, Thielbörger P, Wiegers O, Yaman DK, Zennig K (2023) WorldRiskReport 2023. In: MediaCompany (ed) nativaComm, M Williams-Wendl, L Cohen, S Kirsch (Trans), WorldRiskReport 2023. Bündnis Entwicklung Hilft. https://weltrisikobericht.de/wp-content/uploads/2024/01/WorldRiskReport_2023_english_online.pdf

Carey J (2003) The functions and uses of media during the September 11 crisis and its aftermath. In: Michael Noll A (ed) Crisis communications: lessons from September 11. Rowman & Littlefield Publishers, Oxford, UK, pp 1–16

Chaturvedi S (2022) Training module on management of crowd at large congregations. National Institute of Disaster Management, Delhi, India. p 58

Gilbert A (2014) Top 10 countries with solid disaster management system. https://www.elist10.com/top-10-countries-solid-disaster-management-system/

Kaniasty K (2019) Social support, interpersonal, and community dynamics following disasters caused by natural hazards. Curr Opin Psychol 32:105–109. https://doi.org/10.1016/j.copsyc.2019.07.026

Khan AA (2023) Kashmir floods 2014: recovery to resilience. National Institute of Disaster Management (NIDM), p 158. https://nidm.gov.in/PDF/pubs/KashmirFloodReport_23.pdf

Kumar S, Soni P (2023) Compendium of best practices on women's role in disaster risk management. National Institute of Disaster Management, Delhi, India, p 122. https://nidm.gov.in/PDF/pubs/Women_Compendium_NIDM24.pdf

National Disaster Management Authority (2009) National policy on disaster management. Government of India. https://nidm.gov.in/PDF/policies/ndm_policy2009.pdf

National Disaster Response Force. https://ndrf.gov.in/sites/default/files/NDRF%20Brochure_0.pdf

Parkash S, Sharma R, Menchion WC, Chauhan H, Thapa R, Kaur H, Kathait A, Dogra A (2023) Training module on public health emergency and disaster management-professional development programme (PHEDM-PDP)-Tier II. National Institute of Disaster Management, Delhi. https://nidm.gov.in/PDF/TrgReports/2023/May/Report_30-01May2023sp.pdf

Parkash S, Chandra S, Devada HHK, Batham A, Haldar AL (2023) Documentation on dam leakage, breaching and disaster management at Karam Dam, Madhya Pradesh. National Institute of Disaster Management, Delhi, India. Pages. 70 + xxii. https://nidm.gov.in/PDF/pubs/Karam%20Dam%20NIDM%202023.pdf

Section 67 in The Disaster Management Act, 2005. indiankanoon.org

Slutskiy, P. (2021) Communication art vs communication science. In: Communication and libertarianism. Springer, Singapore. https://doi.org/10.1007/978-981-33-6664-0_1

The Gazette of India (2005) The disaster management act, 2005 (report no. 53 of 2005). Ministry of Law and Justice (Legislative Department), New Delhi. https://ndma.gov.in/sites/default/files/PDF/DM_act2005.pdf

Waldman S (2011) Information needs of communities: the changing media landscape in a broadband age. Diane Publishing

7 Social Media: Potential Communication Tool During Disaster Response—Cyclone Ockhi and Sri Lankan Floods Using AI and ML

N. A. Hareesh and Balamurugan Guru

Abstract

Social media has been used widely during crises and disasters. The data generated by social media is huge and has the potential for deep analysis. The aim of the study is to find out the social media uses in different phases of disaster situations like pre-, during-, and post-. The study will help in understanding the categories for which social media in particular Twitter is used and also to find out the areas in which attention is needed for further effective communication. ML (Machine Learning) algorithms like Decision Tree, Naive Bayes, and IBK are implemented on the categorised data and the classes are predicted by the algorithms. The Twitter data was collected during Sri Lankan Floods and Cyclone Ockhi which occurred during the year 2017. The collected data is then categorised using Keyword analysis and then the data is analysed for trends of usage. Also, this data is given as input to an open-source visualisation tool called WEKA. It was found that social media had political factors influencing the tweet activity during Cyclone Ockhi. The study also identifies the opportunities and challenges faced in disaster situations using social media like information about missing people which is underutilised by the line departments of government and the potential of social media in disseminating search and rescue measures. The study also tries to list out the critical impacts of social media on the user community.

Keywords

Social media · Machine learning · Cyclone Ockhi · Sri Lankan floods · India

7.1 Introduction

Disasters affect every country and effective disaster management is a global challenge. Information tools that can be used during the disaster situation significantly support and properly guide the government agencies, NGOs, and field personnel including the affected communities. Social media has been used efficiently during crises by some developed countries like the United States of America and Europe whereas others especially developing countries have not yet tapped the tool efficiently in a disaster context. The data generated by social media is huge and has a potential for deep analysis during

N. A. Hareesh
Centre for Geoinformatics, Tata Institute of Social Sciences, Mumbai 400088, India

B. Guru (✉)
Department of Geology, Central University of Tamil Nadu, Thiruvarur 610005, India
e-mail: balamurugan.guru@gmail.com

S. Talukdar et al. (eds.), *Communication, Science, Technology and Innovation in Disaster Risk Management*, GIScience and Geo-environmental Modelling, https://doi.org/10.1007/978-3-031-77189-7_7

disasters, especially for rapid action. Recently, there have been various tools used in disaster situations where taking and sharing pictures and video messages as well as text forms. This information is then uploaded to social media using the internet. Soft sensors are defined as conveying that knowledge to an aggregation structure for collaboration, discussion, dissemination, reporting, and decision-making and it may act as an individual or as part of an ordered or unordered (ad-hoc) network (Hall 2008). Social Media is a platform that significantly connects people wise participation, communication, and interaction among them (Kaplan 2010) and it operates at three different levels namely; individual, technological, and sociological. Each level affects one another and only the technological level is involved in making the information exchange and new forms of communication feasible (Mao 2014). There are different types of social media such as bloggers, Twitter, digital content sharing (Flickr, Pinterest, Youtube, and Instagram), discussion forums (Reddit, Quora), social networking (Facebook, Google+, LinkedIn, WhatsApp, and Telegram) and can be accessed by smartphones, laptops, desktops, and tablets.

The features of disasters can be classified based on four structural dimensions such as disaster event, impact units, and its characteristics. The location, hazard specifications, and characteristics are defined as disaster events. Impacts consist of physical, social, economic, and environmental and also losses caused as a result of the disaster. The social units affected by any kind of disaster can range from individual level to entire society in a specific region or across the country (Kreps 1984). Social media is used in disasters by the affected community for a variety of reasons such as getting updates about the event, personal updates related to family, friends and well-wishers, emergency contact numbers, and information about the ongoing relief works (Vieweg 2012). Social media use during disasters can also be classified based on their role in the organisations. Social media information flows from one citizen to another, citizen to organisations, organisations to citizens, and between organisations (Christian Reuter 2011).

The uses of social media data like Twitter analysed scientifically in the various disaster crises situations like technological failure, Hurricanes and flooding (Vieweg et al. 2010). Generally, the Twitter data was analysed based on short messages that were Twittered during the particular crises especially focusing on geographical locations, hot spots, victim groups, level of damages, numbers of responding tweets, retweets, and broadcasts. Retweets act as an evaluation process for critical information in this environment, ensuring that erroneous tweets are corrected and preventing incorrect information from spreading (Vieweg et al. 2010). There are individual Twitter users, who collect valid information from various sources and share it to help victims (Vieweg et al. 2010). Recent studies show that the collective intelligence of the citizens could help to correctly identify the victims because users were concerned about reliable sources in this particular situation. Recently, there was an analysis of social media especially the use of Twitter during the Chennai floods 2015. The study figured out the usage of Twitter into five categories namely Need for help, Relief measures, Express Gratitude, Complaints, and Others. It also used ML (machine learning) algorithms for the analysis of Twitter data (Meera 2017).

According to Guy et al. (2010), the Twitter Earthquake Detector (TED) programme was based on tweets and scans for (already established) crisis-relevant hashtags. Relevant Tweets were then filtered and preserved. The programme was applied during earthquakes to reduce the 20-min delay between the eruption and the release of scientific data. Machines could read standardised hashtag syntax, which could aid in gathering more pertinent data. In order to get information from citizens, Latonero and Shklovski (2011) thought about monitoring a fire department's Twitter account. Their 'eStoryS' project used social software mash-ups to locate photographs from the Flickr API that were relevant to crises and place them on a map. For the prediction of the classes of social media data, a number of machine learning techniques are frequently utilised. Instance-based learner

(IBK), Decision Tree, and Naive Bayes are the machine learning (ML) algorithms that were employed in this work (Khamar 2013; Rossi 2012; Meera 2017).

The scope of this study is to find out the uses of social media before, during, and after disasters and take it forward to further harness the usage through the output of this study. The study will help in understanding the categories for which social media in particular Twitter is used and find out the areas in which attention is needed for further effective communication. This study also uses machine learning algorithms to predict the output of social media data and tries to find out the best possible algorithm for predicting the output from social media data.

To analyse the data shared on social media (Twitter) during the Sri Lankan Floods and Cyclone Ockhi.

- To find out the different categories where social media is used during disaster situations.
- Analyse the timelines of the usage of social media based on the categories.
- To find out the most influencing user on the Twitter community.
- To analyse the Twitter data using Machine Learning Algorithms and find out the best suitable one.
- To find out the possibilities for utilisation of Twitter data during disaster situations.

7.2 Background of the Study

7.2.1 Sri Lankan Floods

In Sri Lanka, floods occur frequently every year, and the resulting damage gets worse each time. The 2017 floods in Sri Lanka affected 15 districts. The Western, Sabaragamuwa, and portions of the Central Provinces of Sri Lanka were severely impacted by the flooding. Kalutara, Matara, and Ratnaputra were the districts most severely impacted. On 25 May 2017, the southwest monsoon began to spread across the nation, bringing with it heavy rainfall that in some places reached 500 mm in 24 h and 300 mm in others. The water level in the Kelani River at the Glencourse gauging site in the upper catchment area had increased to 15.44 m on May 27 at 9:30 am, while other gauging points were buried in the Nitwala River in the southern region of the country. As of 27 May 2017, eight gauge locations indicated flooding. Putupaula (Kalu River) and Baddegama (Gin River) both experienced alarming increases in water levels. Mudslides occurred along with the rains. Mudslides were started in Kalutara as a result of the river Kalu overflowing. Landslides rendered many areas of Kalutara district's Agalawatte town inaccessible.

7.2.2 Cyclone Ockhi

Cyclone Ockhi started as a deep depression in the south of Kanyakumari and west of Sri Lanka over the Comorin area on 28 November 2017. The deep depressions also lead to heavy rainfall and damage to life and property in Sri Lanka on 29 November 2017. Due to favourable conditions and warmer temperatures in the Bay of Bengal, the depression intensified into a Cyclonic storm within 6 h and stuck Kanyakumari on 30 November 2017. The Cyclone then changed its path and headed towards Lakshadweep in the Arabian Sea. It then intensified into a Category 3 equivalent Cyclone on 2 December 2017. The Cyclone was headed towards the Gujarat coast as per forecasts but it soon dissipated near the southern Gujarat coast on 6 December 2017 due to unfavourable conditions. Cyclone Ockhi was an unusual cyclone due to its peculiar path as it originated in the Bay of Bengal and ended up over the Gulf of Cambay. It caused widespread damage in Maldives, Sri Lanka, and India. In India, there was extensive damage in Kerala, Tamil Nadu, and Lakshadweep. In Goa, many beach shacks were hit due to the high tides. Mumbai and Southern Gujarat witnessed a significant amount of rainfall. The study will focus on the tweets shared during this time and analyse the usage of Twitter and the emerging categories for which Twitter was used during and after the cyclone and floods.

7.3 Materials and Methodology

Twitter data was collected for the events of the Sri Lankan Floods and Cyclone Ockhi. For this, Twitter Archiver tool was used in the extraction of Tweets. The hashtags mentioned in Table 7.1 were used in this study based on extensive searches on Twitter and following the popular hashtags used.

The flow chart displayed in Fig. 7.1 shows that the Twitter Archiver tool is used in the present research for data collection. Twitter Archiver is used to save tweets for any keyword or hashtag from Twitter into a Google Spreadsheet. Simple queries and Boolean search are used to pull the data into a spreadsheet. The tool polls Twitter every hour for new tweets. The data collected is categorised using Keyword analysis and given as input to the **Weka** tool to give the required outputs. Also at the same time, the categorised data is analysed for emerging trends in the categories and also a timeline of usage of social media. The Weka tool consists of many machine learning algorithms and visualising tools for analysing the data. This study uses algorithms namely Decision Tree, Naive Bayes, and IBK to predict the classes or categories of the input data (Twitter). The Performance measures of the three algorithms are then compared to find out the best among them for analysing social media data, especially Twitter.

Table 7.1 Hashtags used

S. No.	Hashtags used
1	#CycloneOckhi
2	#FloodSL

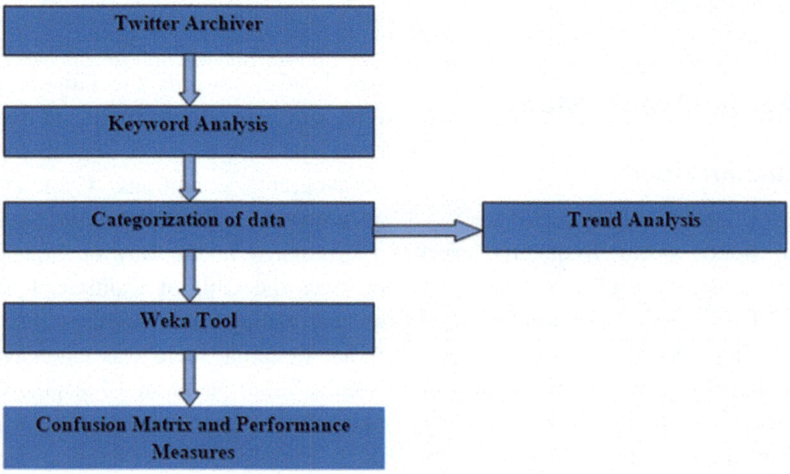

Fig. 7.1 Methodology-flow chart

7.3.1 Details of Information Collected from Twitter

The fields captured from Twitter are as below:

Date—The date and time when the Tweet was made.

Full Name—Full name of the Tweeting person as captured in Twitter.

Screen Name—User name of the Tweeting person.

Tweet Text—The tweet made by the person.

Tweet ID—Unique identifier for every Tweet.

App—The application from which the Tweet was made. E.g. Web client, android, iPhone, etc.

Followers—The number of persons who follow the user.

Follows—The number of persons whom the user follows

Retweets—The number of times the particular Tweet was shared by others.

Favourites—The number of times the particular Tweet was listed as a favourite by other users.

Location—Location from where the tweet was made.

Verified—Whether the user has verified himself/herself on Twitter.

User Since—The date of creation of the Twitter profile.

7.3.2 Methods

7.3.2.1 Weka Tool

For output generation for machine learning algorithms, Weka Tool is used. Weka is a Java-based open-source machine learning programme created by the University of Waikato in New Zealand. The GNU General Public Licence governs the use of this free software. It consists of a set of visualisation tools along with a collection of algorithms for data analysis and predictive modelling. There are many ML algorithms out of which this study focuses on Decision Trees, Instance-based learner (IBK), or K-nearest neighbour algorithm and Naive Bayes algorithm.

7.3.2.2 Decision Tree

A Decision Tree is a type of graph. To show every potential result of a decision, the programme uses a branching mechanism. Recursively and repeatedly dividing is used to build the tree from top to bottom. All the instances or samples are initially placed at the root. They are then recursively partitioned using built-in criteria depending on chosen attributes. There are two nodes: leaf nodes and interior nodes. An internal node contains two or more branches, where each one denotes the test on a particular attribute. The leaf node is the bottom-most node which consists of the resulting class label or decision. This algorithm can handle both categorical and numerical data (Meera 2017). Weka has a Decision Tree algorithm named J48 which is an extension of the C4.5 algorithm. C4.5 has evolved from the ID3 algorithm. ID3 algorithm cannot handle numeric data whereas C4.5 can handle both numerical as well as categorical data. Since Weka is implemented in Java the name was given as J48. The output tree of this algorithm is easy to understand as well as reliable and robust. The categorised Tweets are given as input to the Weka tool and the output is explained later in upcoming sections.

7.3.2.3 Naive Bayes

Naive Bayes is a common classifier that gives much importance to the conditional independence between the predictors. This works based on Bayes theorem. It works on the assumption that the attributes are conditionally independent. Nave Bayes algorithm is useful for high dimensional data sets and these can be built easily without requiring much time. The key feature of this algorithm is it can predict multiple classes. The naive Bayes algorithm works pretty well for categorical data compared to numerical data. The conditional independence assumption works as a limitation for Naive Bayes as it is difficult to get predictors which are completely independent. The categorised tweets are given as input to the Weka tool and the output for this algorithm is explained further in upcoming sections.

7.3.2.4 Instance-Based Learner (IBK) (K-Nearest Neighbours)

This algorithm is called an Instance-based learner with parameter '**k**' or '**k**'-nearest neighbours algorithm (KNN). This algorithm uses '*k*' nearest neighbours to take into account the noisy instances. Unlike the other algorithms, this one does not build a model instead it generates the prediction for every instance under test in less time. For each instance, this algorithm uses a distance measure to locate '**k**' close instances in the data set. Using these selected '**k**' instances, the algorithm makes a prediction. The distance measure by default is calculated using the Euclidean distance which is the sum of squares of the differences between the attribute values. If there is a nominal attribute, the difference is assumed as '1' if the attribute values are different and '0' if they are the same. This algorithm takes more time as it scans the entire data set for every prediction it makes. It assumes that all attributes are equally important. The theoretical assumption in this algorithm is that if the size of the training set or training instances increases, then the value of '*k*' also increases thereby reducing the error in the output. The input file is the categorised tweet data and the output generated is explained further in upcoming sections.

7.4 Data Analysis

The data collected consists of 3777 tweets related to the Sri Lankan floods and 32,897 tweets related to Cyclone Ockhi. The entire data has been categorised using Keyword analysis and divided into categories based on the keywords used in the tweets.

7.4.1 Sri Lankan Floods

The data for Sri Lankan Floods was subdivided into the following 18 distinct categories and Fig. 7.2 shows the total count of Tweets concerning these categories.

1. **Assistance Required**

This category consists of Tweets regarding all kinds of assistance required, like a request for

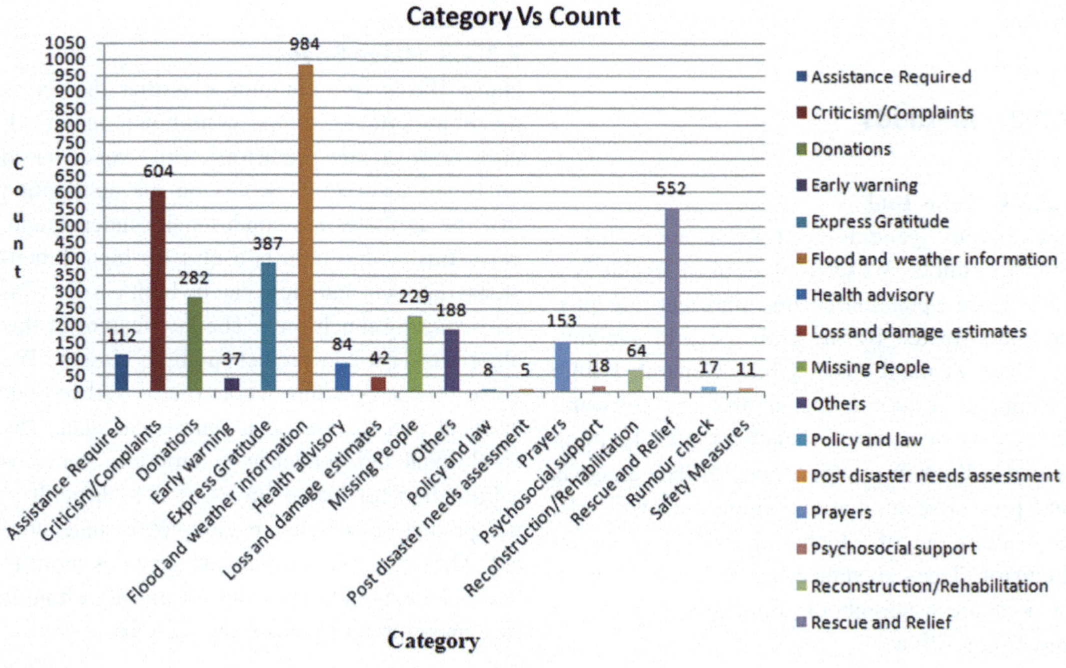

Fig. 7.2 Tweets based on Categories (Sri Lankan Floods)

medicines, water, call for volunteers, navigational assistance, helping in the preparation of a list of relief items, help for providing in kind, and requiring emergency contacts.

2. **Donations**

This category consists of Tweets regarding appeals for donations, and information about an agency/Organisation/individual or country's donations.

3. **Express Gratitude**

This category consists of Tweets that express deep gratitude towards all the persons, countries, agencies, and organisations who wholeheartedly supported the disaster-affected people.

4. **Early Warning**

This category consists of Tweets related to early warning alerts, landslide warnings, tweets with alerts, and flood warnings.

5. **Rescue and Relief**

This category consists of tweets related to rescue and relief measures, updates about the rescue and relief activities, and reviews of rescue activities.

6. **Flood and Weather Information**

This category consists of Tweets related to information regarding water levels, flood updates, weather forecasts and updates, school opening dates, and other information related to making the general public aware of the situation there.

7. **Missing People**

This category consists of Tweets related to information about missing, injured, and deceased people.

8. **Health Advisory**

This category consists of Tweets related to information related to health advisory, disease outbreaks, key needs related to health and hygiene, and priorities of health services.

9. **Safety Measures**

This category consists of tweets related to general safety measures and actions taken or advice related to staying safe.

10. **Rumour Check**

This category consists of tweets related to creating awareness against spreading rumours, informing the public about fake videos, alerting the public about fake accounts asking for donations, and checklists for verifying false claims related to relief.

11. **Loss and Damage Estimates**

This category consists of Tweets related to possible loss and damage estimates due to floods, estimated money required for rectifying the damage, and Sector/Industry-wise loss and damage estimates.

12. **Policy and Law**

This category contains Tweets informing about government policies, new laws being proposed, decisions taken by the government to punish unauthorised builders, and implementation of recommendations.

13. **Post-disaster Needs Assessment**

This category includes Tweets about post-disaster needs assessment (PDNA), and workshops of PDNA.

14. **Prayers**

This category consists of Tweets about prayers to flood victims and people in Sri Lanka, expressions of solidarity, positive thoughts for the victims, and gestures of respect for the people affected.

15. **Psychosocial Support**

This category consists of Tweets related to spreading awareness about Psychosocial Support post-disaster, and mental health support.

16. **Reconstruction/Rehabilitation**

This category consists of Tweets related to rehabilitation and reconstruction activities, funds for rebuilding, and support for rebuilding.

17. **Criticism/Complaints**

This category consists of messages criticising the government's efforts, negative comments about new acts proposed, complaints about Fund agencies/Organisations, corruption complaints, complaints about warning systems, fraud during relief, and other trolls and negative comments.

18. **Others**

This category contains all other messages like general comments, opinions, trends, and general information.

7.4.2 Cyclone Ockhi

The data related to Cyclone Ockhi was categorised into 21 distinct classes. Fourteen categories are common to both Sri Lankan Floods and Cyclone Ockhi. They are from S. No. 1 to 14 in Table 7.3. Figure 7.3 illustrates the various categories along with the count.

The remaining 7 distinct categories are as follows.

1. **Information Dissemination**

This category consists of tweets related to weather forecasts, rain updates, news related to Cyclone Ockhi, visualisation of the cyclone, information related to routes or blockage, school reopening information, and any other information which are pertaining to the cyclone or rallies which were cancelled due to the cyclone. In essence, this category deals with tweets which are intended for general information related to cyclones and informing the public about current happenings that are relevant.

2. **Loss and Damage Assessment**

This category includes Tweets about the losses and damages caused due to the cyclone, Assessment of the damages, loss of property, and loss of livelihood.

3. **Preparedness**

This category consists of Tweets related to appeals for preparedness, contingency plans, and preparedness measures and reviewing of preparedness measures.

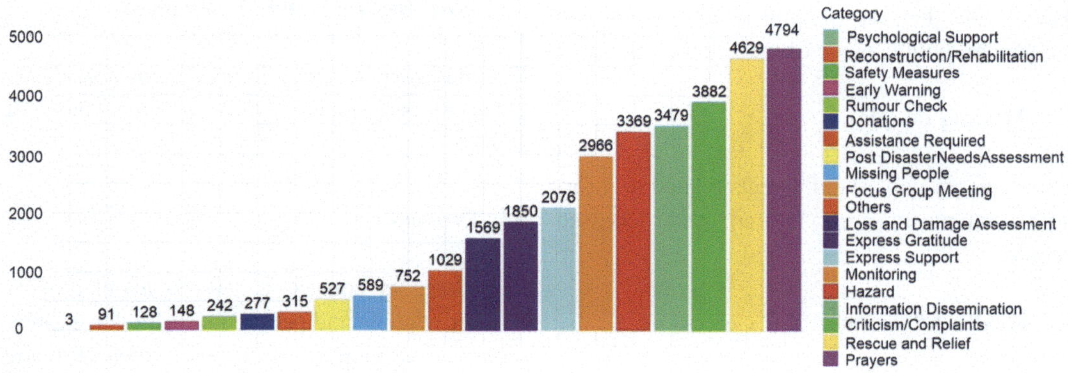

Fig. 7.3 Tweets based on categories (Cyclone Ockhi)

Table 7.2 List of categories

S. No.	Sri Lankan floods	Cyclone Ockhi
1	Assistance required	Assistance required
2	Donations	Donations
3	Express gratitude	Express gratitude
4	Early warning	Early warning
5	Rescue and relief	Rescue and relief
6	Missing people	Missing people
7	Safety measures	Safety measures
8	Rumour check	Rumour check
9	Post-disaster needs assessment	Post-disaster needs assessment
10	Prayers	Prayers
11	Psychosocial support	Psychosocial support
12	Reconstruction/rehabilitation	Reconstruction/rehabilitation
13	Criticism/complaints	Criticism/complaints
14	Others	Others
15	Flood and weather information	Information dissemination
16	Loss and damage estimates	Loss and damage assessment
17	Health advisory	Preparedness
18	Policy and law	Monitoring
19		Hazard related
20		Express support
21		Focus group meeting

4. **Monitoring**

This category consists of Tweets related to monitoring the situation and monitoring activities done by the government and other agencies.

5. **Hazard Related**

This category consists of Tweets which contain hazard-related terms such as landfall, speed and position of the cyclone, disaster, and deep depression.

6. **Express Support**

This category consists of Tweets that express support for the people, assure the support of the government, and express standing along with the public.

7. **Focus Group Meeting**

This category consists of tweets that include meetings with victims of a particular area by political leaders, agencies, and government officers.

Prayers and Rescue and relief had a high percentage of tweets. This was followed by Criticism, Information dissemination, and Hazard-related tweets. The tweets from the Monitoring category followed next along with Expressing Support and Gratitude and Loss and Damage Assessment. All the other categories like Assistance Required, Donations, Early Warning, Focus group meeting, PDNA, Preparedness, Rumour check, and safety measures had comparatively less percentages of tweets. The categories like Reconstruction/Rehabilitation, and Psychosocial support had less than 100 Tweets needing more attention for creating awareness on these topics.

7.5 Results

7.5.1 Sri Lankan Floods

In the first ten days (31 May–9 June 2017) of collection of data, 1 June had the most tweets with 629 and 17% of the tweets were posted on that day. 2 June 2017 had the next highest tweets with 486 tweets (13%). 3 June had 400 tweets amounting to 10.6% of the total tweets collected. 05 and 06 of June had 221 and 200 tweets respectively. These 5 days displayed the highest traffic in terms of tweets in the period of data collected. These 5 days amounted to 51.25% of the total tweets. The first 10 days recorded a tweet count of 2637 which is almost 70% of the tweets. Only 30% of tweets were recorded from 10 June to 08 October. This shows that as the disaster strikes the tweet activity is at its highest and farther the date from the event the tweet activity decreases.

Table 7.3 List of Tweets with high retweet count (Sri Lanka floods)

S. No.	User	Tweet	Retweets
1	India in Sri Lanka	Indian Navy saving precious lives making sure no one is left behind #FloodSL @narendramodi @MEAIndia @SushmaSwaraj @DrSJaisha	714
2	Mohamed Nasheed	Our thoughts are with the victims and displaced families affected by #FloodSL. Commend the ongoing relief effort	245
3	Azzam Ameen	Reached Veyangalle. Agalawatta after a 5 KM boat ride. place still under more than 10 feet underwater #FloodSL	167
4	Azzam Ameen	Inmates of Galle prison join relief effort by preparing food for flood victims, Pics via Manusha Nanayakkara #FloodSL#SriLanka	165
5	UNICEF	Race to rescue flood victims as toll rises. women & children are particularly affected #FloodSL @AJENews	149
6	Logan Amila Gamage	People who wanna put up your hands and help. We need volunteers and donors. Join in. #FloodSL #Flood2017 #SriLanka #RT #Help...	137
7	Sri Lanka Tweet	#PrayForSriLanka 91 Deaths, 110 missing, 300 houses damaged & 20,000 families affected due to Landslides &#FloodSL#SriLanka#LKA	126
8	Atul Keshap	SL FM #RaviKarunanayake, Adm. Wijeygunaratne & I warmly thanked Captain & crew of USS Lake Erie for #FloodSL relief work in #SriLanka	106
9	India in Sri Lanka	2nd Ship from India INS Shardul offloads all the relief material to aid flood relief operations. #FloodSL https://t.co/LyCqn...	106
10	Sri Lanka Tweet	Massive Kindness & Humanity of Sri Lankans! As a Nation Sinhala Tamil Muslim We All Be Can Proud! Respect #FloodSL...	94

During the first 10 days, most of the tweets were related to Flood and weather information (477), Rescue and relief (481). Criticism/Complaints also started during the initial days of data collection. 473 tweets out of 604 Criticism/Complaints related were tweeted in the first 10 days. Most of the donations-related tweets were shared during this period which is 247 (87.5%). Also, most (94%) of the tweets related to Assistance required were tweeted during the initial days (105 tweets out of 112). 188 tweets during this period expressed gratitude towards the people who supported flood victims. The information related to missing people was shared mostly in this period of 10 days with 224 out of 229 tweets shared.

The tweets were analysed for the retweets attribute and then the source of the highly shared contents was listed as shown in Table 7.2. It is found that tweets from individuals were shared more widely than from organisations/ agencies. Here the Tweet handle India in Sri Lanka got the most retweets, 714. The tweet had contents related to rescue and relief efforts in India during the floods. The next highly retweeted message was that of Mohamed Nasheed with 245 retweets. The Tweet expressed solidarity with the victims of families affected by the floods. Next was Azzam Ameen, a reporter who shared updates about the flood situation and also the rescue and relief measures. Only one tweet from Organisation/Agency UNICEF got high retweets (149) which shared the ground situation of the affected people. Another tweet asking for help related to volunteers and donors had the next higher retweet count with 137 shared from the Twitter handle of Logan Amila Gamage. The next widely shared tweet was that of Sri Lanka

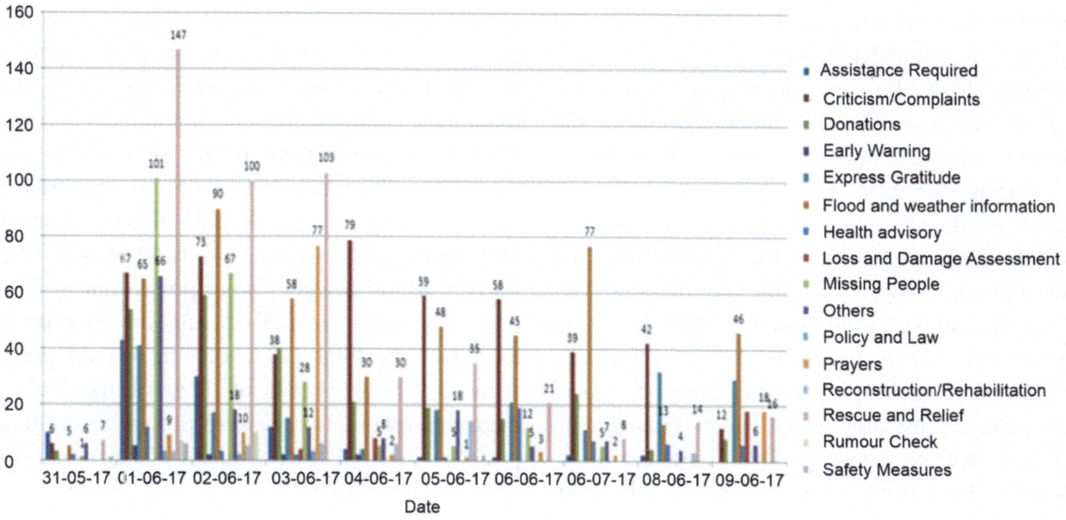

Fig. 7.4 Date-wise Tweet count for first 10 days in Sri Lankan Floods

Tweet with 126 retweets sharing information about the flood situation update with the count of affected people. The next was of Atul Keshap; US Ambassador to Sri Lanka having 106 retweets thanking the crew who helped in rescue and relief measures. Here the individuals had a higher influence on the community than the organisations and the political actors didn't have much influence going by the retweet count (Fig. 7.4).

7.5.2 Cyclone Ockhi

Date-wise analysis of tweet count is given in Fig. 7.5. Maximum tweet activity was observed on 05 December with 12,836 tweets (39%) collected in a single day. The tweet activity decreased gradually after that with 1906 tweets on the next day 06 December 2017. It followed a decreasing trend till 13 December. After this, there was an increase in tweet activity on 14 December with 1759 tweets. After this, the tweet activity gradually declined till 17 December. Again it increased on the 18 and 19 of December due to the visit of the Prime Minister of India to the disaster-affected states of Kerala, Tamil Nadu, and Lakshadweep. During the Prime Minister's visit on 19 December, the tweet count was at 2257. On 20 December, the tweet count was reduced to

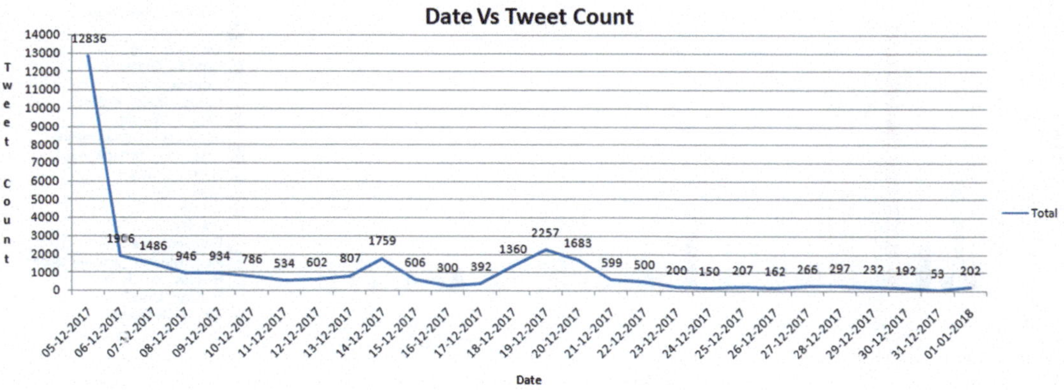

Fig. 7.5 Date-wise Tweet count (Cyclone Ockhi)

1683. Following this, the tweet activity started decreasing gradually. During the period up to 1 January 2018, the tweet activity reached a low of 53 tweets on 31 December 2017 after the search and rescue operations were withdrawn.

During the first 3 days of data collected, most of the Tweets belonged to the prayers category as the Cyclone was approaching and also prayers poured in for the disaster victims in cyclone struck states of Kerala, Tamil Nadu, and Lakshadweep as shown in Fig. 7.6. 3765 tweets belonged to this category. Also, the next highest amount of tweets belonged to Hazard Related category. This contained information about the Cyclone's track, wind speed, landfall path, and other hazard-related terms. As the cyclone was approaching, many people used hazard-related tweets as per the situation and 3208 tweets belonged to this category. All the tweets under the Preparedness category were posted during the initial 3 days indicating that people are aware of the use of social media for Preparedness also. 1652 tweets were posted in criticism or complaints category. The next highest number of tweets belonged to Monitoring Category with 1413 Tweets. Rescue and Relief-related tweets accounted for 1251 as Kerala and Tamil Nadu states still had many fishermen missing and the government was carrying out search and rescue operations. Tweets asking for assistance also were high in this period amounting to 246 out of 315 tweets in this category. 1376 tweets belonged to the Category Information Dissemination informing the public about the possibilities of rainfall, roadblock, news alert, and weather forecasts. People expressed gratitude in 1405 tweets for the cyclone-hit areas and expressed support for the upcoming cyclone in 495 tweets. Almost all tweets (140 out of 143) belonging to the Early warning category were posted during this period. Tweets countering rumours amounted to 238 of the total 242 tweets in this category. Almost 97% (124 out of 128) of the Safety measures-related tweets were posted during this period. The total tweets during the first 3 days were 16,228 (49.3%).

7.5.2.1 Political Factors Influencing Twitter Usage

There was a spike in tweet activity on 14, 19, and 20 December 2017 due to the visits of different National level political leaders to the Cyclone-affected areas. The category-wise tweet count during this period is shown in Fig. 7.7. Most of the tweets during this period belonged to Loss and Damage Assessment and expressed support with 1002 and 913 tweets, respectively. Loss and Damage Assessment contained tweets regarding the visit of the Central team belonging to the National Disaster Management Authority (NDMA) for loss and damage assessment, and estimates of damages. Also during this time, there were tweets from the government assuring

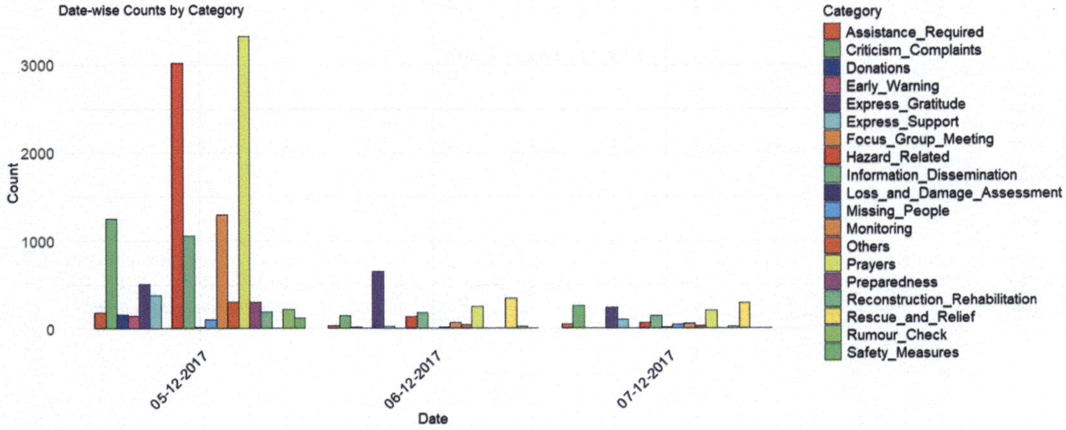

Fig. 7.6 Distribution of categories date-wise

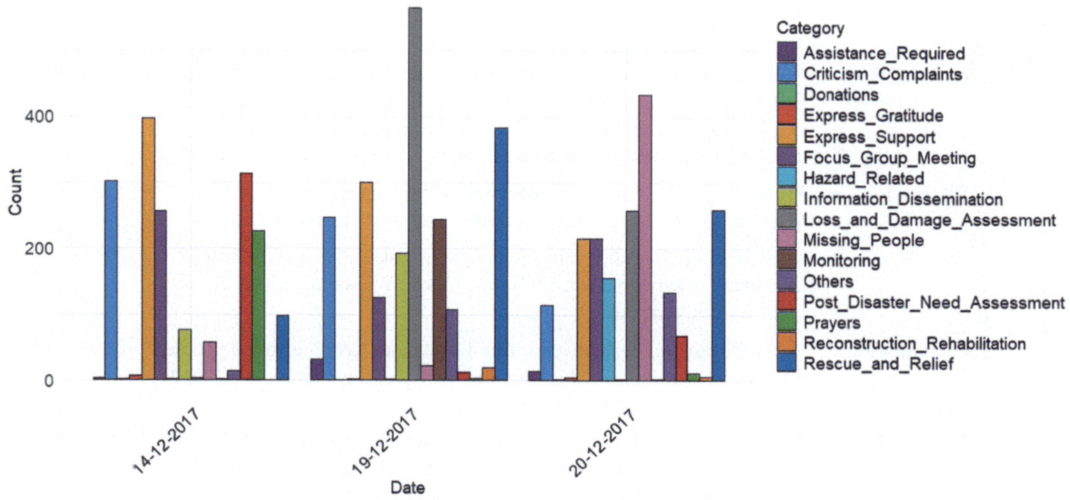

Fig. 7.7 Date-wise category count (middle period)

the state of all the support required for overcoming the damage to negate the criticisms that had been raised. Incidentally, there were 664 tweets pertaining to Criticism/Complaints during this period which was overcome by the support expressed by the leaders and the governments. Focus group meeting-related tweets were 539 in number. This was due to the political leaders' visits and tweets related to meeting the victims of cyclone-hit areas and sharing their grievances in the process. Post-disaster needs assessment-related tweets amounted to 329 in number during this period. There were 740 tweets related to Rescue and Relief with messages related to relief amount, ongoing search and rescue updates, and reviewing of rescue and relief measures by the Prime Minister. Also, there were 379 messages related to Monitoring during this period. The information Dissemination category had 527 tweets about the visits of the leaders and their activities and updates after the cyclone. There were 240 tweets related to prayers for missing and deceased people.

The tweets were analysed based on retweets, the source of these tweets was analysed, and the details are shown in Table 7.4. A source tweet is a tweet from the profile of the user who first posts the tweets. It then gets retweeted and modified based on the content the followers the user has and his/he influence on others. Retweets have **'RT'** at the beginning of the tweet whereas the source doesn't have that. Prime Minister's visit had the highest influence during the period of collection of tweets with his top tweet shared 6379 times during the period of analysis. Congress President Mr Rahul Gandhi's visit also influenced with the highest of 5512 tweets. PMO India the official handle of the PMO office had the highest of 1623 retweets. Next is the official handle of Congress INCIndia with 1069 retweets followed by IAF_MCC with 656 retweets and DDNewsLive with 561 retweets. This shows that the influence of politicians was more during Cyclone Ockhi compared to the Sri Lanka floods.

7.5.2.2 Performance Measures of Machine Learning Algorithms

The categorised data of Sri Lankan floods and Cyclone Ockhi is given as input to Weka. The last field in the file is treated as the class based on which the tool will be predicting the classes for the input. Machine Learning algorithms are inbuilt into the tool and they try to predict the output classes based on the input training and test data. The correctly classified instances are those labels for which the algorithm correctly

Table 7.4 List of Tweets having high retweet count (Cyclone Ockhi)

S. No.	User	Tweet text	Retweets
1	Narendra Modi	With #CycloneOckhi expected to make a landfall in Gujarat, I appeal to @BJP4Gujarat Karyakartas to focus on helping people across the state. Our Karyakartas should devote themselves to providing all possible assistance and stand shoulder to shoulder with fellow citizens	6379
2	Office of RG	My thoughts are with people who will be braving the onslaught of #CycloneOckhi. Urge Congress leaders and workers in Gujarat	5512
3	Office of RG	A salute to bravery shown by the people of Lakshadweep who helped save many fishermen during #CycloneOckhi. We stand with all the victims of this tragedy	5234
4	Narendra Modi	Spent the day in Lakshadweep, Tamil Nadu and Kerala meeting people affected by #CycloneOckhi. I assured them that the entire nation stands firmly with them in these difficult times	4028
5	Narendra Modi	Continuously monitoring the situation arising in various parts of the country due to #CycloneOckhi. Spoke to all relevant authorities and officials. All possible assistance is being provided to those affected	3652
6	Narendra Modi	Since #CycloneOckhi struck, Centre has been monitoring the situation round the clock and ensuring proper rescue and relief operations	3246
7	Narendra Modi	With the influence of #CycloneOckhi waning. I want to compliment Central agencies, various state governments, local administration	2733
8	PMO India	I am here to assure you that India stands shoulder to shoulder with all those affected by #CycloneOckhi. We will leave no stone unturned in the rescue and relief operations: PM @narendramodi to those affected by #CycloneOckhi	1623
9	PMO India	PM @narendramodi met those affected by #CycloneOckhi in Thiru-vananthapuram. https://t.co/abDYUBUQCl	1232
10	INCIndia	The Congress President-Elect and the whole Party stands with those affected by #CycloneOckhi #RahulWithFishermen	1069
11	IAF_MCC	Update on #CycloneOckhi #RescueOperation #Casevac #SavingLives: Fishermens rescued from a capsized boat by the combined effort	656
12	DDNewsLive	PM @narendramodi meets people affected by #CycloneOckhi and reviews progress of relief works in Poonthura. Thiruvananthapuram	561

predicts the class label. Incorrectly classified instances are those whose class names or labels are incorrectly predicted by the algorithm. Three algorithms were used namely Naïve Bayes classifier, Lazy IBK, and Decision Tree algorithm. They are compared based on their Precision, Recall, and *F*-measures.

Table 7.5 illustrates that the Decision Tree is the best among the three classifiers with more than 80% correctly classified instances. The Naive Bayes algorithm correctly classifies 48% of the instances in the case of Sri Lankan floods whereas 78% of the instances were correctly classified in the case of Cyclone Ockhi. This means that when the data set is more Naive Bayes algorithm correctly predicts more instances than when the data set is less in number. IBK algorithm predicts around 50–55% of instances correctly.

The Confusion matrix shows the instances where the algorithms misclassified the class labels to get an idea of the overlapping classes or the details of predicted versus actual classification. Figure 7.8 shows the confusion matrix of the Decision Tree algorithm for the Sri Lankan Floods dataset. Here the misclassified instances were mostly in the category Flood and weather information as this class can cut across other

Table 7.5 Details of correctly and incorrectly classified instances

Prediction of instances	Sri Lankan Floods data set			Cyclone Ockhi data set		
	Naive Bayes	IBK	Decision tree	Naive Bayes	IBK	Decision tree
Correctly classified instances	1826	1977	3110	25,655	18,025	28,736
Incorrectly classified instances	1951	1800	667	7242	14,872	4161
Correctly classified instances in %	48.3	52.3	82.3	77.9	54.8	87.3
Incorrectly classified instances in %	51.7	47.7	17.7	22.1	45.2	12.7

```
   a   b   c   d   e   f   g   h   i   j   k   l   m   n   o   p   q   r   <-- classified as
 130  58   0   0   0   0   0   0   0   0   0   0   0   0   0   0   0   0 |   a = Others
   1 975   0   0   5   0   2   1   0   0   0   0   0   0   0   0   0   0 |   b = Flood and weather information
   0  18  66   0   0   0   0   0   0   0   0   0   0   0   0   0   0   0 |   c = Health advisory
   1 133   0 469   0   0   0   0   1   0   0   0   0   0   0   0   0   0 |   d = Criticism/Complaints
   0 115   0   0 434   0   0   0   0   0   0   3   0   0   0   0   0   0 |   e = Rescue and Relief
   0   5   0   0   0   6   0   0   0   0   0   0   0   0   0   0   0   0 |   f = Safety Measures
   0  89   0   0   0   0 189   0   0   0   0   1   0   3   0   0   0   0 |   g = Donations
   0  46   0   0   0   0   0  66   0   0   0   0   0   0   0   0   0   0 |   h = Assistance Required
   0  51   0   1   0   0   0   0 177   0   0   0   0   0   0   0   0   0 |   i = Missing People
   0   5   0   0   0   0   0   0   0  12   0   0   0   0   0   0   0   0 |   j = Rumour check
   0  17   0   0   0   0   0   0   0   0  20   0   0   0   0   0   0   0 |   k = Early warning
   0  52   0   0   0   0   0   0   0   0   0 335   0   0   0   0   0   0 |   l = Express Gratitude
   0   6   0   0   0   0   0   0   0   0   0   0   2   0   0   0   0   0 |   m = Policy and law
   0  22   0   0   1   0   3   0   0   0   0   0   0  38   0   0   0   0 |   n = Reconstruction/Rehabilitation
   0  19   0   0   0   0   0   0   0   0   0   0   0   0 134   0   0   0 |   o = Prayers
   0   5   0   0   0   0   0   0   0   0   0   0   0   0   0  37   0   0 |   p = Loss and damage estimates
   0   2   0   0   0   0   0   0   0   0   0   0   0   0   0   0  16   0 |   q = Psychosocial support
   0   1   0   0   0   0   0   0   0   0   0   0   0   0   0   0   0   4 |   r = Post disaster needs assessment
```

Fig. 7.8 Confusion matrix of decision tree (Sri Lankan Floods)

classes with flood-related words common across the classes.

Figure 7.9 shows the Confusion Matrix of the IBK algorithm for the Sri Lankan Floods data set and can be observed that the misclassified instances are spread across all the classes indicating that the accuracy of this algorithm is very less and not quite suitable for this kind of data.

Figure 7.10 shows the Confusion matrix of the Naive Bayes algorithm. Here also the misclassified instances are spread across most of the class labels indicating that the accuracy or Precision of this algorithm is less.

Figure 7.11 illustrates the Confusion Matrix of the Decision Tree algorithm for the Cyclone Ockhi data set. Here it can be inferred that most of the misclassified instances belonged to the class label prayers. The diagonal elements of the matrix refer to the correctly classified instances in each class. The prayers category has data which are overlapping with other class labels and hence the algorithm cannot differentiate between the correct and incorrect instances. This can happen as the data from social media contains mixed responses and words cutting across different categories. The prayers for the people also may contain information to the public or criticism at first and hence the misclassification by the algorithm. To make the algorithm refine its precision, large amount of data set may be required to train the algorithm related to the class labels. Still with the huge data and a large number of classes, this precision of more than 88% for this algorithm is far better as the

a	b	c	d	e	f	g	h	i	j	k	l	m	n	o	p	q	r	<-- classified as
74	47	0	15	19	1	7	6	4	0	1	10	0	2	2	0	0	0	a = Others
48	619	14	65	73	1	41	13	16	3	11	47	2	13	7	7	4	0	b = Flood and weather information
1	21	38	2	11	0	2	1	1	0	0	2	0	0	0	0	5	0	c = Health advisory
12	87	1	386	43	1	23	9	7	2	2	12	3	5	8	3	0	0	d = Criticism/Complaints
18	84	8	51	264	2	34	10	28	3	1	27	0	6	13	3	0	0	e = Rescue and Relief
1	1	0	1	2	1	1	0	3	0	1	0	0	0	0	0	0	0	f = Safety Measures
6	55	1	22	35	0	106	10	11	3	0	17	0	6	10	0	0	0	g = Donations
5	21	5	13	13	0	17	28	2	3	0	1	1	0	3	0	0	0	h = Assistance Required
11	44	1	11	32	2	13	3	100	3	1	6	0	1	1	0	0	0	i = Missing People
0	5	0	0	1	0	1	2	4	4	0	0	0	0	0	0	0	0	j = Rumour check
1	18	0	2	2	2	0	0	1	0	7	2	0	0	2	0	0	0	k = Early warning
11	47	3	27	23	1	18	2	5	1	0	237	0	5	7	0	0	0	l = Express Gratitude
0	4	0	1	1	0	1	0	0	0	0	1	0	0	0	0	0	0	m = Policy and law
3	22	0	5	9	0	4	1	3	0	0	6	0	9	0	2	0	0	n = Reconstruction/Rehabilitation
4	16	0	17	17	0	10	4	6	0	3	2	0	0	74	0	0	0	o = Prayers
0	9	1	8	2	0	0	0	0	0	0	1	0	1	0	20	0	0	p = Loss and damage estimates
0	5	4	0	0	0	0	0	0	0	0	0	0	0	1	0	8	0	q = Psychosocial support
0	2	0	1	0	0	0	0	0	0	0	0	0	0	0	0	0	2	r = Post disaster needs assessment

Fig. 7.9 Confusion matrix of IBK algorithm (Sri Lankan Floods)

a	b	c	d	e	f	g	h	i	j	k	l	m	n	o	p	q	r	<-- classified as
56	50	0	19	5	3	23	2	14	6	4	3	2	1	0	0	0	0	a = Others
12	601	10	90	19	23	86	3	29	47	27	12	2	7	0	13	2	1	b = Flood and weather information
1	24	10	7	2	2	18	1	2	4	3	4	2	1	1	0	2	0	c = Health advisory
6	96	0	373	0	12	43	6	13	30	2	3	3	1	6	10	0	0	d = Criticism/Complaints
8	109	7	51	201	7	70	1	25	51	2	8	6	0	3	3	0	0	e = Rescue and Relief
1	2	0	0	1	0	2	0	2	2	0	0	0	0	0	1	0	0	f = Safety Measures
1	39	3	18	2	2	172	2	8	20	0	5	5	0	0	5	0	0	g = Donations
0	16	1	11	6	1	27	6	16	24	1	0	1	0	0	1	1	0	h = Assistance Required
1	32	4	13	3	6	27	1	129	9	0	3	1	0	0	0	0	0	i = Missing People
0	3	0	2	0	0	2	0	3	7	0	0	0	0	0	0	0	0	j = Rumour check
0	10	2	4	0	1	2	0	4	0	13	0	1	0	0	0	0	0	k = Early warning
4	67	8	31	18	4	39	0	6	10	2	189	1	2	1	4	1	0	l = Express Gratitude
0	1	0	1	0	1	2	0	1	1	0	0	0	0	0	1	0	0	m = Policy and law
2	19	2	5	1	3	13	0	4	3	0	1	1	4	1	5	0	0	n = Reconstruction/Rehabilitation
4	25	1	11	3	0	19	3	12	11	11	1	2	0	49	0	1	0	o = Prayers
1	21	0	12	1	0	1	0	0	0	0	0	0	0	0	6	0	0	p = Loss and damage estimates
0	9	1	0	0	0	0	0	0	0	1	0	0	0	0	0	7	0	q = Psychosocial support
0	1	0	0	0	0	0	0	0	0	1	0	0	0	0	0	0	3	r = Post disaster needs assessment

Fig. 7.10 Confusion matrix of Naive Bayes algorithm (Sri Lankan Floods)

a	b	c	d	e	f	g	h	i	j	k	l	m	n	o	p	q	r	s	t	<-- classified as
3225	0	0	0	0	0	0	0	0	0	3	0	0	0	0	141	0	0	0	0	a = Hazard Related
0	1793	0	0	0	0	1	0	0	0	0	0	0	0	0	56	0	0	0	0	b = Express Gratitude
0	0	241	0	0	0	0	0	0	0	0	0	0	0	0	36	0	0	0	0	c = Donations
0	0	0	205	0	0	0	0	0	0	0	0	0	0	0	37	0	0	0	0	d = Rumour Check
0	0	0	0	499	0	0	1	0	0	0	0	0	0	0	89	0	0	0	0	e = Missing People
0	0	0	0	0	2930	0	0	0	0	0	0	0	0	0	36	0	0	0	0	f = Monitoring
0	1	0	0	0	0	1899	0	0	0	3	0	0	0	0	173	0	0	0	0	g = Express Support
0	0	0	0	1	0	0	2876	0	0	1	0	1	0	0	1003	0	0	0	0	h = Criticism/Complaints
0	0	0	0	0	0	0	0	80	0	0	0	0	0	0	48	0	0	0	0	i = Safety Measures
0	0	0	0	0	0	0	0	0	150	0	0	0	0	0	37	0	0	0	0	j = Preparedness
3	0	0	0	0	0	3	1	0	0	2351	1	0	0	0	1122	1	0	0	0	k = Information Dissemination
0	0	0	0	0	0	0	0	0	0	1	4050	0	0	0	578	0	0	0	0	l = Rescue and Relief
0	0	0	0	0	0	0	1	0	0	0	0	489	0	0	539	0	0	0	0	m = Others
0	0	0	0	0	0	0	0	0	0	0	0	0	113	0	30	0	0	0	0	n = Early Warning
0	0	0	0	0	0	0	0	0	0	0	0	0	0	263	52	0	0	0	0	o = Assistance Required
0	0	0	0	0	0	0	0	0	0	0	0	0	0	0	4794	0	0	0	0	p = Prayers
0	0	0	0	0	0	0	0	0	0	1	0	0	0	0	59	1509	0	0	0	q = Loss and Damage Assessment
0	0	0	0	0	0	0	0	0	0	0	0	0	0	0	31	0	60	0	0	r = Reconstruction/Rehabilitation
0	0	0	0	0	0	0	0	0	0	0	0	0	0	0	50	0	0	701	1	s = Focus Group Meeting
0	0	0	0	0	0	0	0	0	0	0	0	0	0	0	19	0	0	0	508	t = Post Disaster Needs Assessment

Fig. 7.11 Confusion matrix of decision tree algorithm (Cyclone Ockhi)

complexities of social media data are high and the tweets are not structured and they express various emotions and meanings which are difficult to predict.

Figure 7.12 shows the confusion matrix of the IBK algorithm for the Cyclone Ockhi data set. We can observe that the misclassified instances for this algorithm are spread across the categories and hence low accuracy for this algorithm.

Figure 7.13 shows the Confusion Matrix of Naïve Bayes algorithm for the Cyclone Ockhi data set. Here the misclassified instances mostly belong to the classes Criticism/Complaints, Information Dissemination, Rescue and Relief, and Post-Disaster Needs Assessment. This has misclassified instances spread across more classes when compared to Decision Tree where only one class had misclassified instances in the majority. Comparing the three algorithms Decision Tree has a less spread Confusion Matrix in terms of misclassified instances and can be fine-tuned further if a comparatively higher volume of data is given in the training set and some categories can be combined and refined further so the overlapping can be reduced and accuracy increased.

7.5.2.3 Precision of Classification

Figures 7.14 and 7.15 display the performance results of the three algorithms for the data sets Sri Lankan Floods and Cyclone Ockhi respectively. The performance parameters are based on Precision, Recall, and *F*-Measure. Precision

a	b	c	d	e	f	g	h	i	j	k	l	m	n	o	p	q	r	s	t	<-- classified as
1901	64	19	5	18	688	67	105	8	20	129	86	10	14	14	138	58	3	12	10	a = Hazard Related
47	852	10	1	6	237	36	38	0	3	17	65	10	4	7	459	43	0	12	3	b = Express Gratitude
16	8	87	0	3	14	9	12	1	0	40	68	2	2	0	1	6	1	7	0	c = Donations
8	1	1	136	0	1	6	31	3	0	27	4	5	0	2	16	1	0	0	0	d = Rumour Check
17	3	5	0	275	8	48	61	0	4	64	56	4	1	10	17	12	1	1	2	e = Missing People
580	162	10	0	6	1503	118	33	1	8	80	136	38	7	0	18	228	0	37	1	f = Monitoring
48	28	10	6	57	103	816	115	2	10	99	132	102	6	8	159	179	10	115	71	g = Express Support
116	41	16	38	59	61	139	2370	5	7	383	222	78	12	40	163	28	10	35	59	h = Criticism/Complaints
11	1	1	2	0	3	1	11	48	6	24	9	6	2	0	3	0	0	0	0	i = Safety Measures
21	1	0	2	4	11	10	7	2	51	44	8	2	4	3	15	1	0	0	1	j = Preparedness
146	27	42	21	67	75	132	407	26	51	1644	446	55	15	20	136	60	14	53	45	k = Information Dissemination
82	53	59	4	46	125	117	197	8	12	420	3200	28	14	14	57	86	26	67	14	l = Rescue and Relief
24	13	2	9	3	25	104	86	7	1	74	42	587	2	8	19	17	1	4	1	m = Others
20	4	1	0	1	13	3	6	2	6	25	13	1	41	0	4	2	0	1	0	n = Early Warning
14	4	0	1	16	3	18	39	0	5	19	16	5	0	143	26	2	0	0	4	o = Assistance Required
175	460	1	12	20	23	236	177	2	19	150	56	11	5	38	3150	4	1	145	109	p = Prayers
37	42	11	0	9	291	168	28	1	0	57	108	22	2	3	2	756	3	26	3	q = Loss and Damage Assessment
2	1	1	0	3	1	10	7	0	1	14	22	0	0	0	1	4	23	1	0	r = Reconstruction/Rehabilitation
14	12	5	0	3	40	76	21	0	0	52	65	4	2	0	122	18	3	281	34	s = Focus Group Meeting
6	5	0	0	1	2	59	42	0	1	41	10	1	0	2	129	1	0	66	161	t = Post Disaster Needs Assessment

Fig. 7.12 Confusion matrix of IBK algorithm (Cyclone Ockhi)

a	b	c	d	e	f	g	h	i	j	k	l	m	n	o	p	q	r	s	t	<-- classified as
2941	5	3	16	4	23	5	95	11	36	60	90	6	28	1	25	4	0	0	16	a = Hazard Related
10	1649	2	9	7	15	2	34	1	5	4	41	35	9	0	17	0	2	0	8	b = Express Gratitude
1	0	148	1	0	0	0	10	1	2	18	73	1	0	0	4	0	0	0	18	c = Donations
0	0	1	185	0	0	0	29	3	1	10	8	2	1	0	1	1	0	0	0	d = Rumour Check
0	0	1	4	309	0	1	67	1	9	52	69	15	10	9	2	0	0	1	39	e = Missing People
13	1	2	3	0	2769	1	22	3	5	18	51	7	8	0	4	3	0	4	52	f = Monitoring
3	0	0	15	13	26	1305	156	2	12	73	165	21	16	3	35	18	4	1	208	g = Express Support
5	0	1	32	10	2	4	3015	13	29	135	220	82	51	3	12	2	2	0	264	h = Criticism/Complaints
6	0	0	4	0	0	1	23	45	4	15	17	1	7	0	2	0	0	0	3	i = Safety Measures
1	0	0	0	1	0	0	13	1	118	19	27	0	3	0	3	0	0	0	1	j = Preparedness
16	1	10	64	26	5	13	469	28	68	1505	596	66	79	5	18	7	2	9	495	k = Information Dissemination
3	1	1	8	10	10	9	96	11	5	227	3955	41	10	2	8	19	1	15	197	l = Rescue and Relief
3	0	0	22	2	11	6	128	9	22	15	59	641	25	1	9	12	2	1	61	m = Others
3	0	0	2	0	0	1	6	0	3	10	22	0	94	0	2	0	0	0	0	n = Early Warning
0	0	0	2	1	0	1	27	3	2	12	17	3	2	218	5	0	0	0	22	o = Assistance Required
13	0	0	6	20	3	8	115	2	3	49	39	71	10	0	4310	1	0	3	141	p = Prayers
5	1	0	0	6	11	3	17	0	0	40	81	9	1	0	0	1382	2	0	11	q = Loss and Damage Assessment
0	0	0	0	0	0	1	8	0	0	10	42	8	1	0	0	1	7	0	13	r = Reconstruction/Rehabilitation
1	0	0	0	2	1	3	8	0	0	15	59	6	0	0	1	0	0	550	106	s = Focus Group Meeting
1	0	0	0	1	0	2	0	0	0	1	6	3	0	0	1	0	0	3	509	t = Post Disaster Needs Assessment

Fig. 7.13 Confusion matrix of Naive Bayes algorithm (Cyclone Ockhi)

Decision Tree			IBK			Naive Bayes			
Precision	Recall	F-Measure	Precision	Recall	F-Measure	Precision	Recall	F-Measure	Class
0.985	0.691	0.813	0.379	0.394	0.386	0.577	0.298	0.393	Others
0.602	0.991	0.749	0.559	0.629	0.592	0.534	0.611	0.570	Flood and weather information
1.000	0.786	0.880	0.500	0.452	0.475	0.204	0.119	0.150	Health advisory
0.998	0.776	0.873	0.616	0.639	0.627	0.576	0.618	0.596	Criticism/Complaints
0.986	0.786	0.875	0.483	0.478	0.480	0.767	0.364	0.494	Rescue and Relief
1.000	0.545	0.706	0.091	0.091	0.091	0.000	0.000	0.000	Safety Measures
0.974	0.670	0.794	0.381	0.376	0.379	0.315	0.610	0.415	Donations
0.985	0.589	0.737	0.315	0.250	0.279	0.240	0.054	0.088	Assistance Required
0.994	0.773	0.870	0.524	0.437	0.476	0.481	0.563	0.519	Missing People
1.000	0.706	0.828	0.182	0.235	0.205	0.031	0.412	0.058	Rumour check
1.000	0.541	0.702	0.259	0.189	0.219	0.194	0.351	0.250	Early warning
0.988	0.866	0.923	0.639	0.612	0.625	0.825	0.488	0.614	Express Gratitude
1.000	0.250	0.400	0.000	0.000	0.000	0.000	0.000	0.000	Policy and law
0.927	0.594	0.724	0.188	0.141	0.161	0.250	0.063	0.100	Reconstruction/Rehabilitation
1.000	0.876	0.934	0.578	0.484	0.527	0.803	0.320	0.458	Prayers
1.000	0.881	0.937	0.571	0.476	0.519	0.122	0.143	0.132	Loss and damage estimates
1.000	0.889	0.941	0.471	0.444	0.457	0.500	0.389	0.438	Psychosocial support
1.000	0.800	0.889	1.000	0.400	0.571	0.750	0.600	0.667	Post disaster needs assessment
0.888	0.823	0.831	0.520	0.523	0.520	0.565	0.483	0.494	Weighted Avg.

Fig. 7.14 Precision, recall, and *F*-measure for three algorithms (Sri Lankan Floods data set)

Decision Tree			IBK			Naive Bayes			
Precision	Recall	F-Measure	Precision	Recall	F-Measure	Precision	Recall	F-Measure	Class
0.999	0.957	0.978	0.579	0.564	0.571	0.972	0.873	0.920	Hazard Related
0.999	0.969	0.984	0.478	0.461	0.469	0.995	0.891	0.940	Express Gratitude
1.000	0.870	0.931	0.310	0.314	0.312	0.876	0.534	0.664	Donations
1.000	0.847	0.917	0.574	0.562	0.568	0.496	0.764	0.602	Rumour Check
0.998	0.847	0.916	0.461	0.467	0.464	0.750	0.525	0.617	Missing People
1.000	0.988	0.994	0.466	0.507	0.485	0.963	0.934	0.948	Monitoring
0.998	0.915	0.955	0.376	0.393	0.384	0.955	0.629	0.758	Express Support
0.999	0.741	0.851	0.625	0.611	0.618	0.695	0.777	0.734	Criticism/Complaints
1.000	0.625	0.769	0.414	0.375	0.393	0.336	0.352	0.344	Safety Measures
1.000	0.802	0.890	0.249	0.273	0.260	0.364	0.631	0.462	Preparedness
0.996	0.675	0.805	0.483	0.472	0.478	0.658	0.432	0.522	Information Dissemination
1.000	0.875	0.933	0.672	0.691	0.681	0.702	0.854	0.771	Rescue and Relief
0.998	0.475	0.644	0.605	0.570	0.587	0.630	0.623	0.626	Others
1.000	0.790	0.883	0.308	0.287	0.297	0.265	0.657	0.378	Early Warning
1.000	0.835	0.910	0.458	0.454	0.456	0.901	0.692	0.783	Assistance Required
0.537	1.000	0.699	0.680	0.657	0.668	0.967	0.899	0.932	Prayers
0.999	0.962	0.980	0.502	0.482	0.492	0.953	0.881	0.916	Loss and Damage Assessment
1.000	0.659	0.795	0.240	0.253	0.246	0.318	0.077	0.124	Reconstruction/Rehabilitation
1.000	0.932	0.965	0.326	0.374	0.348	0.937	0.731	0.822	Focus Group Meeting
0.998	0.964	0.981	0.311	0.306	0.308	0.235	0.966	0.378	Post Disaster Needs Assessment
0.932	0.874	0.882	0.550	0.548	0.549	0.823	0.780	0.789	Weighted Avg.

Fig. 7.15 Precision, recall, and *F*-measure for three algorithms (Cyclone Ockhi data set)

is out of the positively labelled instances what percentage is positive. Precision refers to accuracy. Recall is the completeness. Out of the positive instances what percent did the classifier labelled as positive is called Recall. *F*-Measure is usually the harmonic mean of Precision and Recall. The weighted average of all class labels for Recall for Decision Tree, IBK, and Naive Bayes for Sri Lankan Flood's data set is 0.823, 0.523 and 0.483 which means that Decision Tree identified 82.3% of positive instances, IBK 52.3 and Naïve Bayes 48.3%. The weighted average of all the class labels of Precision for Decision Tree, IBK and Naive Bayes is 0.888, 0.520, and 0.565, respectively. This means out of the positively identified instances 89% of the instances were positive for Decision Tree, 52% for IBK, and 56% for Naive Bayes.

Figure 7.15 shows the performance results of the three algorithms for the Cyclone Ockhi data set. The weighted average for Recall of all the class labels for Decision Tree, IBK, and Naïve Bayes is 0.874, 0.548, and 0.780, respectively, which means that Decision Tree identified 87.4% of all the positive instances, IBK 54.8% and Naïve Bayes 78%, respectively. The weighted average for Precision of all the class label instances of Decision Tree, IBK and Naïve Bayes is 0.932, 0.550, and 0.823, respectively which means the accuracy of predicted positive classes is 93.2% for Decision Tree, 55% for IBK and 82.3% for Naïve Bayes. From both the tables we can conclude that among the three algorithms, Decision Tree is more suitable and accurate. For social media data we can use the Decision Tree algorithm with the huge data set and still the algorithm will predict the classes with greater accuracy if the overlapping between different classes is reduced to a certain extent.

7.6 Discussion

7.6.1 Critical Impact of Social Media

The advent of smartphones has led to more use of social media than before as access to the Internet and smartphones has increased quite drastically. This can be observed in the data also with 81% of the tweets being tweeted through mobile phones in the case of Cyclone Ockhi. In the case of Sri Lankan Floods, the use is almost 50% through mobile phones. The variation is due to the difference in connectivity, level of damage, access to the Internet and smartphones, and also the advent of competitive Telecom companies like Jio. The impact made by social media in disaster situations is huge with the rapid spread of critical information across the community. It helps in sharing emergency numbers and information which is picked up by the people who require it and also the volunteers take this further in reaching the affected people. Situational updates were shared more in case of a disaster which means that the ground situation was being updated frequently and the outside world got to know the impact of the disaster and its severity almost immediately. This helps in the flow of aid materials and donations from humanitarian actors. This is very crucial because for any help to be received some amount of information is needed and this is a tool where information, though not always accurate is spread rapidly even in remote locations where traditional media reaches late.

Establishing connectivity has been a problem during any kind of disaster. Even when network connectivity has been down and power lines have been disrupted, Internet connectivity has helped people to communicate through social media with the outside world. This has been observed during Chennai Floods 2015, Sri Lanka Floods, and Cyclone Ockhi to name a few. Support pours in from the outside world and people come together and help each other using social media even before other Humanitarian Organisations and Government agencies act. The main impact it has created among the community is to own responsibility and has given a sense of power in their hands to prepare, mitigate, and respond to a disaster. It has helped in mobilising the support and resources required based on the tweets and requests received through social media. Fundraising no longer needs to go through a lengthy process nowadays. It is being done through appeals in social media and the information passes to the larger community within seconds. The requests from social media can be tracked if proper hashtags are used and the information on the rescue measures is updated daily as seen in the case of Cyclone Ockhi with precise information on the location of search and rescue and the count of people rescued.

Social media has become a tool where problems can be identified and the gaps in the response measures can be immediately identified without needing extensive review of policy documents and Initial reports. It helps in providing a rapid situational update of the current scenario and the needs arising currently such as medical kits, food, water, sanitation, shelter, clothing, and so on. Keyword analysis helps in tracking these needs based on the keywords emerging and also with previous knowledge of similar events. These needs and issues can be immediately addressed by tracking the trends and issues being shared by the people and keeping a positive sentiment among the public by sending assurance and support messages and actions without delaying further. Grievance redressal mechanisms can be created much in advance and tracked regularly and verified so that no false complaints or rumours are spread and also genuine complaints are addressed immediately.

7.6.2 Challenges and Opportunities

The information generated by social media is wealthy and yet to be tapped by developing countries such as India and Sri Lanka in the context of disasters. As the data is huge, the issue of classification of relevant tweets and cleaning of the data also arises. The categories emerging from the social media data change for every event or hazard and hence the different categories for different hazards. The challenge of data validation and reliability makes government agencies and renowned Humanitarian organisations take a step back in using social media for real-time communication during disasters or using social media platforms for information gathering to ascertain the initial situation on the ground. They prefer to wait for the official count and data which is released at least a month after the event. Cyclone Ockhi created a devastating impact on the minds of the fishermen and farmers in Tamil Nadu and Kerala due to its sudden onset and failure of adequate warnings and preparedness from the governments. The information related to missing fishermen was shared on social media regularly but it was not taken further in verifying with actual data on the ground Flash reports during disaster situations require a rapid account of the situation on grounded and precision is not warranted at that stage. Social media can be a potential tool in this aspect. Search and rescue measures were shared regularly by government agencies through social media to disseminate the information as to larger audience and positive image of the government and a positive atmosphere. This shows that social media as a communication tool during disasters has a huge potential.

7.7 Conclusion

Social media if properly utilised can be an asset for the Emergency Managers. Through this research, the use of social media has been analysed during Sri Lankan Floods and Cyclone Ockhi. It has been observed that proper grievance redressal channels need to be established during disasters to overcome the negative emotions arising from the people. Machine learning algorithms have been analysed and found that the Decision Tree algorithm works better for social media data and the classes need to be refined further for more accuracy. A skeleton framework has been arrived at for the usage of social media during disasters or emergencies with the Twitter data collected. Apart from the framework mentioned proper training and education need to be given to the volunteers and emergency responders about the usage of social media and about effectively using and monitoring hashtags. This helps in tracking and analysing the requests and responses. Further work needs to be done on how to integrate the data from different social media platforms and provide it to the public so that redundancy is avoided.

The focus needs to be shifted to the usage of social media during preparedness so that people are aware of preparedness measures well in advance and this can be easily achieved through social media due to the connectivity nowadays and the accessibility. This research was a step in the direction of analysing how people are using social media in a disaster situation and how algorithms implementing machine learning can be used to classify and predict the data generated from social media. Future work may include integrating the data generated from social media before, during and after the disaster and providing a platform to share the same in real-time.

Acknowledgements We would like to thank the social media platforms such as Twitter and online developer communities who contributed to the data analysis tools for using the social media data. We are also grateful to the Twitter users who shared their honest and informative views on several issues on Twitter during Ockhi Cyclone and Sri Lankan Floods.

References

Christian Reuter AM (2011) Social software as an infrastructure for crisis management—a case study about current practice and potential usage. In: Proceedings of the 8th international ISCRAM conference, Lisbon, pp 1–11

Guy M, Earle P, Ostrum C, Gruchalla K, Horvath S (2010) Integration and dissemination of citizen reported and seismically derived earthquake information via social network technologies. In: Cohen PR, Adams NM, Berthold MR (eds) Advances in intelligent data analysis IX. IDA 2010. Lecture Notes in Computer Science, vol 6065. Springer, Berlin, Heidelberg. https://doi.org/10.1007/978-3-642-13062-5_6

Hall DL, Llinas J (2008) A framework for dynamic hard/soft fusion. In: 11th international conference on information fusion. IEEE, Cologne, pp 1–8

Kaplan AM, Haenlein M (2010) Users of the world, unite! The challenges and opportunities of social media. Business Horizons, 59–68

Khamar (2013) Short text classification using kNN based on distance function. Int J Adv Res Comput Commun Eng: 1916–1919

Kreps G (1984) Sociological Inquiry and disaster research. In: Annual review of sociology, 309–330

Latonero M, Shklovski I (2011) Emergency management, Twitter, and social media evangelism. Int J Inf Syst Crisis Resp Manage: 1–16

Mao J (2014) Social media for learning: a mixed methods study on high school. Comput Human Behav: 213–223

Meera GP (2017) Usage and analysis of Twitter during the 2015 Chennai flood towards disaster management. Procedia Comput Sci: 350–358

Palen L, Hughes AL (2018) Social media in disaster communication. In: Rodrigues H, Trainor JE, Donner W, Guzman AP (eds) Handbook of disaster research, 2nd edn. Springer, pp 497–518

Rossi PF (2012) Inductive model generation for text categorization using a bipartite heterogeneous network. In: IEEE 12th international conference on data mining. IEEE

Statista (n.d.) Mobile social media. Retrieved 15 Jan 2018 from Statista. https://www.statista.com/topics/2478/mobile-social-networks/

Vieweg S (2012) Situational awareness in mass emergency: a behavioral and linguistic analysis of microblogged communications (Doctoral Dissertation). Qatar Computing Research Institute

Vieweg S, Hughes AL, Starbird K, Palen L (2010) Microblogging during two natural hazards events: what twitter may contribute to situational awareness. In: Proceedings of the ACM 2010 conference on computer-human interaction, pp 1079–1088. https://doi.org/10.1145/1753326.1753486

Administrative Response to Cyclone Nilam: A Case Study of Tiruvallur District Administration

8

Shailendra Rai

Abstract

Effective disaster management heavily relies on administrative response, yet it remains inadequately documented and researched in literature. Documenting best practices, challenges, and enhancing coordination between departments and states is essential for comprehending the intricacies of disaster situations. This will serve as a valuable resource for future administrators, practitioners, policymakers, and stakeholders in this field. The study aims to explore the admirative responses to Cyclone by coinciding with a case study of cyclone Nilam. The case is structured as a cliff-hanger, appraising an eye to the different facets of administrative response to cyclone. In terms of methodology, the research utilized a combination of questionnaire surveys and interviews to delve deeply into disaster management processes, such as coordination and information management, aiming to pinpoint areas of improvement and challenges. A substantial aspect of the research methodology comprised qualitative methods. These qualitative approaches encompassed various data collection techniques, including field visits, interviews, observations, document analysis, and examination of visual records. The study showed that district administration encountered significant difficulties in evacuating fishermen, constituting a major challenge. Subsequently, assessing damages and implementing relief measures posed another major hurdle for the district administration. The study will be useful in disaster risk reduction course and particularly for administrators, planners, practitioners, and other relevant stakeholders.

Keywords

Administrative response · Cyclone Nilam · Disaster readiness · Coordination · Information management

S. Rai (✉)
Natural Resources Institute Finland (LUKE), Paavo Havaksen Tie 3, 90570 Oulu, Finland
e-mail: rajanshailendrarai@gmail.com; ext.shailendra.rai@luke.fi

University of Jyväskylä, Kokkola University Consortium Chydenius, P.O. Box 567, 67701 Kokkola, Finland

8.1 Introduction

Tropical cyclones are prevalent disaster occurrences along the coastal regions of India. The Indian subcontinent is one of the worst cyclone-prone areas after the tropical western northern

S. Talukdar et al. (eds.), *Communication, Science, Technology and Innovation in Disaster Risk Management*, GIScience and Geo-environmental Modelling, https://doi.org/10.1007/978-3-031-77189-7_8

Pacific Ocean. More than 102 very severe cyclonic storms have originated and dissipated in the Bay of Bengal since 1986 and have impacted the subcontinent countries.

In literature, numerous studies conducted on understating the response experiences to tropical cyclones. As, Paul and Routray (2011) examined how households in coastal Bangladesh reacted to cyclones, while Alam and Collins (2010) investigated the experiences of individuals dealing with cyclone hazards. Paul and Routray (2013) focused on the lack of response to cyclone warnings and forecasts, drawing on indigenous knowledge systems in Bangladesh. Sharma et al. (2009) looked into public reactions to cyclone warnings on the eastern coast of India. Paul (2014) delved into the factors influencing evacuation responses during cyclones. Tatham et al. (2012) analyzed cyclone responses from a military planning standpoint. However, a limited study explored the admirative response to tropical cycle (Pal et al. 2017). The present study is going to understand the admirative responses considering the case of Cyclone Nilam. Further, the study discussed the administrative challenges and lesion learning from the event. The current research aims to comprehend the administrative responses, specifically focusing on the case of Cyclone Nilam. Additionally, the study addresses the administrative challenges encountered and explores the lessons learned from the event.

8.1.1 Cyclone Nilam

During the early years of the twenty-first century, Cyclone Nilam emerged as a formidable cyclone striking the Coromandel Coast of India. This cyclone followed a distinctive path characterized by frequent directional changes. Initially, it moved westward, then stalled near the coast of Sri Lanka before veering north-northwestward until landfall. Subsequently, it moved west-northwestward over land, reaching south interior Karnataka, then shifted northwest and northward. The residual low-pressure system continued northeastward across Andhra Pradesh.

Doppler radar first detected Cyclone Nilam 500 km southeast of Chennai on October 29, 2012. To forecast its trajectory and intensity, various numerical weather prediction models and IMD's global and mesoscale models were employed. Cyclone warning bulletins were issued by the Cyclone Warning Division on December 28, 2012, and subsequently, every three hours until October 30, 2012, disseminated to national disaster management agencies and the states of Tamil Nadu, Pondicherry, and Andhra Pradesh via email, fax, SMS, and personal briefings.

Formed on October 28, 2012, Cyclone Nilam dissipated on November 1, 2012, impacting primarily India and Sri Lanka. Originating from a low-pressure area in the Bay of Bengal, it developed into a weak depression 550 km northeast of Trincomalee, Sri Lanka. It intensified into a deep depression and later a cyclonic storm by October 30, 2012, making landfall at Mahabalipuram on October 31, 2012, with peak winds of 85 km/h. Strong winds on October 31, 2012, caused sand to pile up along the shore, with seawater reaching nearly 100 m (330 ft) inland. Cyclone Nilam's landfall near Mahabalipuram prompted the evacuation of over 3000 people, and Chennai declared school holidays until November 1, with 282 schools converted into relief centers.

In Tamil Nadu, due to flash floods (rain (mm) > 100 mm per day), 17 deaths were reported, 298 livestock was lost; paddy crops of 4646 hectares were damaged in Tiruvallur districts; 242 km of roads; 415 huts were fully damaged; 3283 was partly damaged, and 5000 electric poles were reported damaged. The cyclone's impact in Tiruvallur was relatively mild, resulting in three deaths, damage to 670 huts, and minor agricultural and infrastructure damage. Even neighboring countries like the Maldives were affected, with 142 households severely impacted in Hoarafushi Harbour and Nolhivaranfaru. Another 27 households were damaged in 22 islands, and displaced around 600 individuals. In Sri Lanka, the North and North-East regions bore the brunt of the cyclone and floods. According to Sri Lanka's National

Disaster Management Centre, 68,217 individuals from 18,187 families were affected, with three reported deaths and 7,434 people displaced. The impacts of cyclone Nilam are shown in Table 8.1.

8.2 Methodology

The current studies draw upon field-based observations conducted during the post-disaster phase, supplemented by face-to-face interviews with disaster managers and telephonic discussions with field responders in the region. A significant portion of the research methodology involved qualitative approaches, including the analysis of publicly available secondary data. Qualitative research methods encompassed data collection through field visits, interviews, observations, document analysis, and visual records. Through qualitative analysis of primary and secondary data, this research aims to comprehensively understand the institutional framework and the roles and functions of the various stakeholders involved in disaster management. The data analysis focused on identifying patterns and themes, allowing for a nuanced understanding of human behavior in response to disaster events, and capturing the complexities and unexpected challenges that communities face.

8.2.1 Disaster Readiness

Planning is a crucial aspect of the Disaster Preparedness cycle, encompassing measures aimed at anticipating, preventing, mitigating, and responding to disasters to minimize their impact. Effective disaster preparedness facilitates coordinated planning, minimizes redundant efforts, and enhances the effectiveness of household and community response efforts. Integrating risk reduction measures into preparedness activities can prevent disasters and save lives and livelihoods during crises, enabling affected populations to recover swiftly.

According to NDMA IRS Guidelines, Incident Response Teams (IRTs) are to be established at various administrative levels under the direction of District Magistrates. These teams, organized into Planning, Operations, and Logistics sections, coordinate response efforts. The District Emergency Operations Centre (DEOC) serves as the central coordination point, ensuring communication, decision support, and alert systems are operational during crises.

Response planning involves swift and disciplined incident assessment to enable a scalable, adaptable, and flexible response. It adheres to national and state response doctrines, delineating roles and responsibilities across government and the private sector. Before initiating response activities, the District Magistrate, acting as the Incident Commander, convenes a meeting to assess the situation, mobilize resources, and assign tasks. The Incident Action Plan guides response efforts, with the DM appointing an Operation Section Commander based on incident type and issuing procedural guidelines.

Response planning entails developing and analyzing various courses of action. This process combines scenario-based, functional, and capabilities-based planning to ensure

Table 8.1 Impact of Cyclone Nilam

Country	Affected	People affected	Death	Injured	Missing	IDPs	House damaged and destroyed
Maldives	24 Islands	14,152	0	0	0	600	169/0
Sri Lanka	16 islands	167,466	10	15	0	13,865	4843/335
India	2 Islands	150.00	28	0	2	68,000	766/480
Total		331,618	38	15	2	82,465	5778/815

Source Emergency and Humanitarian Action (EHA) Unit, Regional Office for South-East Asia (SEARO), World Health Organization (WHO) situation November 2, 2012

comprehensive solutions are generated. Typically, it follows a structured series of steps:

a. Establish the timeline.
b. Identify and depict decision points.
c. Identify and depict operational tasks.
d. Identify resources.
e. Identify information needs.
f. Assess progress.
g. Incident Specific SOP.

In Tiruvallur, the District Emergency Operations Centre (DEOC) operated continuously, and the Toll-Free Number 1077 was fully operational. Complaints received via the Toll-Free Number were promptly relayed to the relevant authorities for immediate action.

As per the circular instructions issued from the Office of the Commissioner of Revenue Administration, the rainfall and the damage data such as human loss, cattle loss, and hut damages was uploaded in the revenue intra-site before 8.30 A.M. daily. Further, rainfall data was also uploaded in the evening at 4.00 PM daily. That apart, information is sent to Personal Secretary of Relief and Rehabilitation Department through WhatsApp in case of heavy rainfall/or in any emergency in the district. The Taluk EOC was also functioning at each Taluk office with sufficient staff.

In case of line departments, coordination meeting was conducted at divisional, district, and taluk levels. For proper coordination, mock exercises and training were conducted for effective implementation of plan during crisis. The planning is of utmost importance in case of preparedness before implementation, the district also has disaster management plan which focuses on district profile, hazard analysis, and risk assessment. But it is a general practice observed that disaster management plan was not properly used or implemented during crisis/ disaster.

The Relief Operations in Tiruvallur were overseen by Revenue Divisional Officers, designated as Divisional Zonal Officers, with each Taluk Officer supervised by a Deputy Collector, known as an Additional Zonal Officer. These officers, operating from Taluk Headquarters, worked closely to manage disaster response during the Northeast Monsoon period. Control rooms were established at Divisional offices, staffed round the clock on rotating shifts, maintaining registers to record flood/cyclone-related messages for immediate dissemination to relevant departments.

Throughout October, November, and December, a Deputy Collector was assigned to each Taluk, with Tahsildars, Zonal Deputy Tahsildars, and Revenue Inspectors instructed to identify disaster-prone areas and potential shelter locations in public buildings. They were tasked with inspecting and ensuring the suitability of these buildings for use as shelters, informing owners of private buildings that could serve as emergency shelters.

Taluk Tahsildars were responsible for utilizing police VHF sets for urgent communications, while Revenue Inspectors were placed under the supervision of special Tahsildars or Panchayat Union Commissioners during disaster operations.

In municipal areas, the Municipal Commissioner acted as the Relief Officer, coordinating with the Collector and mobilizing manpower for immediate response. Executive Officers in town Panchayats also served as Relief Officers, overseeing relief operations and ensuring public health measures to prevent epidemics, including maintaining clean drinking water supplies.

Coastal districts were equipped with VHF communication systems for effective communication during the Northeast Monsoon (Fig. 8.1). Revenue officials in these districts were directed to utilize these systems extensively. However, a significant challenge was the lack of adequate training for the population. Additionally, a roster of individuals and organizations possessing boats, pump sets, generators, power saws, and other relief and rescue equipment was compiled to facilitate immediate mobilization during emergencies. Evacuation Shelter homes for the victims in the Coastal Taluks and Low-lying areas in the District were identified and manned by the respective Taluk Officials in

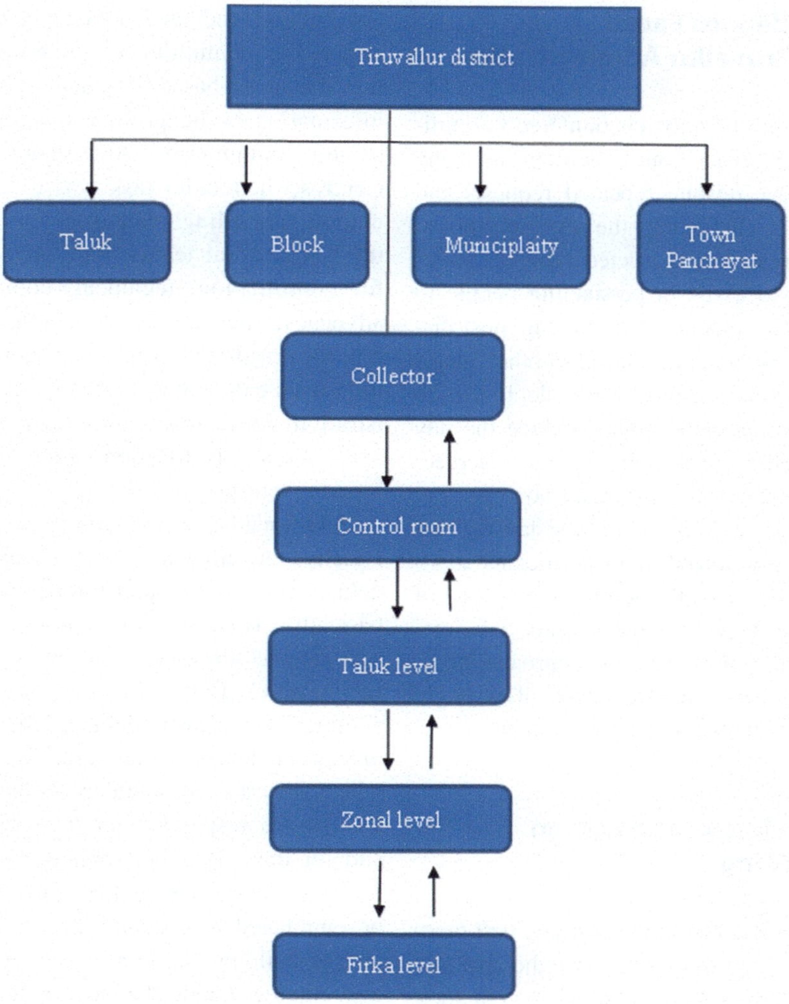

Fig. 8.1 Key stakeholders and communication flow

coordination with fire, co-operative, health, and Police Departments. Effective steps were taken to ensure the availability of boats, inflatable rubber boats, and vital equipment by the Fire and Fisheries Department, and availability of panel of swimmers was also ensured by the Fire and Fisheries Department. Early Warning Team was formed comprising 12 members with Personal Assistant (General) to Collector as the head.

The district's coastline comprised 35 fishing villages, each overseen by a Village Officer acting as the Nodal Officer. Two villages were supervised by a Revenue Inspector. Cyclone shelters, both permanent and temporary, were organized, with each under the control of a Village Administrative Officer (VAO) and monitored by a Taluk Supply Officer at the Taluk level. Fifteen evacuation camps were established, and officers were tasked with coordinating transport from transport depots to these shelters. Arrangements for generators and drinking water were also made. The health department was instructed to be prepared for immediate action. Approximately 1500 coastal residents were relocated to these evacuation camps, receiving an allowance of INR 1000 per family and free clothing, including Dhotis for men and Sarees for women.

8.3 Challenges Faced by Tiruvallur Administration

The district administration encountered a significant obstacle in evacuating fishermen, as many refused to move despite repeated requests and announcements, believing the cyclone posed no serious threat. Local elected representatives were enlisted to assist in persuading people to relocate to safe shelters. Conducting post-disaster damage assessments posed another major challenge. Special teams were deployed to assess damages in each fishing village the day after the cyclone. Some false claims emerged, with individuals who suffered no cyclone-related damage seeking relief assistance. This issue became politicized, with politicians advocating for relief to all victims regardless of actual damage. However, the revenue administration adopted a more cautious approach, highlighting the adverse consequences of wrongly including individuals for relief assistance.

8.4 Conclusion and Lesson Learning

We can always learn from the past and best practices can be used to further improve the response and recovery. Let us take a case when civil society organization, Society for Nutrition, Education and Health Action (SNEHA) and South Indian Federation of Fishermen Societies (SIFFS) initiated NGOs Coordination and Resource Centres (NCRC) to have better coordination among local NGOs in the aftermath of Indian Ocean Tsunami, 2004. Senior officials from Tamil Nadu government have also partnered with NCRC to improve coordination and information exchange among the affected communities, government, and NGOs. NCRC was involved in three stages—(a) front office, (b) Village Information Centres, and c. Sectoral Teams.

In front office, they were responsible for linking affected communities to NGOs and government and they advocated on behalf of affected communities to government regarding the need-based demand. There was an information exchange in vernacular languages to the communities. At Village Information Centres, their main task was to collect information regarding damages and compensation; at sectoral teams, they were responsible for coordination, technical, communications, advocacy, and policy development in community-prioritized sectors. These types of initiatives can be considered by district administration while responding to a disaster, but every disaster is different which requires need-based response.

Additionally, accountability is essential for effective governance. There should be clearly defined roles and responsibilities which address basically two crucial questions—"who is accountable for what" and "to whom they are accountable". Even when the roles and responsibilities are clearly defined, there were significant challenges to initiating the activities for which they are accountable. Authorities do not want to be responded for their wrong actions and for that, they fear initiating actions which leads to delay in response. The fear needs to be eliminated to expedite the response process and in building the confidence of the authorities. Another major challenge is the capacity of the government to respond and the functioning resources available in the time of crisis. The capacity can be enhanced by properly designed training and mock exercises which can be tailor made depending upon the context and event. Good Governance often calls for participation in decision making and ensuring public participation can be challenging but fruitful for solving real-life challenges and for ensuring public participation digital tools such as Public Participation Geographic Information System (PPGIS) can be used but all citizens may not be technologically advanced to participation in PPGIS which can be a challenge for inclusive participation.

References

Alam E, Collins AE (2010) Cyclone disaster vulnerability and response experiences in coastal Bangladesh. Disasters. https://doi.org/10.1111/j.1467-7717.2010.01176.x

Pal I, Ghosh T, Ghosh C (2017) Institutional framework and administrative systems for effective disaster risk governance—perspectives of 2013 Cyclone Phailin in India. Int J Disaster Risk Reduct. https://doi.org/10.1016/j.ijdrr.2017.01.002

Paul SK (2014) Determinants of evacuation response to cyclone warning in coastal areas of Bangladesh: a comparative study. Orient Geographer 55(01 & 02)

Paul SK, Routray JK (2011) Household response to cyclone and induced surge in coastal Bangladesh: coping strategies and explanatory variables. Nat Hazards. https://doi.org/10.1007/s11069-010-9631-5

Paul SK, Routray JK (2013) An analysis of the causes of non-responses to cyclone warnings and the use of indigenous knowledge for cyclone forecasting in Bangladesh. In: Climate change management

Sharma U, Patwardhan A, Parthasarathy D (2009) Assessing adaptive capacity to tropical cyclones in the East coast of India: a pilot study of public response to cyclone warning information. Clim Change. https://doi.org/10.1007/s10584-009-9552-z

Tatham P, Oloruntoba R, Spens K (2012) Cyclone preparedness and response: an analysis of lessons identified using an adapted military planning framework. Disasters. https://doi.org/10.1111/j.1467-7717.2011.01249.x

Part III
Science and Technology in Disaster Risk Management

Assessing Dynamics of Exposed Elements to Natural and Anthropogenic-Induced Landslide Hazard Model for Land Use Policy

9

Somnath Bera, Shivam Priyadarshi, Pritha Ghosh, Guru Balamurugan, and Atish Prashar

Abstract

Understanding the dynamic risks of landslides is imperative for land use policies in mountainous settlements. The aim of this research is to develop a framework for hazard modeling of different types of landslides, such as those induced by natural slopes and human-modified slopes. The study, first, generates landslide susceptibility using Geospatial Artificial Intelligence (GeoAI) techniques, such as the Random Forest algorithm. Next, it estimates the temporal probability of landslide over three different time periods using the Poisson distribution model. Subsequently, the study integrates landslide susceptibility with temporal probability for hazard estimation. The modeling approach is applied in the Kalimpong region located in the Eastern Himalayas of India. Further, the study analyzes changes of buildings, roads, and agricultural lands between 2008 and 2016 under different hazard scenarios. The research identifies notable distinctions between scenarios in which landslides are induced along natural and anthropogenic slope in similar environmental conditions. Moreover, the study exposes a concerning pattern: an increase in the construction of roads and buildings as well as a decrease in agricultural land, especially in high-hazard regions. These findings indicate the increasing risk in the study area.

Keywords

Landslides · Integrated hazard model · GeoAI · Poisson distribution model · Exposed elements · Dynamic scenarios

9.1 Introduction

Landslide occurrences are increasing globally, resulting in more than 4000 fatalities each year (Haque et al. 2019). The Himalayan region witnesses the highest number of fatal landslide

The original version of the chapter has been revised. A correction to this chapter can be found at https://doi.org/10.1007/978-3-031-77189-7_17

S. Bera (✉) · S. Priyadarshi · P. Ghosh
Department of Geography, Central University of South Bihar, Gaya, India
e-mail: somnath@cusb.ac.in

A. Prashar
Department of Mass Media and Communication, Central University of South Bihar, Gaya, India

G. Balamurugan
Department of Geology, School of Earth Sciences, Central University of Tamil Nadu, Thiruvarur, India

S. Talukdar et al. (eds.), *Communication, Science, Technology and Innovation in Disaster Risk Management*, GIScience and Geo-environmental Modelling, https://doi.org/10.1007/978-3-031-77189-7_9

events worldwide, owing to its susceptibility to landslides due to high seismic activity and intense rainfall (Froude and Petley 2018). Despite this inherent risk, the population in the Himalayas has surged from 19.9 million to 53.8 million over the past five decades (1961–2011) (Apollo 2017). Notably, the condition becomes worse when the human settlements and construction activities is closely associated with this population growth and severity of landslides (Bera et al. 2021). Collectively, it is restricting the carrying capacity of the hilly region and, simultaneously, generating a scarcity of safe spaces for future habitation.

Evaluating landslide hazard is crucial for effective land use planning in any area (Nefeslioglu and Gorum 2020). For the first time, Varnes (1984) proposed hazard as a potentially damaging phenomenon which was brought by integrating both spatial and temporal dimensions. The spatial aspect, the likelihood of landslides within a given area termed as susceptibility. It is typically determined through relationship between previous landslides and associated geoenvironmental factors. In the last two decades, there has been a significant progress in landslide susceptibility evaluation through numerous methodological approaches (Chen et al. 2019; Gao et al. 2022). Recent trends in machine learning include the use of methods like MaxEnt, support vector machine (SVM), logistic regression, ensemble machine learning, artificial neural network, and deep learning (Merghadi et al. 2020; Nurwatik et al. 2022).

On the other, the temporal probability means the landslide's frequency occurrence in a given time. All the past landslides occurred in different periods, providing a trend of future landslide occurrences in a given time series. Usually, the landslide's temporal probability can be described in frequency, exceedance probability, or return period. The frequency means, in a specific time interval, the number of landslides that occurred (e.g., annual frequency). The average time it takes for a landslide of a certain magnitude to occur is known as the return period. On the other hand, exceedance refers to a landslide occurring with a magnitude equal to or greater than a specific value in a definite period. Landslide exceedance probability may be estimated using statistical methods like Poisson model or binomial (Crovelli 2000) or power-law distribution. Many literatures have estimated the temporal likelihood based on the rainfall threshold method where rainfall triggers the rate of landslides (Jaiswal and van Westen 2009; Ghosh et al. 2012). It assumes that landslides will also occur in the future, but the frequency will remain the same. Assessing the impact of dynamic phenomena, such as climate change or land-use land-cover change, on an area is challenging (Jaiswal and van Westen 2009). Accurate information regarding the specific dates of landslide events and the characteristics of rainfall is necessary. In regions with limited data, this specific type of information may not always be accessible. In order to address this constraint, the temporal probability is calculated by taking into account the average rate of the landslide, like the Poisson probability (Coe and Michael 2000; Guzzetti et al. 2005). Based on historical landslide data, the frequency of future landslides is expected to remain constant. This model assumes that it can estimate the chance of one or more landslides occurring at any given period. Thus, it is not necessary to deal triggering events.

In hazard assessment, the selection of spatial units is essential for mapping temporal probability. Das et al. (2011) proposed an automated homogenous susceptible unit using a region-growing algorithm. On the other hand, Ghosh et al. (2012) analyze the temporal probability using terrain unit. The sections were delineated according to the land use pattern and local topography for analyzing temporal probability. Many studies consider regular grid- or pixel-based analysis (Gaprindashvili and Van Westen 2016).

In general, literature indicates a growing focus on landslide hazard studies in recent years (Jaiswal and van Westen 2009; Das et al. 2011; Ghosh et al. 2012; Gaprindashvili and Van Westen 2016; Fu et al. 2020). However, a key observation is that many studies primarily address landslide susceptibility, which may not fully justify hazard as previously defined. Moreover, despite there has been an increase in evaluating model performance, assessments of model uncertainty remain limited. Few studies explore the geomorphological significance of the data (Reichenbach et al. 2018), such as the differing responses of landslides along natural slopes versus cut slopes. Yet, there's a gap in past modeling frameworks when it comes to assessing landslide hazard by differing between these various types of landslides. Finally, there's a need to harness the potential of these models to inform better land use policies, an area that has been relatively underexplored in the literature.

This present study proposes the methodological framework of the novel hazard model for three different landslides: natural-slope-induced debris slides (NSDS), cut-slope-induced debris slides (CSDS), and rock fall-rock slides (RFRS). At first, the study assesses the three sets of landslide susceptibility applying random forest. Second, this study estimates the temporal probability using Poisson distribution model, for three returning periods of every individual model. Last, we are integrated with landslide susceptibility with the temporal probability. We are considering the annual exceedance of landslide probability rather than the rainfall threshold technique because of the unavailability of detailed data about the exact rainfall date and occurrence of landslides. The study selected the third-order catchments as a spatial unit of the temporal probability analysis.

9.2 Study Area

The study area lies within the hilly terrain of Kalimpong, part of the Darjeeling-Sikkim Himalaya region (see Fig. 9.1). To the west flows the Teesta River, while the Relli and Reshi Rivers mark the eastern boundary of the study area. It covers an area of 238.7 square kilometers with altitudes ranging from 250 to 3000 m and slopes up to 76°. Numerous small streams cross the area, which makes landslides more common. Seismic activity is frequently caused by the Main Central Thrust (MCT). The area has dry winters and rainy summers, with 250–300 cm of rain falling on average per year. Numerous landslides, such as rock slides, debris slides, and rock falls are frequently made worse by intense rains during the monsoon season. Landslides are also caused by human activity, especially when slope cutting occurs alongside a road. The region has a history of fatal landslide disasters dating back to 1899. Particularly disastrous years include 1950, 1968, 1980, 1996, 2007, and 2015, leading to considerable loss of lives and property damage. Landslides remain a perennial issue, causing damage to houses, roads, and agricultural land yearly (Bera et al. 2020).

9.3 Data Sources and Methodology

The inventory of landslides from year 2002 to 2017 was taken to assess temporal probability. The inventory was drawn up based on satellite images, Google Earth images, field surveys, and various reports. The resulting susceptibility maps of NSDS, CSDS, and RFRS were used to input spatial probability.

Landslide hazard (H) was calculated by integrating temporal probability (PT) and spatial probability (PS) of a region following Eq. 9.1.

$$H = \mathrm{PT} \times \mathrm{PS} \tag{9.1}$$

9.3.1 Landslide Susceptibility

Random Forest was applied to generate the spatial probability of landslides. The interaction of geoenvironmental conditions (see Fig. 9.2) and past landslides was considered to find probability. The Random Forest (RF) technique, an ensemble machine learning approach, is utilized in constructing susceptibility models. RF excels at capturing intricate systems and their interdependencies, a task that can pose challenges for traditional global regression models. The RF technique is suitable for regression or classification tasks involving both discrete and continuous datasets due to its capability to generate appropriate functions. The model's performance is evaluated by analyzing the AUC-ROC curve. The final susceptibility maps of NSDS, CSDS, and RFRS are categorized into four groups (i.e., low, moderate, high, and very high).

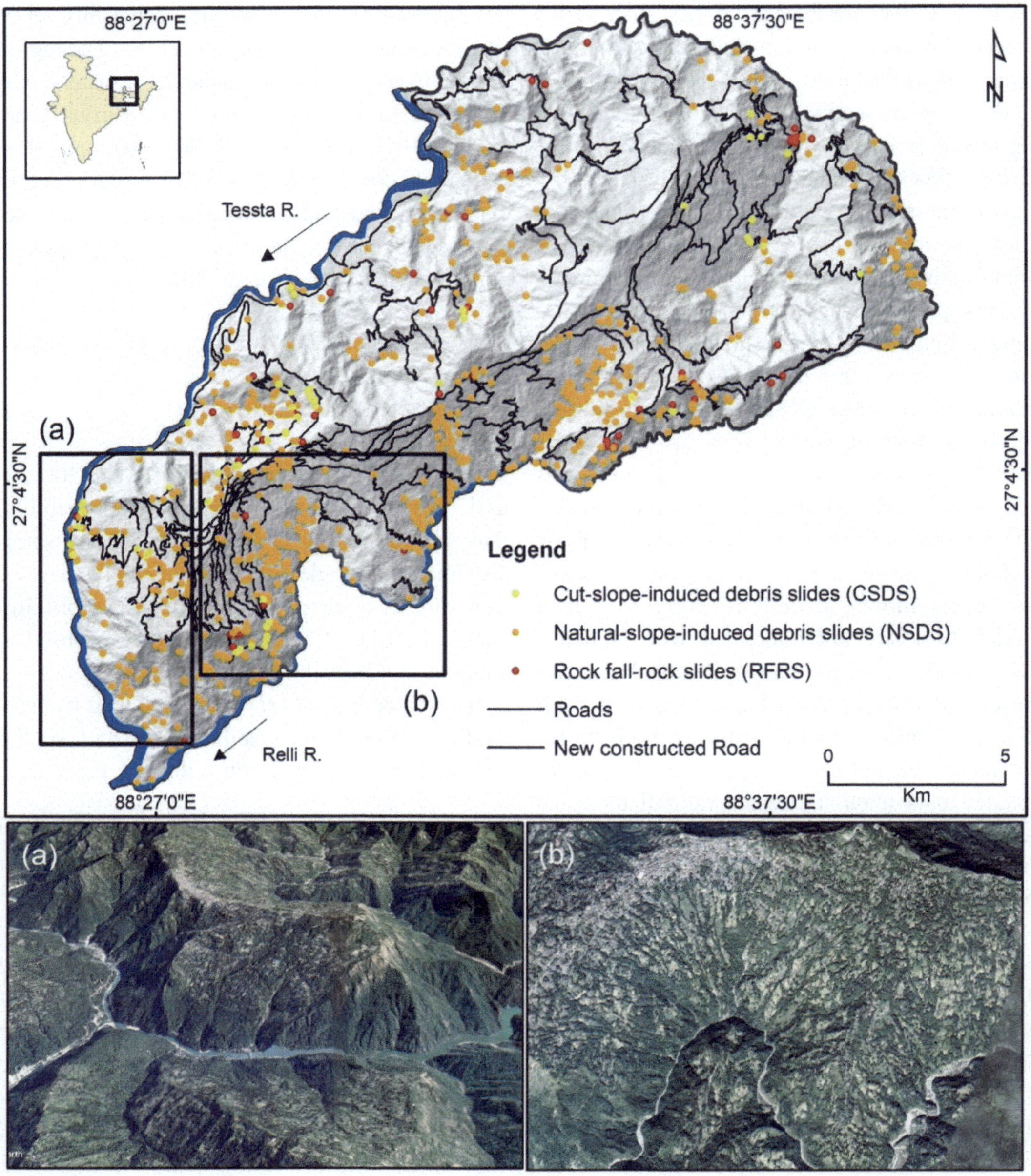

Fig. 9.1 Landslide distribution in study area. **a** Landscape in the western side; **b** landscape in the eastern side of the Kalimpong ridge

9.3.2 Temporal Probability

Temporal probability shows the frequency of landslides within a particular timeframe. In this study, we used the Poisson probability model to determine the exceedance probability of landslides (Crovelli 2000; Coe and Michael 2000). In this model, during the time "t", the exceedance probability is described below:

$$P\{N(t) > 1\} = 1 - e^{-t/\mu} \tag{9.2}$$

where N is the number of landslides, P is the probability, and μ is the future average recurrence interval between the successive events.

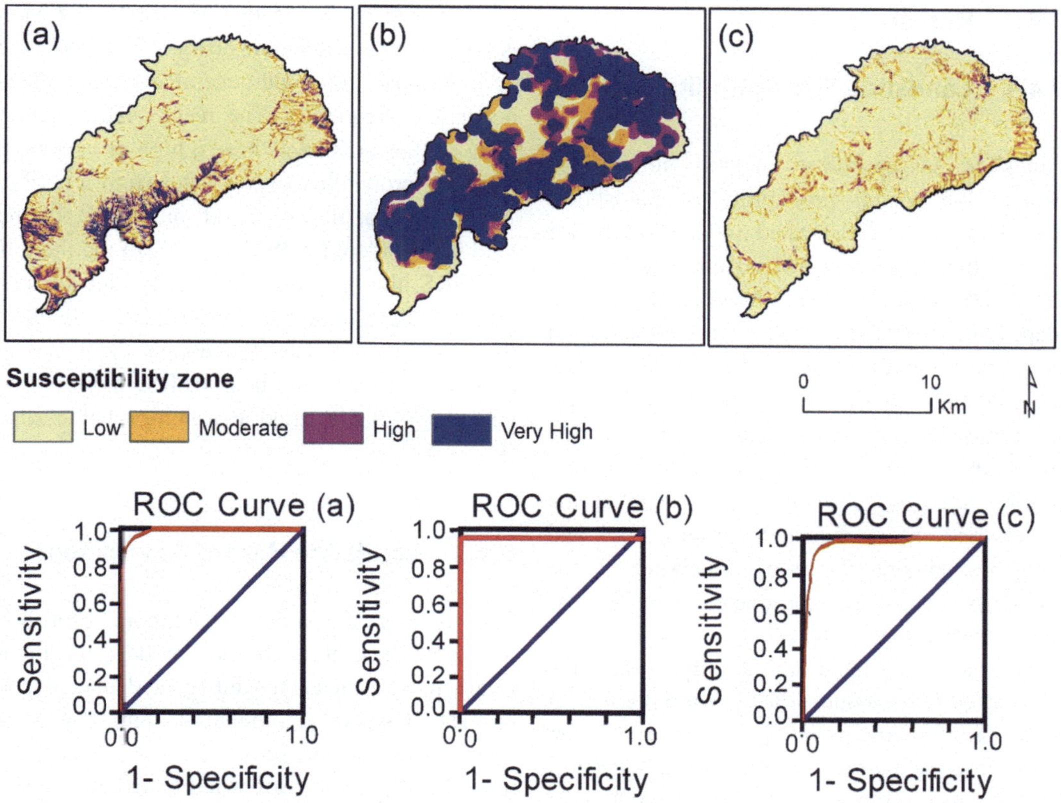

Fig. 9.2 Landslide susceptibility zonation map and ROC curve: **a** NSDS, **b** CSDS, and **c** RFRS

We calculated the temporal probability for return periods of 10, 5, and 3 years using the Poisson probability model to analyze landslide exceedance probabilities in each catchment. This analysis covered the 17-year period from 2002 to 2018, assuming a consistent frequency of landslides. All models have a final temporal probability value that falls between "0" and "1". Greater than 0.75 in the probability value was treated as the highly high zone. In contrast, less than 0.25 (above "0") was considered the low zone. Similarly, when the result falls between 0.25 and 0.50 and 0.50 and 0.75, we evaluated the moderate and high likely zones. Due to no prior landslides, the basin's temporal probability was determined to be 0.00.

9.3.3 Assessment of Hazard and Exposed Elements

The hazard assessment involves: (1) generating a landslide inventory data set, (2) developing a landslide susceptibility index layer, (3) identifying catchments as mapping units, (4) computing temporal probability for each catchment, (5) developing landslide hazard scenarios by combining susceptibility and temporal probability, (6) constructing geospatial database of exposed elements (i.e., roads, buildings and agricultural land) for 2006 and 2018 from Google Earth Images, and (7) deriving changes of the exposed elements to different hazard scenarios.

9.4 Result

9.4.1 Landslide Susceptibility

Landslide susceptibility mapping in Fig. 9.2 depicts the spatial probability of landslides. The maps were constructed independently for each of the three types of landslides: (a) NSDS, (b) CSDS, and (c) RFRS. Associated with the map, ROC AUC shows that the models achieved excellent accuracy of greater than 90%; hence, we used these models for further hazard analysis. Figure 2a reflects the spatial pattern of the susceptible zone of NSDS. The very high susceptible area is generally concentrated at the ridge's bottom, eastern and western sides and along the streams. The top of the ridge in the middle portion of the map comes under the low susceptibility zone of NSDS.

On the other hand, the highly susceptible zone of CSDS is concentrated randomly, i.e., it does not follow the topographic pattern. The top of the ridge and some portions of the bottom of the hill are prone to high CSDS. Next, Fig. 2c shows an entirely different pattern of high RFRS susceptibility that is limited and distributed sparsely in some pockets of the study region.

9.4.2 Temporal Probability

The overall result demonstrates a general pattern whereby the probability values for all landslide models rise with an increasing return duration (Fig. 9.3). The most excellent chance of an NSDS is 0.8 during a 3-year return time. When the return period is 5 or 10 years, the highest temporal probability rises to 0.93 and 0.99, respectively. As a result, when return durations rise, so does the coverage area of the highly high probable area. Very high probability area of NSDS only covers 0.69% of the area with a 3-year return time and climbs to 16.3% with a 10-year return period, according to the results (Figs. 9.3a1–a3, and Fig. 9.4).

The probable value in simulations of CSDS is "0" for most catchments (Fig. 9.3b1, 3b2, and 3c3) because cut-slope-generated debris slides occur less frequently. The results indicate that none of the catchments have a high or very high temporal probability over a 3-year return period. At the 10-year return period, only 3.6% of the land is considered high (Fig. 9.4). In the models of RFRS, the proportion of highly high temporal probability regions is likewise low (Fig. 9.3). Since only a few catchments in the research region experience frequent RFRS, most of the land (67%) is classified as a non-probable zone (Fig. 9.4).

9.4.3 Landslide Hazard Assessment

The study used spatial and temporal probability to produce three hazard models for three return periods of each kind of landslide. In the context of hazard computation, temporal probability rises with the return duration. Hence, the hazard index value rises when landslides' return times increase. Therefore, we utilized the quantile approach for categorizing the hazard index to compare changes. At first, the hazard classification breaking value for the three-year return period was noted. Later, the hazard map for the five- and ten-year return periods was further categorized using its breaking value. The four groups of naturally occurring slope-induced debris slides that were reclassified are incredibly high, high, moderate, and low (Fig. 9.5a). However, we divided CSDS and RFRS into three categories: low, high, and very high (Fig. 9.5b, c). Due to the moderate class's inability to reliably show differentiation, we combined it with the low-hazard class.

The hazard zone distributions of different models differed across the area (Fig. 9.7). The result demonstrates the significant change in landslide hazards (Fig. 9.6). In the NSDS hazard model of return periods 5 and 10, the region of a low class is decreased concerning the return

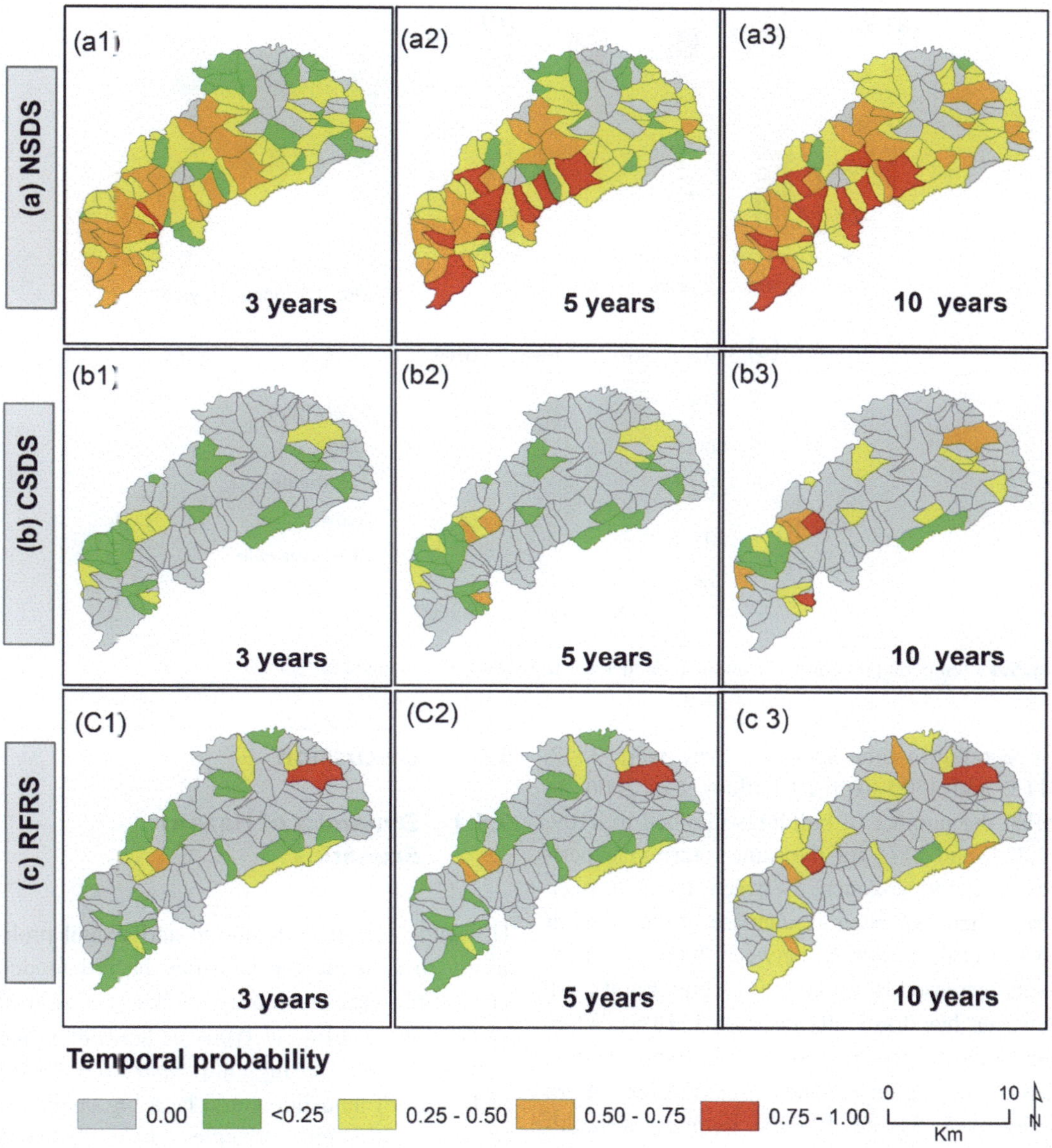

Fig. 9.3 Temporal probability of **a** NSDS, **b** CSDS, **c** RFRS for recurrence periods of **a** 3 years, **b** 5 years, and **c** 10 years

period of 3 years. At the same time, the highly hazardous areas increased from 3.4 to 6.2% for the five and ten-year return periods, respectively, compared to the three-year return period's hazard model. Similarly, the low-hazard class area decreases while the very high-hazard class area increases in the CSDS hazard model and the RFRS hazard model.

9.4.4 Exposed Elements

The changes of exposed elements for the returning periods of 3, 5, and 10 years, are shown in Figs. 9.8, 9.9 and 9.10. Figures 9.11 and 9.12 depict the significant changes in buildings, roads, and agricultural land in the NSDS between 2006 and 2018. For the hazard zone of a 3, 5, and

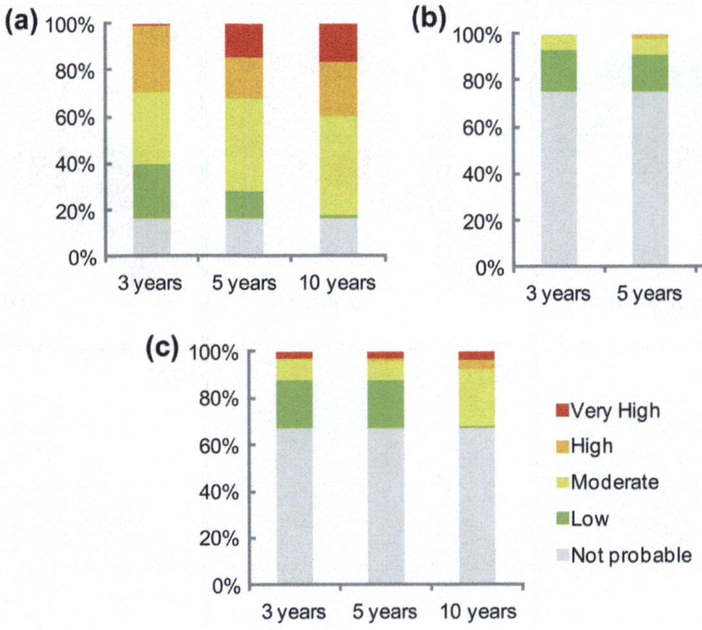

Fig. 9.4 Temporal probabilities classification includes **a** NSDS, **b** CSDS, and **c** RFRS

10-year return time scenario, there were 3542, 4112, and 4500 exposed buildings in 2006. In 2018, the number of buildings climbed to 4109, 47,734, and 5174 for the same return year periods' hazardous zones. Between these thirteen years, there has been an increase in buildings of 15% overall. Similarly, between 2006 and 2018, roads increased by about 22%. Agricultural land, however, has drastically decreased (18%). When analyzing the growth pattern, it becomes evident that as the region expands over different return periods—3, 5, and 10—the number of buildings and road length both increase. The changes to buildings and roads in the CSDS scenarios are identical to those in the NSDS scenarios. Agricultural land in CSDS fell 5% compared to 18% in an NSDS hazard scenarios. The exposed maps of RFRS show that buildings increased by about 20%. The length of the road has not changed during this time. The average expansion of agricultural land is 1.8% (Fig. 9.12). The all-hazard scenario with a 3, 5, and 10-year return period revealed a rising tendency in the amount of agricultural land. This pattern is dissimilar to the cases of CSDS and NSDS.

9.5 Discussion

9.5.1 Dynamics of Hazard Scenarios

The study integrates temporal and spatial probability to generate the landslide hazard model. In multiple cases, a hazard model was created by combining different types of landslides into a single dataset. Getting unique hazard maps for each form of landslide could be a big challenge because mitigating strategies differ between them (Ajin et al. 2022). In this study, we create three distinct hazard models for integration and examine the markedly varied patterns of the hazardous region resulting from various landslides. The highly prone areas of NSDS are located along the streams and at the bottom of the ridge. This region comes under the village located surrounding Kalimpong town. The region's NSDS conditions are favorable due to thick soil and stream erosion. According to the RFRS model, areas with steep slopes and pockets of exposed rock are the most hazardous. Landslides frequently happen here due to the favorable

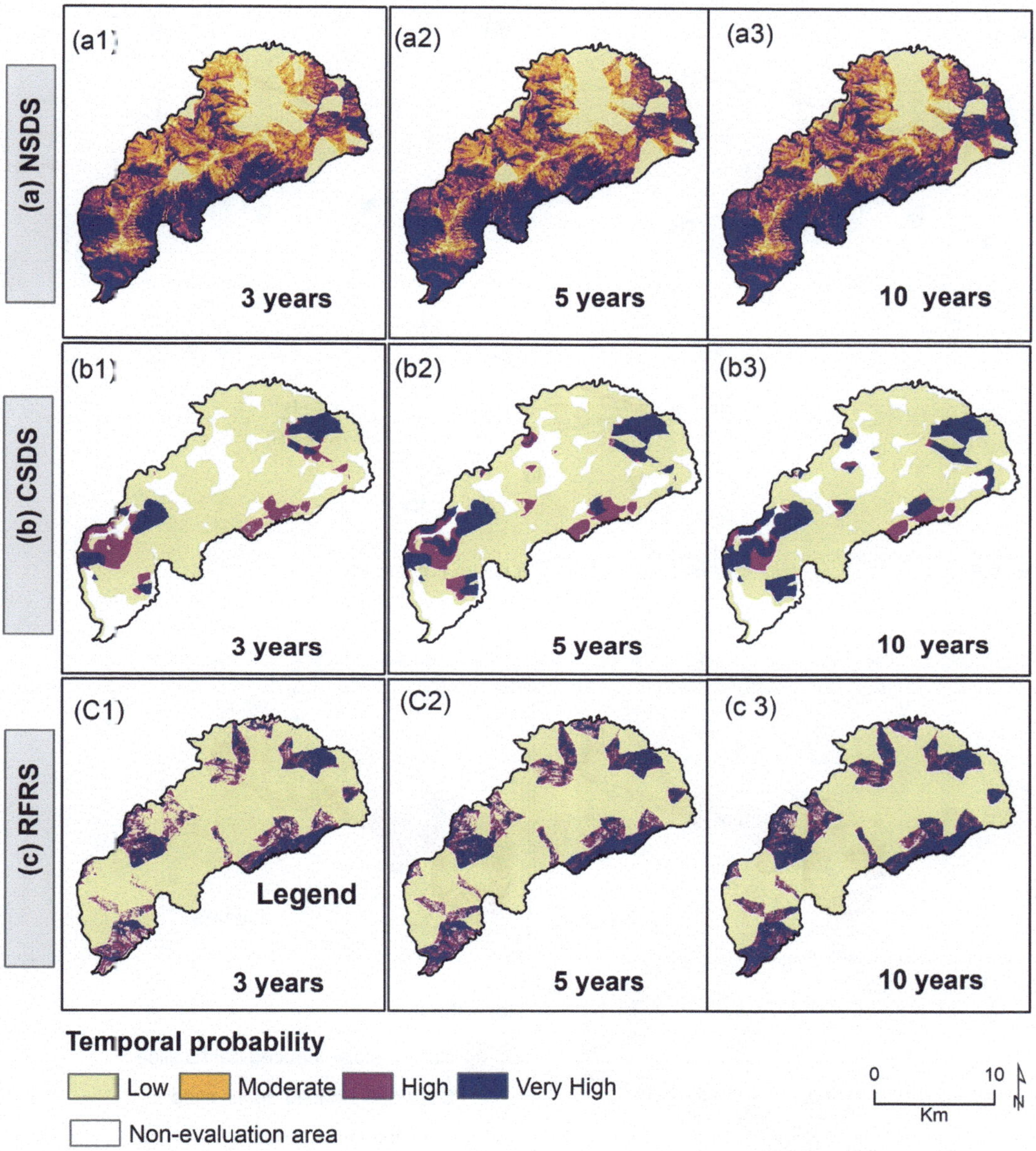

Fig. 9.5 Hazard scenarios of **a** NSDS, **b** CSDS, and **c** RFRS

geoenvironmental circumstances. The majority of the area is covered with debris, which reduces the temporal probability and low susceptibility of RFRS, with the exception of this little area of extremely steep terrain. As a result, there is often little chance of RFRS throughout the area. The hazard zone of CSDS is spectacularly found around the newly cut-slope area that is not always favorable for NSDS. The differences are also associated with the process of hazard generation. Such as stream erosion is the dominant factor of NSDS in the region as usual in literature (Tsai et al. 2022; Cebulski 2022). On the other, intersection of roads and streams is closely correlated with CSDS that is also frequently observed in the study region. It reflects the potentiality of anthropogenic activities in hazard interference.

The dynamics of hazard scenario also show the intertwining effect of susceptibility and

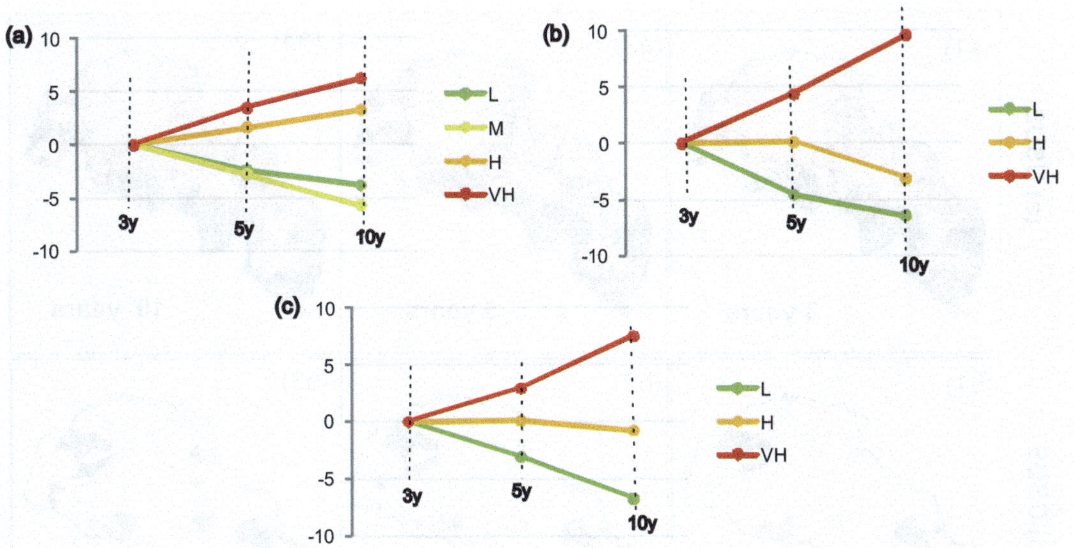

Fig. 9.6 Area of the hazardous zone varying with different recurrence intervals

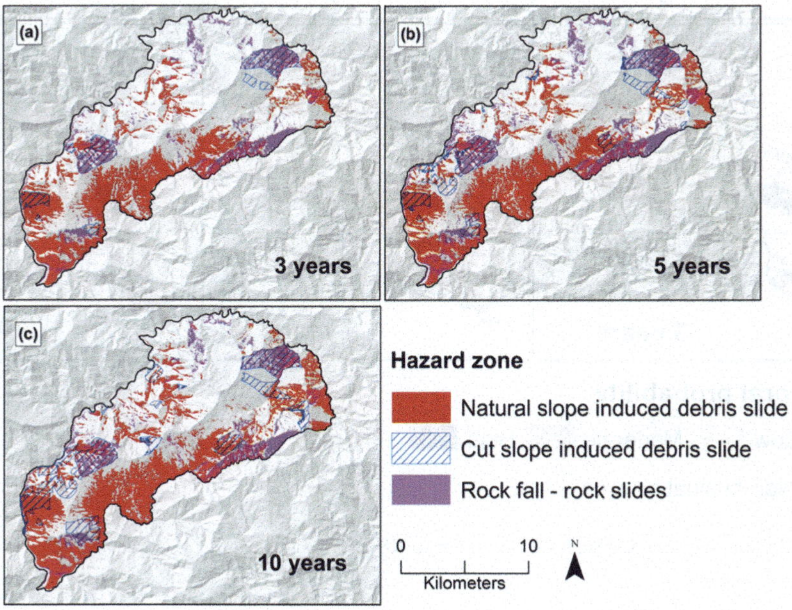

Fig. 9.7 Hazard zones with different recurrence intervals

temporal probability. In NSDS and RFRS, high susceptible area and high temporal probable have close association. It reflects that high susceptible area has repeatedly experienced landslide occurrences in the past. This similar pattern was also found in the studies of Budha et al. (2016), Li and Chen (2020), Mersha and Meten (2020). Interestingly, at the same time, we have also noticed that relatively highly susceptible areas are not always highly hazardous, especially in CSDS. It reflects that CSDS although not occurred in the past significantly,

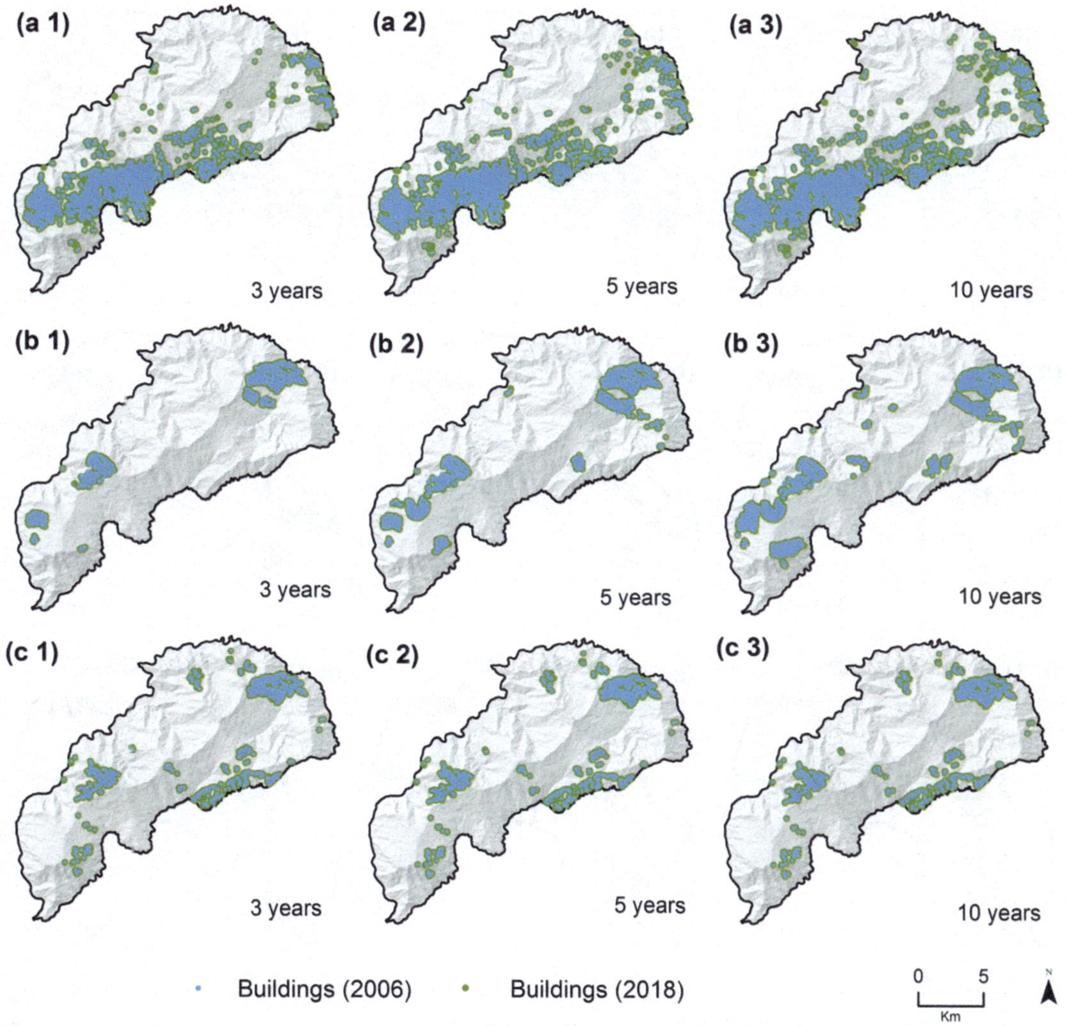

Fig. 9.8 Building exposed to the hazards of NSDS, CSDS, and RFRS with 3, 5, and 10-year return periods, respectively

new cut-slope activities make a favorable condition for occurring landslides. Hence, this hazardous area is controlled by both the landslide's periodic occurrence and the terrain's susceptible condition (Jaiswal et al. 2010; Lombardo et al. 2020).

9.5.2 Dynamics of Exposed Elements

The trend of exposed buildings, roads, and agricultural land to CSDS is quite similar to NSDS. The favorable region for NSDS is predominantly agricultural. Therefore, the continuous trend of NSDS may have caused to the decline of agricultural land between 2006 and 2018. However, the same region has experienced a simultaneous increase in the construction of buildings and roads, which poses a growing risk to both the people and infrastructure in the villages of the research area (Rumbach and Follingstad 2019). Urban expansion and the construction of infrastructure on agricultural land may be a significant contributing element to the decline in available agricultural land, a trend that has

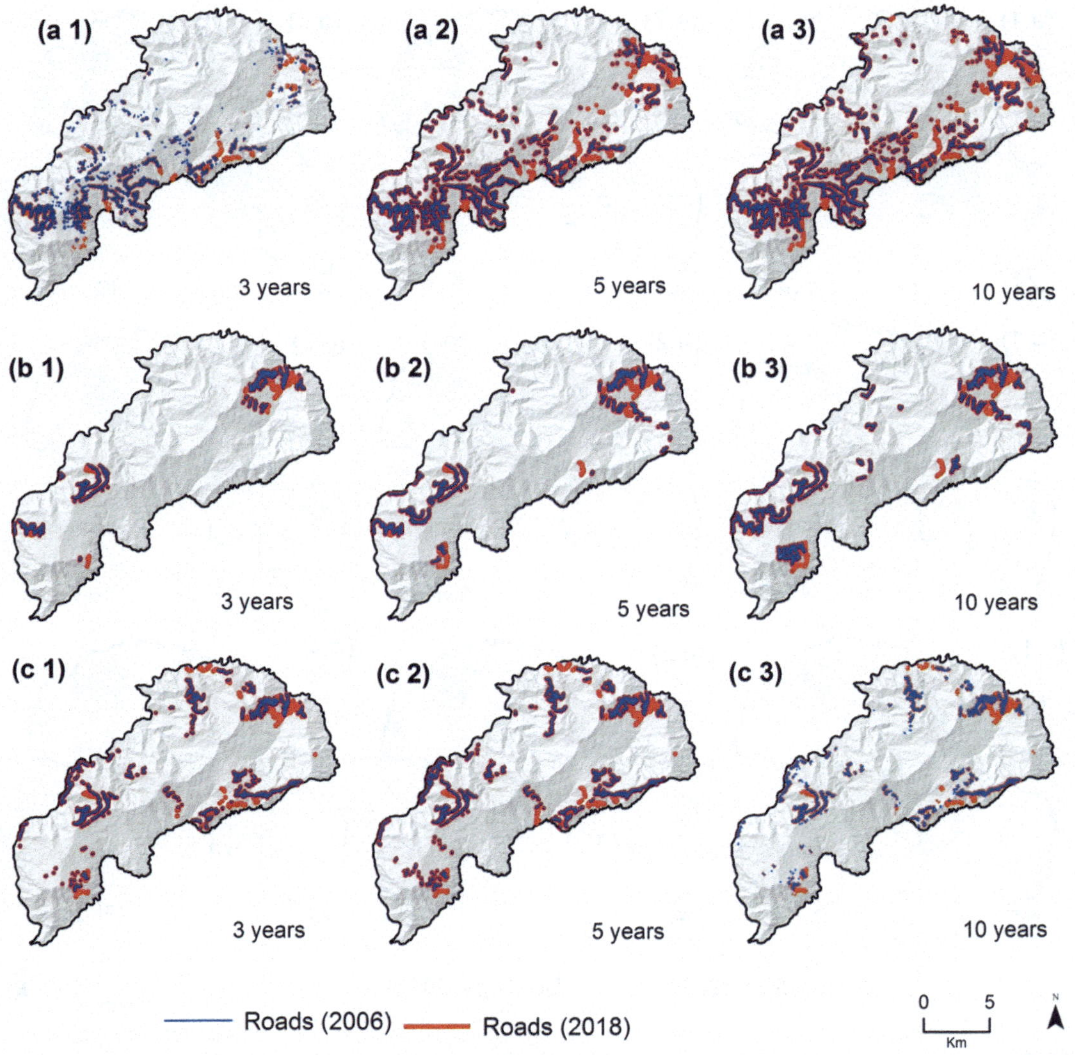

Fig. 9.9 Exposed roadways at the risk scenario of **a** NSDS, **b** CSDS, and **c** RFRS with 3, 5, and 10-year return periods

been shown in several studies (such as Pandey and Seto 2015; Song et al. 2016; Beckers et al. 2020). However, spatial pattern of these exposed elements is rarely overlapped among CSDS and NSDS scenarios. For instance, habitation areas, particularly below summit of ridge and middle of the ridge are at more risk to CSDS due to prevalent cut-slope activities. Conversely, the bottom of the ridge is more risk to NSDS where cut-slope activities are less frequent. Furthermore, an increasing trend of exposed buildings and roads is observed in RFRS, akin to NSDS, and CSDS.

Generally, the condition of RFRS is associated with rock-exposed surface, open land, thin soil, and steep slope that are unfavorable for agricultural activities. Therefore, we found the exposed agriculture land is lower to RFRS than NSDS in terms of its spatial coverage. The increasing trend in such a condition indicates the conversion of open land to agricultural land. It is possibly associated with coupling dynamics of population

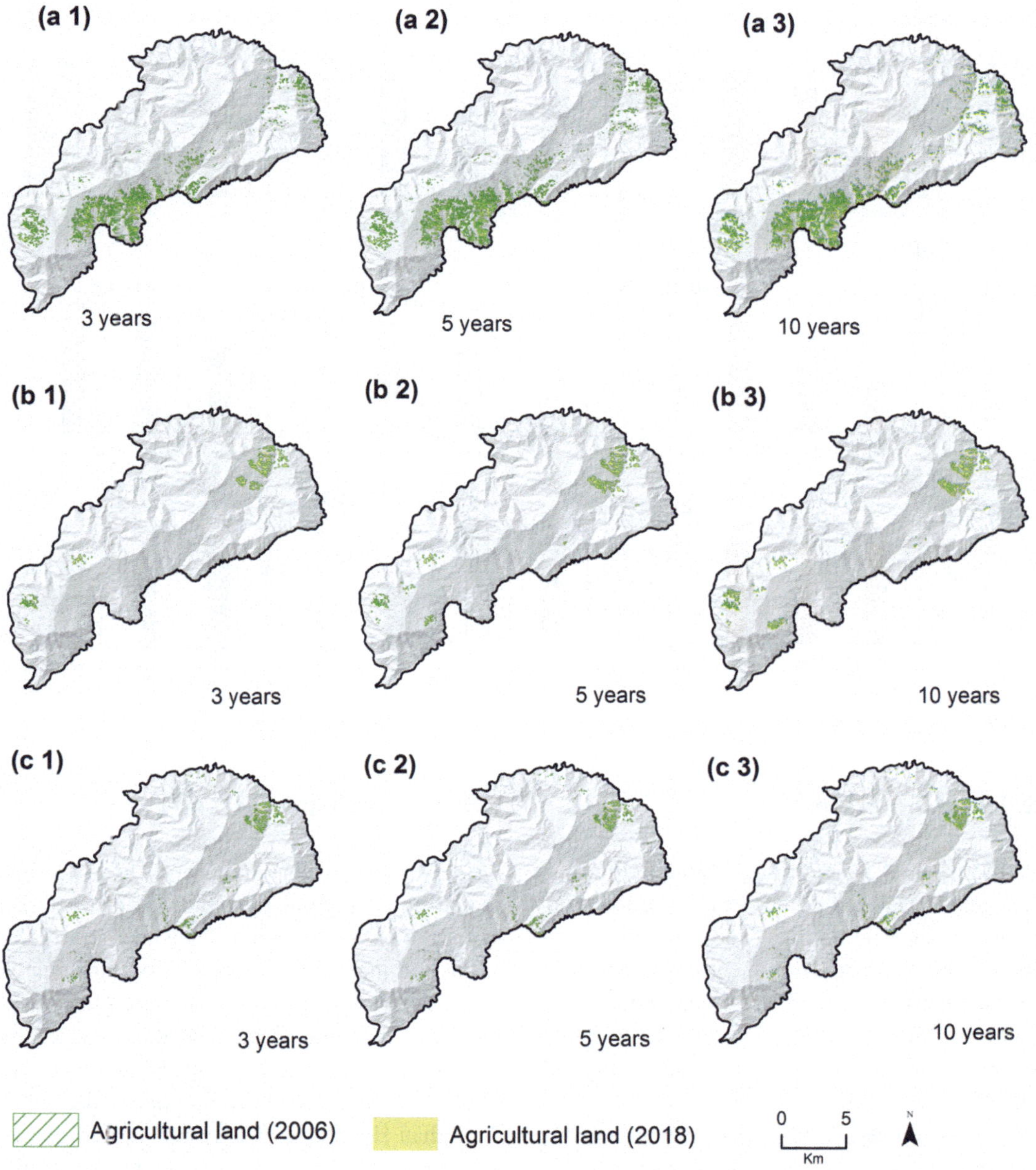

Fig. 9.10 Exposed agricultural land under the hazard scenarios of **a** NSDS, **b** CSDS, and **c** RFRS with 3, 5, and 10-year return periods

growth and damaging agricultural land observed in NSDS and CSDS. However, the opposite scenario, namely an increasing trend in agricultural land, is observed within RFRS. Typically, RFRS conditions include rock-exposed surfaces, open land, thin soil, and steep slopes, generating them unsuitable for agricultural activities. Hence, the coverage of exposed agricultural land is lower in RFRS compared to NSDS. The increasing pattern in these circumstances indicates a transformation of open land into agricultural use, potentially influenced by the interaction of population increase and the negative effect on agricultural land found in NSDS and CSDS scenarios.

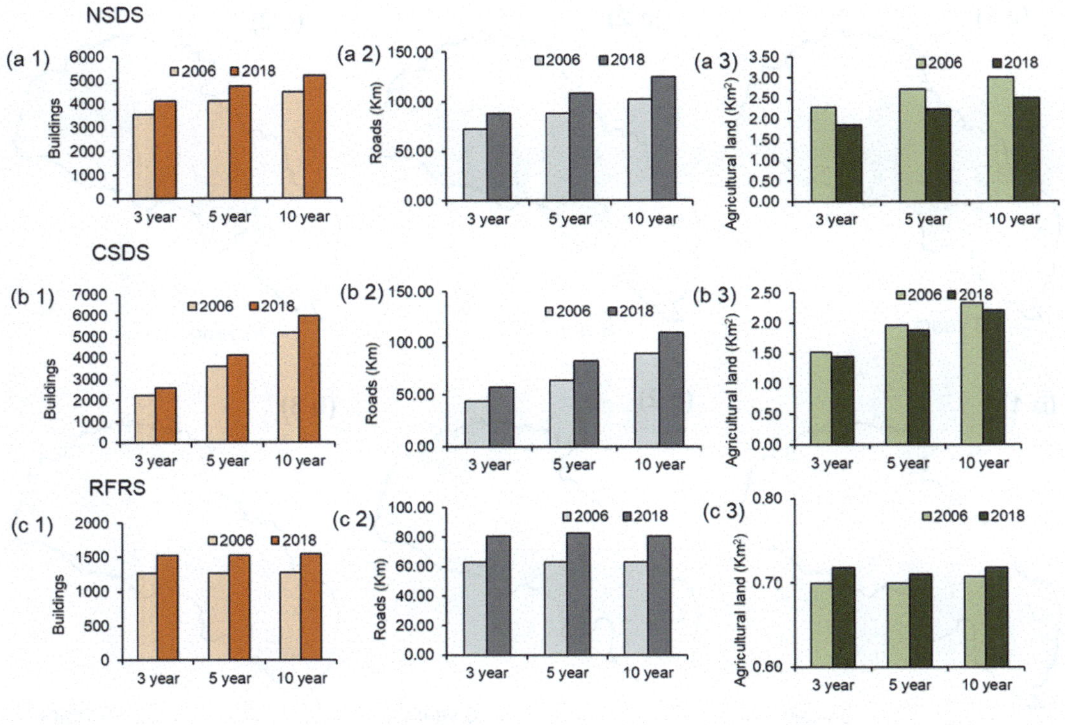

Fig. 9.11 Variations in exposed elements among NSDS, CSDS, and RFRS hazard scenarios

9.5.3 Limitations

At the regional scale, these research findings are reasonable (Jaiswal et al. 2011). Yet, the study considered the limitations and uncertainty of the model. In these models, uncertainty may arise in the debris slides of cut-slope induced. As anthropogenic activities control the landslides of cut-slope induced, temporal probability calculation is complicated based on past landslide occurrences. In respect of actual landslide occurrences, we also use limited inventory data from 2002 to 2018. This study omits extreme event landslides like 1950, 1968, or 1996. This limited event data may enhance the uncertainty between the temporal probability variations. In our hazard model, landslide size data are limited, and the volumes of different periods also do not allow the magnitude probability to be evaluated.

9.6 Conclusion

We have developed a framework to assess the spatial and temporal dynamics of distinct risk features exposed to various landslide hazard scenarios. An innovative component of our research is the capacity to distinguish between landslides initiated by natural slopes and those caused by human activities within a spatial context. We conducted our study in the Kalimpong district of the Indian Himalayas. The results of our investigation revealed the complex relationship and differences between spatial and temporal landslide maps. Our observation indicates that the differences between situations involving natural and anthopogenic slope-induced landslides are more significant than the similarities, even when the environmental conditions are identical. In addition, our study has revealed a concerning pattern: a rise in the development of infrastructure such

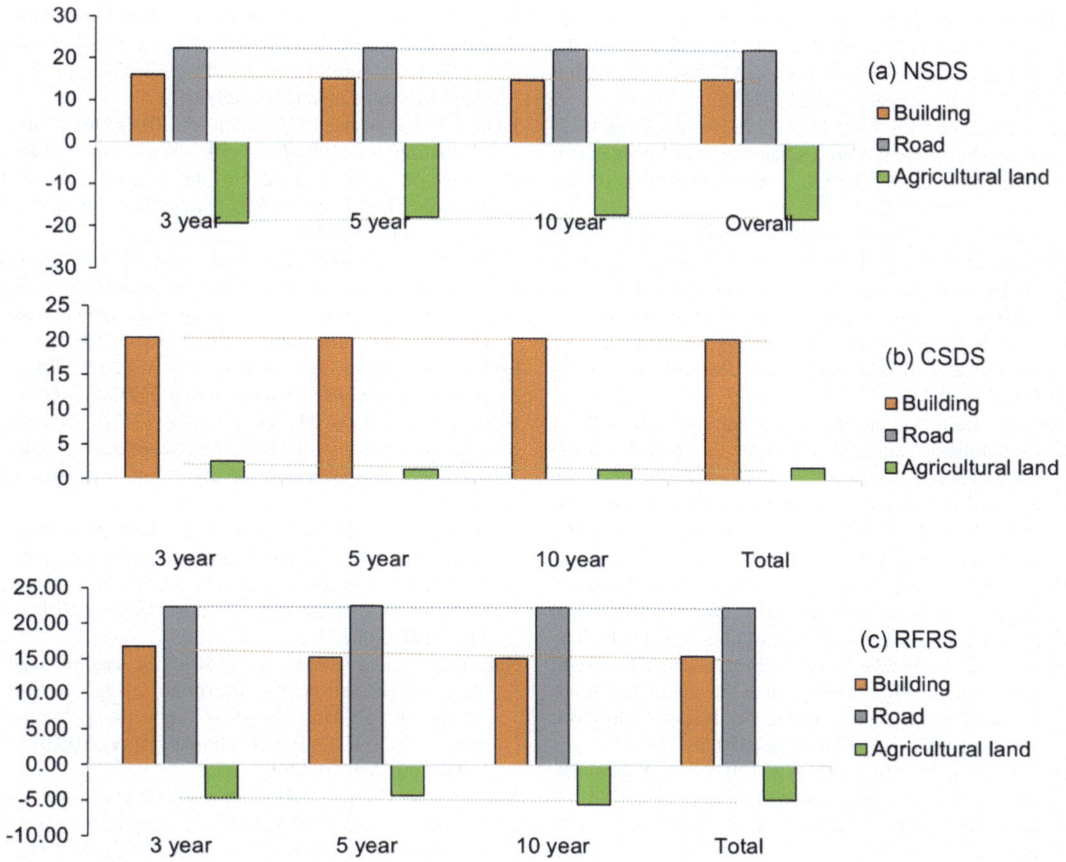

Fig. 9.12 Changing trend in exposed elements among NSDS, CSDS, RFRS hazard scenario

as buildings and roads, along with a decrease in agricultural land, especially in areas with high levels of risk. These data highlight the increasing risk in the study area. Overall, our framework is adaptable and can be applied elsewhere, providing valuable insights for land use policy.

Acknowledgements The research was supported by Junior Research Fellowship of the University Grant Commission (India) (S. Bera; UGC-Ref. No.: 3181) and the Digital Globe Foundation, USA (S. Bera; Sales Order Number: 0057506095)

References

Ajin RS, Nandakumar D, Rajaneesh A et al (2022) The tale of three landslides in the Western Ghats, India: lessons to be learnt. Geoenvironmental Disasters. https://doi.org/10.1186/s40677-022-00218-1

Apollo M (2017) The population of Himalayan regions—by the numbers: past, present and future. In: Contemporary studies in environment and tourism

Beckers V, Poelmans L, Van Rompaey A, Dendoncker N (2020) The impact of urbanization on agricultural dynamics: a case study in Belgium. J Land Use Sci. https://doi.org/10.1080/1747423X.2020.1769211

Bera S, Guru B, Chatterjee R, Shaw R (2020) Geographic variation of resilience to landslide hazard: a household-based comparative studies in Kalimpong hilly region. Int J Disaster Risk Reduct. https://doi.org/10.1016/j.ijdrr.2019.101456

Bera S, Melo R, Guru B (2021) Assessment of exposed elements in a changing built environment by using an integrated model of debris flow initiation and runout (Kalimpong region, Himalaya). Bull Eng Geol Environ. https://doi.org/10.1007/s10064-021-02352-w

Budha PB, Paudyal K, Ghimire M (2016) Landslide susceptibility mapping in eastern hills of Rara Lake, western Nepal. J Nepal Geol Soc. https://doi.org/10.3126/jngs.v50i1.22872

Cebulski J (2022) Impact of river erosion on variances in colluvial movement and type for landslides in the Polish Outer Carpathians. CATENA. https://doi.org/10.1016/j.catena.2022.106415

Chen CW, Tung YS, Liou JJ et al (2019) Assessing landslide characteristics in a changing climate in northern Taiwan. CATENA. https://doi.org/10.1016/j.catena.2018.12.023

Coe JA, Michael JA, Crovelli RA, Savage WZ (2000) Preliminary map showing landslide densities, mean recurrence intervals, and exceedance probabilities as determined from historic records, Seattle, Washington

Crovelli R (2000) Probability models for estimation of number and costs of landslides. US Geol Surv Open-File Rep

Das I, Stein A, Kerle N, Dadhwal VK (2011) Probabilistic landslide hazard assessment using homogeneous susceptible units (HSU) along a national highway corridor in the northern Himalayas, India. Landslides 8:293–308. https://doi.org/10.1007/s10346-011-0257-9

Froude MJ, Petley DN (2018) Global fatal landslide occurrence from 2004 to 2016. Nat Hazards Earth Syst Sci. https://doi.org/10.5194/nhess-18-2161-2018

Fu S, Chen L, Woldai T et al (2020) Landslide hazard probability and risk assessment at the community level: a case of western Hubei, China. Nat Hazards Earth Syst Sci. https://doi.org/10.5194/nhess-20-581-2020

Gao Z, Ding M, Huang T et al (2022) Landslide risk assessment of high-mountain settlements using Gaussian process classification combined with improved weight-based generalized objective function. Int J Disaster Risk Reduct. https://doi.org/10.1016/j.ijdrr.2021.102662

Gaprindashvili G, Van Westen CJ (2016) Generation of a national landslide hazard and risk map for the country of Georgia. Nat Hazards. https://doi.org/10.1007/s11069-015-1958-5

Ghosh S, van Westen CJ, Carranza EJM et al (2012) Generating event-based landslide maps in a data-scarce Himalayan environment for estimating temporal and magnitude probabilities. Eng Geol 128:49–62. https://doi.org/10.1016/j.enggeo.2011.03.016

Guzzetti F, Reichenbach P, Cardinali M et al (2005) Probabilistic landslide hazard assessment at the basin scale. Geomorphology. https://doi.org/10.1016/j.geomorph.2005.06.002

Haque U, da Silva PF, Devoli G et al (2019) The human cost of global warming: deadly landslides and their triggers (1995–2014). Sci Total Environ. https://doi.org/10.1016/j.scitotenv.2019.03.415

Jaiswal P, van Westen CJ (2009) Estimating temporal probability for landslide initiation along transportation routes based on rainfall thresholds. Geomorphology 112:96–105. https://doi.org/10.1016/j.geomorph.2009.05.008

Jaiswal P, van Westen CJ, Jetten V (2010) Quantitative landslide hazard assessment along a transportation corridor in southern India. Eng Geol. https://doi.org/10.1016/j.enggeo.2010.09.005

Jaiswal P, Van Westen CJ, Jetten V (2011) Quantitative estimation of landslide risk from rapid debris slides on natural slopes in the Nilgiri hills. Nat Hazards Earth Syst Sci. https://doi.org/10.5194/nhess-11-1723-2011

Li Y, Chen W (2020) Landslide susceptibility evaluation using hybrid integration of evidential belief function and machine learning techniques. Water (Switzerland). https://doi.org/10.3390/w12010113

Lombardo L, Opitz T, Ardizzone F et al (2020) Space-time landslide predictive modelling. Earth-Sci Rev

Merghadi A, Yunus AP, Dou J et al (2020) Machine learning methods for landslide susceptibility studies: a comparative overview of algorithm performance. Earth-Sci Rev

Mersha T, Meten M (2020) GIS-based landslide susceptibility mapping and assessment using bivariate statistical methods in Simada area, Northwestern Ethiopia. Geoenviron Disasters. https://doi.org/10.1186/s40677-020-00155-x

Nefeslioglu HA, Gorum T (2020) The use of landslide hazard maps to determine mitigation priorities in a dam reservoir and its protection area. Land Use Policy. https://doi.org/10.1016/j.landusepol.2019.104363

Nurwatik N, Ummah MH, Cahyono AB et al (2022) A comparison study of landslide susceptibility spatial modeling using machine learning. ISPRS Int J Geo-Inf. https://doi.org/10.3390/ijgi11120602

Pandey B, Seto KC (2015) Urbanization and agricultural land loss in India: comparing satellite estimates with census data. J Environ Manage. https://doi.org/10.1016/j.jenvman.2014.05.014

Reichenbach P, Rossi M, Malamud BD et al (2018) A review of statistically-based landslide susceptibility models. Earth-Sci Rev

Rumbach A, Follingstad G (2019) Urban disasters beyond the city: environmental risk in India's fast-growing towns and villages. Int J Disaster Risk Reduct. https://doi.org/10.1016/j.ijdrr.2018.11.008

Song J, Ye J, Zhu E et al (2016) Analyzing the impact of highways associated with farmland loss under rapid urbanization. ISPRS Int J Geo-Inf. https://doi.org/10.3390/ijgi5060094

Tsai YJ, Chen YJ, Weng CH et al (2022) Application of stream conductivity to activity of potential large-scale landslide. Front Earth Sci. https://doi.org/10.3389/feart.2022.759556

Varnes D (1984) Landslide hazard zonation: a review of principles and practice. Nat Hazards

Cyclone Induced Hazard Assessment Using Geospatial Technology in East Coast of India

10

Toushif Jaman, K. Dharanirajan, and S. V. Shivaprasad Sharma

Abstract

Cyclone is the most hazardous phenomenon all over the globe and its effect can be found in the coastal regions of the world. Cyclone causes flooding, storm surge, landslides, heavy rain, and huge property and environmental damage. The intensity and number of cyclones are triggering due to climate change and its relative impact over ocean and land surface temperature in recent decade. The purpose of the study is to assess cyclone induced hazard. The study adopts Geographic Information Systems (GIS)-based Analytical Hierarchy Weightage matrix of Bhadrak district and proposes strategic measures for sustainable cyclone disaster management of that area. Land use/land cover (LULC), geomorphology, lithology, elevation, slope, drainage density, topographic wetness index (TWI), and rainfall are considered important cyclone hazard factors. The overall accuracy and Kappa coefficient statistics are 90% and 87.50%, respectively. The resultant cyclone induced hazard zonation map of the study area reveals that the total area is subjected to four categories such as very high—7.80% (18,564.28 ha), high—40.24% (95,890.43 ha), moderate—31.48% (74,974.42 ha), and low—20.48% (48,821.54 ha)—cyclone induced hazard area of Bhadrak district. This study will help the decision-makers and district authority to yield good disaster mitigation measure for the study area.

Keywords

Cyclone · GIS · Analytical Hierarchy Process (AHP) · Kappa coefficient statistics · Hazard zonation

T. Jaman (✉)
North Eastern Space Application Centre (NESAC), Department of Space, Government of India, Shillong, Meghalaya 793103, India
e-mail: toushifphd@gmail.com

T. Jaman · K. Dharanirajan
Department of Coastal Disaster Management, Pondicherry University, Puducherry, India

S. V. Shivaprasad Sharma
Disaster Management Support Division, National Remote Sensing Centre (ISRO), Hyderabad, India

10.1 Introduction

Tropical cyclone is one of the destructive natural calamities in coastal regions (Jaman et al. 2021). It also causes other hazards such as storms, heavy storm surge, and coastal inundation in the low-lying area that is common in most of the coastal province of the world. Cyclone induced hazard includes all these hazards that

S. Talukdar et al. (eds.), *Communication, Science, Technology and Innovation in Disaster Risk Management*, GIScience and Geo-environmental Modelling, https://doi.org/10.1007/978-3-031-77189-7_10

are associated with cyclonic event. Study shows that these hazards are affecting millions of people every year. Its risk is shaped by the physical characteristics and socioeconomic variables of the region. Especially, unscientific development and climate change are becoming a threatening factor to intensify the risk. Therefore, researchers have given importance to spatial planning for disaster preparedness, especially since it may aid in lengthy urgent situation prevention (Basawaraja et al. 2011; Steinberg and Lindfield 2012; Wapwera and Egbu 2013; Watson 2009; UNDP 2015). In the recent era by the European Union and other countries, "Disaster Risk Reduction" approach has been adopted to minimize the impact of the hazardous events and to save human lives. To put it another way, effective disaster goals are to reduce harm to society and assets just before catastrophic event, but its effectiveness in disaster prevention and response necessitates a high degree of expertise and governmental collaboration and integration, as well as an adherence from many other stakeholders involved as a partnership in environmental sustainability (Onyenechere 2010; Sutanta 2012; Watson 2009).

The cyclone induced hazard zonation mapping is widely acknowledged for spatial planning (Dewan 2013; Schmidt-Thome 2006; Dewan 2013; Khailani and Perera 2013). The method of zoning involves integrating data components with high levels of similarity and isolating data units with low levels of similarity, all in accordance with the assessment criteria. A zoning system is able to be employed to recognize spatial trends and to analyze a phenomenon's underlying causes (Zheng et al. 2005). Geographic information systems (GIS) and remote sensing technologies are essential tools for preparing for cyclone induced hazard mapping (Dewan and Yamaguchi 2009; Manfreda et al. 2011; Suriya and Mudgal 2011; Albano et al. 2014; Samela et al. 2016, 2017).

In Indian sub-continent, Bay of Bengal is highly favorable for tropical cyclone formation. The eastern coastal region, especially Odisha Coast of India, has faced numerous cyclones in past history (Table 10.1). This study evaluated analytical hierarchy process-based cyclone hazard induced zonation map utilizing remote sensing and GIS tools along the Odisha coast, India. The cyclone induced hazard includes other associated hazards such as storm surge, soil erosion, and high wind speed. Annual rainfall, elevation, slope, drainage network (Density), lithology, land use/land cover, geomorphology, and TWI are all variables that contribute to cyclone induced hazard. These variables were taken into account when determining the severity of the hazard using consolidated matrix value (Saini et al. 2012). Cyclone induced hazard zonation will help the executives and policymakers in identifying the hazardous areas and emphasizing cyclone prevention and mitigation activities.

Table 10.1 Tropical cyclone along Odisha Coast, India

Sl. No	Name of the cyclone	Month and Year	Intensity (km/h)	Pressure gradient [hPA (mbar)]
1	Super cyclone	October, 1999	260	912
2	Phailin	October, 2013	230	936
3	Hudhud	October, 2014	215	950
4	Titli	October, 2018	195	944
5	Fani	May, 2019	215	900
6	Bulbul	November, 2019	155	990
7	Amphan	May, 2020	260	920
8	Yaas	May, 2021	140	970
9	Gulab	September, 2021	110	976
10	Jawed	December, 2021	65	1000

Sources Meteorological Department (IMD)

10.2 Study Area

Bhadrak district is one of the cyclone-affected districts in Odisha covering 2505 km^2 (Fig. 10.1). It has seven tahasils namely Bant, Bhadrak, Dhamanagar, Chandabali, Tihidi, Basudevpur, and Bhandaripokhari. The average daily temperature ranges between 24.6 °C and 36.4 °C. Annual average rainfall of this region is 1427 mm. The region is severely affected by high rainfall during south-west monsoon, wherein 70% of rainfall occurs within this season. Bhadrak district is frequently affected by cyclonic storms originating in the Bay of Bengal during the monsoon, leading to heavy rainfall and strong winds in the region. The cyclone in this region causes consequent disaster such as flood and soil erosion.

10.3 Data and Methods

10.3.1 Description of Data and Database Generation

Sentinel-2 images from 2020 and an ESRI virtual raster overlaid on a vector layer at a scale of 1:50,000, and rainfall data were also used in the research (2000–2019) for the study region from the Tropical Rainfall measuring mission (TRMM), a digital elevation model (ALOS PALSAR) of the province from Alaska Satellite Facility (ASF), lithology and geomorphology map from Bhukosh, human population, and cyclone data of Digital satellite data (Monkkonen 2008; Gondo and Zibabgwe 2010), questionnaire survey, and legal research were used to collect relevant data for cyclone induced hazard analysis (Table 10.2).

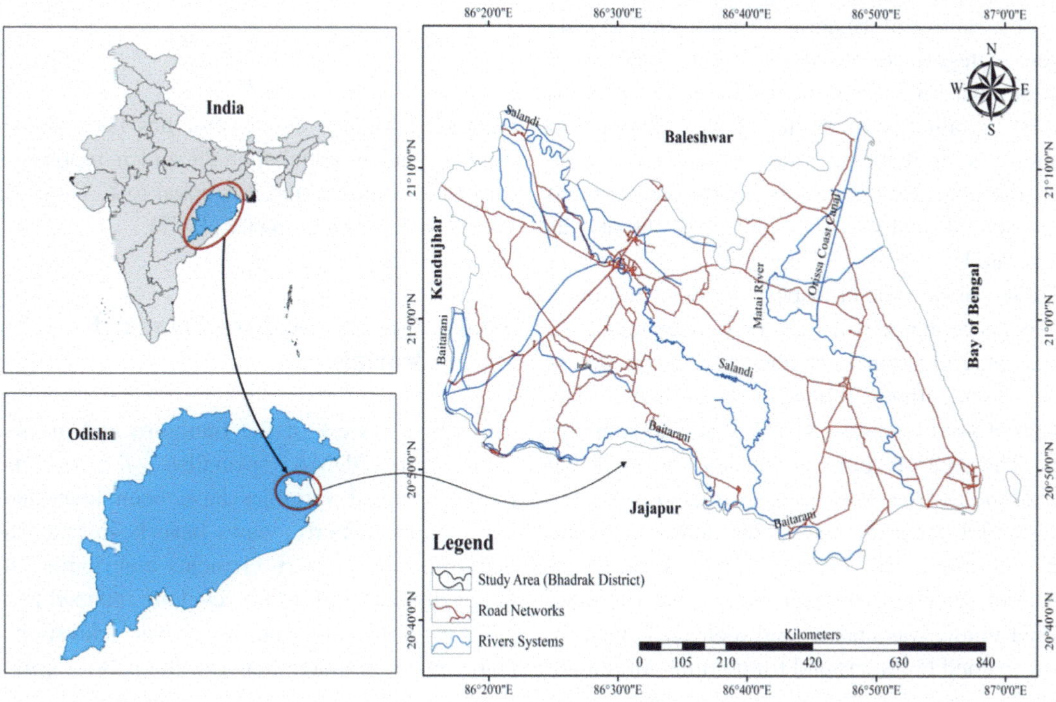

Fig. 10.1 Locational overview of study area

Table 10.2 Material used for research

Satellite/Sensor	Resolution (m)	Acquisition time	Sources
Sentinel 2A	10	2020	Sentinel Copernicus
DEM ALOS PALSAR	12.5	–	Alaska Satellite Facility (ASF)
Rainfall	5 × 20	2000–2019	TRMM
Geomorphology/Lithology	1:50,000	2020	Bhukosh

Both primary and secondary data have processed on geospatial software (ArcGIS-Vs10.4). Surface modeling function has produced a map of the research area's relative relief. The river network data have been extracted by applying the hydrological model tools on the ArcGIS platform using Alos Palsar DEM. The secondary data are spatially merged with the block level database of Bhadrak district, and all data were categorized based on hazard categories. After that, the Analytical Hierarchy Process (AHP) qualitative techniques have been used to determine influencing value of each criterion. All the processed data was finally incorporated into spatial analysis tools and final output was generated.

The most crucial aspect in determining a map's acceptance is its accuracy assessment. No data categorization is deemed successful unless the classification's reliability is evaluated. The method used to assess accuracy is Kappa analysis. The area of land use and land cover in the Bhadrak district has been estimated for each class after accuracy assessment and adjustment. To determine the classification's accuracy, a random selection of pixel value from the classified image was chosen and their class identity was likened to the ground truth points. In remote sensing, assessing the quality of a classification result is crucial because it shows how successfully the classifier is able to separate the targeted objects from the image.

10.3.2 Application of AHP

Analytical Hierarchical Process (AHP) is a qualitative decision-making judgmental tool. All expertise judgmental weights have been consolidated (based on geometrical mean) into a single matrix (consolidated matrix) to develop the pairwise comparisons matrix. The pairwise comparison matrix weights are determined based on each variable's priority and relative importance with cyclone hazard. After that, the normalized Eigenvector Matrix has been generated by dividing each column's corresponding sum values. Next, the Lambda (λ) max has been calculated by integrating the pairwise comparisons matrix's column sum value and normalized Eigenvector Matrix's normalized weights value. This Lambda (λ) max value has been used to generate the consistency ratio (CR). The CR value provides the likelihood of the matrix's judgments.

$$\mathrm{CI} = \frac{\lambda_{\max} - n}{n - 1} \tag{10.1}$$

where the highest eigenvector has been shown by λ_{max} and is equal to 8.29 and n represents the order of matrix and it is equal to 8. These RI values were given by Saaty (1977).

10.4 Cyclone Induced Hazard Variables

Through a review of the literature and discussions with DDMA specialists, the cyclone induced hazard variables have been identified. Ten cyclone experts' views have been taken to determine the primary elements contributing to the cyclone hazard. Main "cyclone induced hazard" variables are land use/cover, elevation", slope angle, drainage network (density), rainfall (Precipitation), lithology, geomorphology, etc. (Fig. 10.2). Every variable has its own importance in cyclone induced hazard zonation, which is discussed.

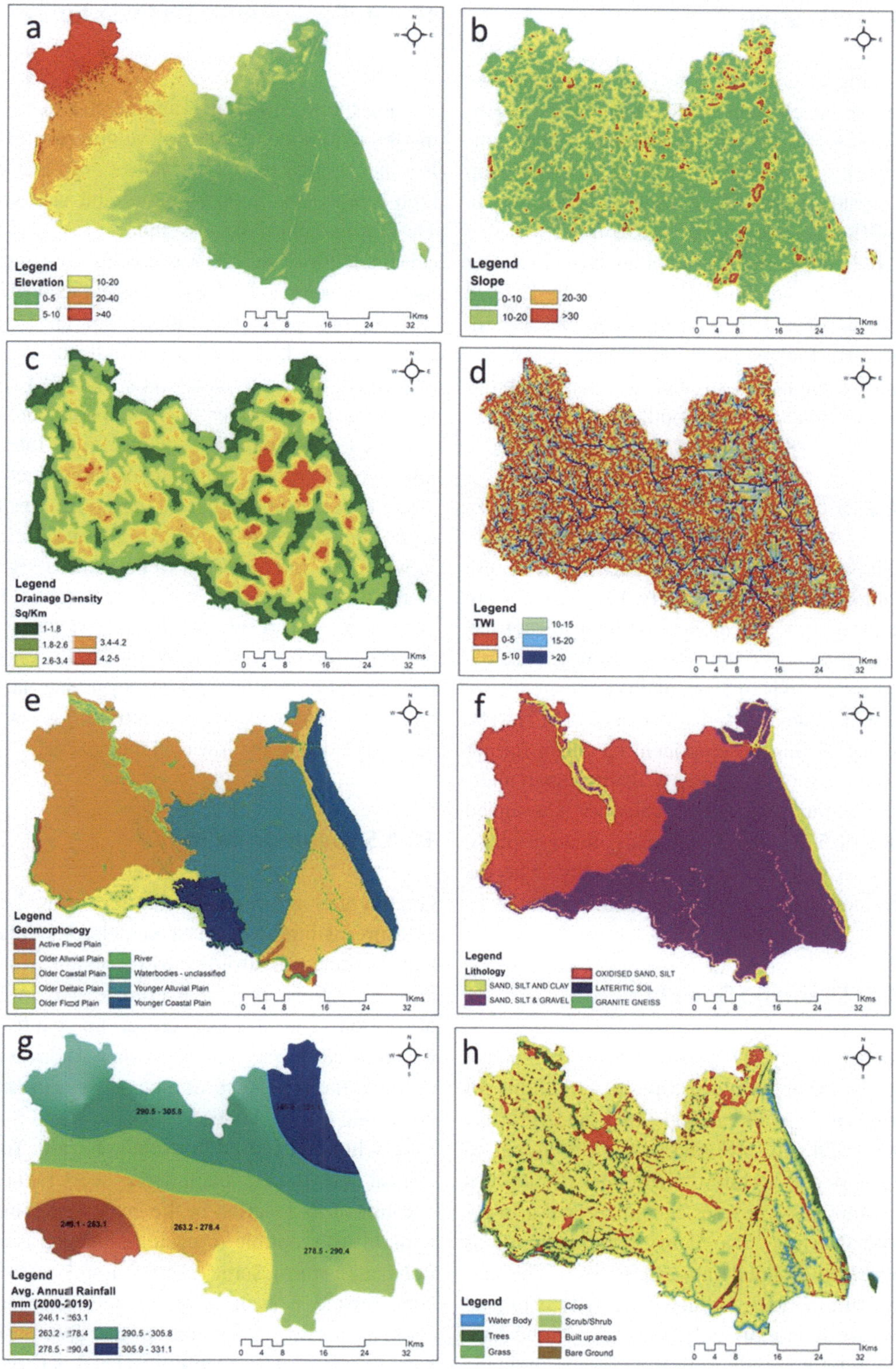

Fig. 10.2 Thematic layers of cyclone induced hazard: **a** Elevation, **b** Slope, **c** Drainage density, **d** Topographic Wetness Index, **e** Geomorphology, **f** Lithology, **g** Rainfall, **h** LULC

10.4.1 Elevation

According to Gigović et al. (2017), elevation is an important element in determining the orientation of water movement as well as the complexity of sea level. The elevation raster map was created using Equal Interval technique in Arc GIS. The elevation was further divided into five subgroups. Throughout this clustering algorithm, the smallest altitude attribute is rated on a scale of 5 due to its high cyclone hazard, whereas the highest elevation category is given a score of 1 due to its extremely low susceptibility zone to cyclone induced flood.

10.4.2 Slope

The slope has a significant impact on the length and amount of stream velocity. Flat surfaces are more prone to water inundation than steeper terrain because water flows more slowly, receives for longer periods of time, and is deposited there (Gigovi'c et al. 2017; Rimba et al. 2017). The slope digital image component has been further classified into five segments using equal interval. The minimum gradient category is assigned a value of 5 throughout this classification phase due to its high cyclone susceptibility, whereas the greatest slope group is assigned a value of 1.

10.4.3 Drainage Density

Drainage density (DD) is the ratio of drainage length to basin area. Absorption, erodibility of external surfaces, flora, slope, and time all affect drainage density. Infiltration is proportional to drainage density. More drainage density implies more runoff for the river catchment and a lower risk of flooding. The density component has been further divided into five groups. The category with the lowest drainage density is ranked 5 because it is extremely high to cyclone hazards, while the group with the greatest drainage network is ranked 1.

10.4.4 Topographic Wetness Index (TWI)

Among the most crucial variables for determining the likelihood of runoff development is TWI. In other words, high TWI values correlate with high runoff generation potential and vice versa. The Topographic Wetness Index (TWI), a crucial predictor of runoff, was calculated using the method developed by Kirkby and Beven (1979).

TWI represents the capacity of an area to store water. A high TWI score indicates a great propensity to accumulate water. The TWI raster layer has been further classed using common classification techniques, such as Equal Interval, into five subgroups. This categorization methodology divides the number of input variables into equivalent sub-ranges, with Arc Layout deciding in which the breaks ought to be and reassigning distinct beliefs in order of cyclone potential hazard score. As a result, the highest value (> 20) classification is given a value of 5 in the clustering algorithm due to the higher cyclone hazard, while the lowest value (0–5) attribute is set to a value of 1 due to the low cyclone hazard.

10.4.5 Geomorphology

Geomorphology also offers a tool for analyzing and forecasting how river channels, floodplains, and catchment hydrology will respond to long-term alterations in global climate and in land utilization. The classes were grouped into five different categories based on their influence to cyclone hazard. Accordingly, river and water body, one of the classes, has been ranked 5 as it is very highly exposed to cyclone hazard. Young coastal and deltaic plain have been ranked 4 because of its location in the coastal area and it is highly susceptible to cyclone induced coastal flood and storm surge. Active flood plain has been ranked 3 as it is moderately exposed to cyclone hazard. Older coastal and deltaic plain have been 1 as its susceptibility level is very low to cyclone hazard.

10.4.6 Lithology

Lithology, which regulates the pace of water infiltration, has a substantial effect on modifying the terrestrial river flow. The lithological classes for the study region were taken. Later, the lithological map was transformed into a 30 m raster. Lithology is significant mainly in connection with permeability. In this study, granite and gneiss lithology class has been ranked 5 because granite and gneiss have low porosity and permeability and it is highly exposed to cyclone. The permeability of whole rock ranged from about 10^{-6} to $10-^{7}$darcies. Laterite soil is ranked 4 due to its level of permeability. The permeability of laterite soil helps to determine the partitioning of runoff and interflow. The rank 3 has been given to oxidized sand, silt, and clay with caliche nodules because this class is moderately exposed to cyclone hazard. Sand, silt, and gravel is ranked 1 due to its high permeability and porosity.

10.4.7 Rainfall

Changes in terrestrial weather patterns and precise meteorological variables, as well as changes in coastline and storm surge, can all influence cyclones (Santato et al. 2013; Jha et al. 2012). Storms, particularly flash floods, are becoming more common as cyclone induced rainfall intensifies (Guo et al. 2004; Few et al. 2004).

In order to obtain point precipitation data for the years 2000–2019, the watershed area has received data from TRMM. The gross estimated precipitation for each year at each location, as well as the mean of the last 20 years (2000–2019) for every depot, has been estimated using the data obtained and then overlaid to Inverse Distance Weight in the spatial database. It had been transformed into a digital image stratum to maintain a consistent digital image rainfall dataset for the river catchment and its surroundings. Eventually, using Equal Interval, this was divided into five categories. Due to its significant contribution to cyclone risk, the category with the highest rainfall received a ranking of 5 in this category. The minimum rainfall group, on the other hand, is given a ranking of 1, as there is very little cyclone hazard.

10.4.8 Land Use/Land Cover

The area's land use (2020) criteria have been again divided into five broad groupings based on one's potential to raise or reduce the incidence of cyclones, from extremely particularly prone to very low exposed. As a result, the water body received a rating of 5. Settlement (Built-up) area land use/land cover has been given weight of 4 due to its high cyclone danger susceptibility. The value of 3 was given to the land use/cover type categorized as "cultivated" because it has a moderate cyclone hazard. The value of 2 was given to the scrub/shrub use/land cover type because it poses a low cyclone hazard. The value of 1 was given to the vegetation/grassland use/land cover category because the risk of cyclones is very low.

10.5 Results

10.5.1 Cyclone Induced Hazard

10.5.1.1 Contributing Variables

One of the variables in the Bhadrak district's cyclone threat that was taken into consideration was land utilization and land cover. Accuracy in the classification process has been evaluated utilizing the statistical method known as the Kappa coefficient. For cyclone hazard land use, vegetation/grassland, scrub/shrub, agricultural lands, built-up (settlement) location, and sources of water received very low, low, moderate, high, and very high ratings, respectively. Another category of extremely significant cyclone induced hazard variables is geomorphology. The cyclone induced hazard ratings for rivers and ponds, younger coastal and deltaic plains, active flood plains, younger alluvial and older flood plains, and older coastal and deltaic plains were assigned as very low, low, moderate, high, and very high. Because lithology regulates porosity and permeability, it is one of the most valuable significant variables. Cyclone hazard lithology categories for sand, silt, and gravel, sand, silt, and clay, oxidized sand, silt, and clay with caliche nodules, lateritic soil, and granite/granite gneiss are very low, low, moderate, high, and very high, respectively.

Elevation was thought to be one of the most important cyclone hazard contributing variables of Bhadrak districts. The watershed's highest elevation is > 40 m, and its lowest height is 0–5 m. The highest altitude category (30–40 m) was classified as having very low cyclone threat, while the lowest altitude category (0–5 m) had very high cyclone hazard. Slope was thought to be one of the Bhadrak district's cyclone hazard contributing variables. Slope has been divided into percentage-based categories. The research area has a slope of roughly 30% at its greatest point and 0% at its lowest. The greatest slope category (20–30%) was evaluated as having very low cyclone hazard, whereas the lowest slope group (0–5%) had extremely high cyclone induced hazard. One factor that contributed to the cyclone induced hazard zonation in the research area was drainage density. The watershed's highest drainage density is 5 km/sq., while its smallest is 0 km/sq. The greatest drainage network (density) group was classified as having very low cyclone threat, while the lowest drainage density group had very high hazard. Another factor that affects the hazard is the topographic wetness index (TWI), which regulates how much water can accumulate on the ground. While the lowest TWI for the research area is 0–5, the highest TWI for the Bhadrak district is > 20, which was evaluated as a very less cyclone hazard group.

Rainfall has been considered one of the most vital variables in the study area that contributed to cyclone threat. The district experiences yearly rainfall ranging from 246.1 to 263.1 mm on average, with 305.9 to 331.1 mm being the highest. While the greatest rainfall category (305.9–331.1 mm) was grouped as having a very high cyclone hazard region, the smallest precipitation category (246.1–263.1 mm) was classified as having a very low cyclone hazard category (Fig. 10.2g).

10.5.1.2 Cyclone Induced Hazard Mapping

In this research, the outcomes obtained through the application of AHP in cyclone hazard zonation have been represented well. In AHP analysis technique, a CR value below 0.10 indicates that the pairwise comparison matrix exhibits acceptable consistency and the value greater than 0.10

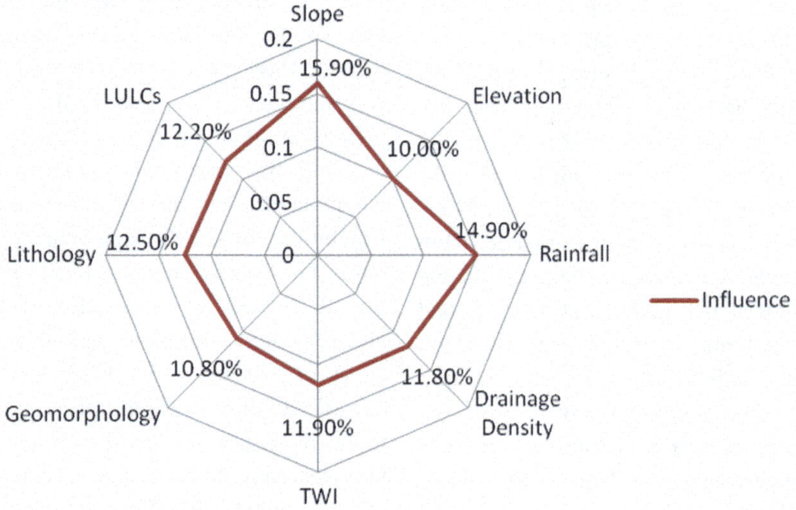

Fig. 10.3 Weightage matrix of evaluation criteria

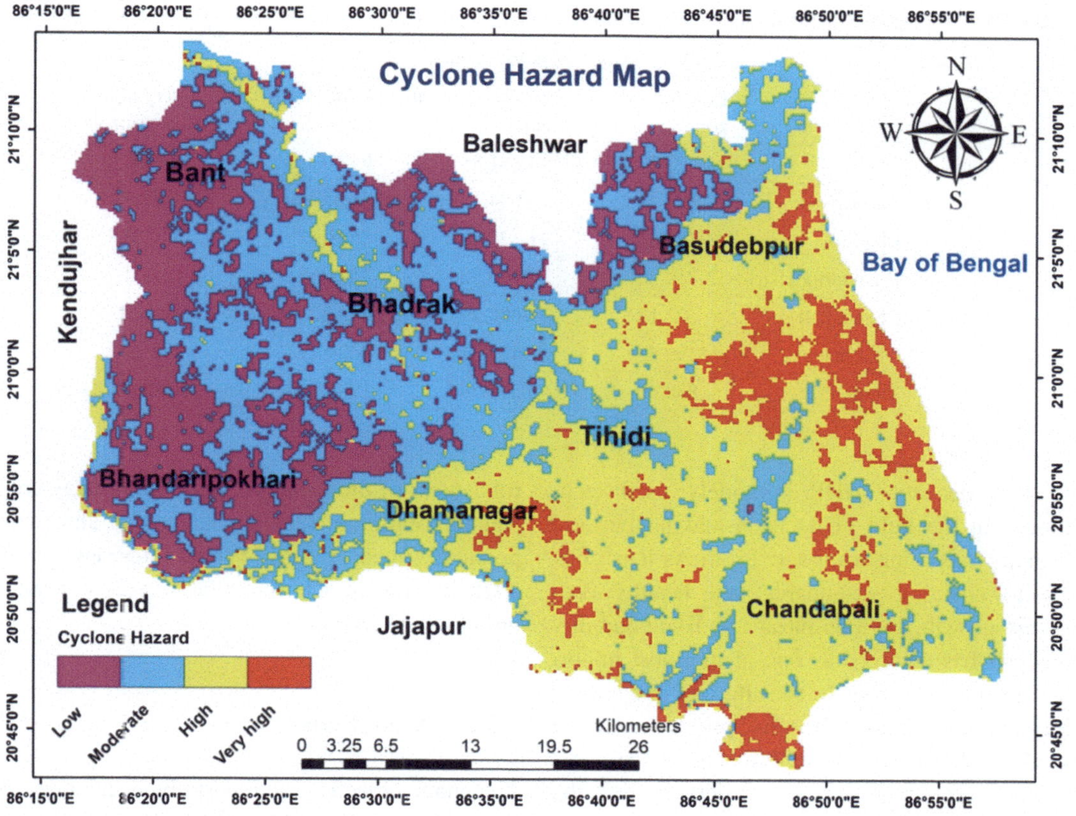

Fig. 10.4 Cyclone induced hazard zonation

demonstrates unacceptable consistency in decision-making. In this study, the CR value is 0.037, which revealed very good consistency. Results show that very high, high, moderate, and low hazard levels have covered 7.80% (18,564.28 ha), 40.24% (95,890.43 ha), 31.48% (74,974.42 ha), and 20.48 % (48,821.54 ha) of total area, respectively (Fig. 10.4; Table 10.3). According to this, a larger section of the study region must be a high- and moderate-hazard area (71.72%). Additionally, the Bhadrak district contains areas with high and medium cyclone induced hazard rates of 40.24% (95,890.43ha) and 31.48% (74,974.42ha), respectively (Table 10.3).

Table 10.3 Cyclone hazard area distribution

Sl. No	Cyclone hazard value	Area (ha)	Percentage (%)
1	Low	48,821.54	20.48
2	Moderate	74,974.42	31.48
3	High	95,890.43	40.24
4	Very high	18,564.28	7.80

10.6 Discussion

A cyclone is one of the most common types of natural disasters in the world. The thorough research of the study area depicts that the area adjacent to the Bay of Bengal is highly vulnerable to cyclone and its landfall. Several weather dynamics which help to form depression over the ocean exist over Bay of Bengal (Table 10.4). The study revealed that the coastal area of Bhadrak district is falling under very high cyclone hazard zone. Due to cyclone, the coastal low-lying area is highly sensitive to storm surge and coastal flooding. Several physical and hydro-meteorological parameters were taken and "Analytical Hierarchy Process (AHP)" has been applied

Table 10.4 Block-wise distribution of cyclone hazard

Hazard zonation Block-wise			Cyclone hazard areas in hectares			
			Low hazard zone	Moderate hazard zone	High hazard zone	Very high hazard zone
Block name	1	Bant	14,073.27	10,861.85	949.04	106.98
	2	Basudebpur	5123.29	9084.13	21,550.64	9582.57
	3	Bhadrak	9841.28	21,327.06	2518.63	12,704.46
	4	Bhandaripokhari	12,719.20	8109.80	1071.65	128.04
	5	Chandabali	27.17	7608.29	41,882.29	6776.25
	6	Dhamnagar	5399.68	9433.31	8229.49	958.29
	7	Tihidi	838.66	8554.79	19,488.18	2124.15

to take decision. AHP is widely used techniques in the field of decision-making because it gives maximum accuracy. From As it is critical to ensuring sustainable and healthy economic growth, cyclone hazard and vulnerability analysis has emerged as one of the most important priorities in remote sensing and geographic information systems (GIS) disciplines around the world (Yahaya et al. 2010; Zou et al. 2013).

As significant hazard variables for the Bhadrak district, this research examines land use/land cover, elevation, slope angle, drainage network (density), geomorphology, lithology, TWI, and rainfall. Several more development-related researchers have shown that as an area's municipal development, impassable cover, and forestation decrease, stream intensity increases. The topographic wetness index (TWI) is another factor that increases the proneness of cyclone induced hazard. A higher TWI score indicates a higher cyclone risk location because it increases the ability of the ground to accumulate water. Existing cyclone risk may be exacerbated by observed and changing climatic patterns.

The results of the study indicate that Bhadrak district is significantly prone to cyclone hazards, with approximately 48.04% of its total area falling within the high to very high hazard categories. This highlights the urgent need for comprehensive disaster management planning and mitigation measures. The accuracy of the hazard zonation map, as measured by overall accuracy and Kappa coefficient statistics (90% and 87.50%, respectively), underscores the robustness of the methodology used. GIS is crucial for real-time tracking of cyclone and their potential impact areas. When integrated with AHP, it can help determine the severity of warning based on changing condition.

10.7 Conclusion

The main objective of the study is to analyze the cyclone induced hazard in Bhadrak district from a Geographic Information Systems (GIS)-based analytical hierarchy analysis and to suggest strategic actions for long-term cyclone and associated hazard management. The cyclone hazard variables were identified through a scientific report, field impression, and discussions with specialists as well as locals. As a result, the Bhadrak district's main cyclone hazard variables LULC, elevation (orientation), slope (angle), drainage network (density), geomorphology, lithology, TWI, and Precipitation (rainfall) have been taken into consideration. The AHP method was used in the metrics map competitive equilibrium for cyclone hazard zonation mapping. Land usage and land cover for the year 2020 have been assessed using Kappa coefficient statistics. The figures for the overall accuracy and the Kappa coefficient are 90 and 87.50%, respectively.

According to the cyclone hazard results in the research region, the Bhadrak district has cyclone hazard levels of 7.80% (18,564.28 ha), very high, 40.24% (95,890.43 ha), high,

31.48% (74,974.42 ha), moderate, and 20.48% (48,821.54 ha), low, respectively. According to this, a larger section of the study region must be a high- and moderate-hazard area (71.72%). Additionally, the Bhadrak district contains areas with high and moderate cyclone hazard rates of 40.24% (95,890.43 ha) and 31.48% (74,974.42 ha), respectively.

This research provides valuable insights into cyclone hazard assessment and zonation in Bhadrak district, Odisha. The resulting hazard zonation map can serve as a crucial tool for decision-makers and local authorities in formulating effective disaster management plans and mitigation strategies. Given the increasing frequency and intensity of cyclones in the region, such proactive measures are essential to safeguard the lives and property of the community while promoting sustainable development in the face of climate change challenges.

Acknowledgements The authors would like to thank Training Division, NRSC (National Remote Sensing Centre) for providing the necessary satellite data and the opportunities to carry out the research work. We would also like to acknowledge Pondicherry University for its selfless support. The first author is very much grateful to Kajal Kumar Mandal for his helpful suggestions.

References

Albano R, Sole A, Adamowski J, Mancusi L (2014) A GIS-based model to estimate flood consequences and the degree of accessibility and operability of strategic emergency response structures in urban areas. Nat Hazards Earth Syst Sci Discuss 14:2847–2865

Basawaraja R, Chari KB, Mise SR, Chetti SB (2011) Analysis of the impact of urban sprawl in altering the land-use, land-cover pattern of Raichur City, India, using geospatial technologies. J Geogr Reg Plan 4(8):455–462

Dewan AM, Yamaguchi Y (2009) Land use and land cover change in Greater Dhaka, Bangladesh: using remote sensing to promote sustainable urbanization. Appl Geogr 29:390–401

Dewan AM (2013) Floods in a megacity: geospatial techniques in assessing hazards, risk and vulnerability. Springer Dordrecht and Springer Heidelberg, New York and London

Few R, Ahern M, Matthies F, Kovats S (2004) Floods, health and climate change: a strategic review. Tyndall Centre Working Paper No. 63. Tyndall Centre for Climate Change Research, Norwich

Gigović L, Pamučar D, Bajić Z, Drobnjak S (2017) Application of GIS-interval rough AHP methodology for flood hazard mapping in urban areas. Water 9(360):1–26

Gondo T, Zibabgwe S (2010) GIs solutions and land management in urban Ethiopia: perspectives on capacity, utilization, and transformative possibilities. Manag Res Pract 2(2):200–216

Guo E, Zhang J, Ren X, Zhang Q, Sun Z (2014) Integrated risk assessment of flood disaster based on improved set pair analysis and the variable fuzzy set theory in central Liaoning Province, China. Nat Hazards 74:947–965

Jaman T, Dharanirajan K, Shivaprasad Sharma SV (2020) Assessment of impact of cyclone hazard on social vulnerability of Bhadrak District of Odisha State during Phailin Cyclone in 2013 and Titli Cyclone in 2018 using multi-criteria analysis and geospatial techniques. Int J Disaster Risk Reduc 53:101997. https://doi.org/10.1016/j.ijdrr.2020.101997

Jaman T, Dharanirajan K, Rana S (2022) Land use and land cover change detection and Its environmental impact on South Andaman Island, India using kappa coefficient statistical analysis and geospatial techniques. Int Res J Eng Technol (IRJET) 9(4)

Jha AK, Bloch R, Lamond J (2012) Cities and flooding: a guide to integrated urban flood risk management for the 21st century. The World Bank, Washington, DC

Khailani DK, Perera R (2013) Mainstreaming disaster resilience attributes in local development plans for the adaptation to climate change induced flooding: a study based on the local plan of Shah Alam City, Malaysia. Land use policy 30(1):615–627

Kirkby M, Beven K (1979) A physically based, variable contributing area model of basin hydrology. Hydrol Sci J 24:43–69

Manfreda S, Di Leo M, Sole A (2011) Detection of flood-prone areas using digital elevation models. J Hydrol Eng 16:781–790

Marchi L, Borga M, Preciso E, Gaume E (2010) Characterization of selected extreme flash floods in Europe and implications for flood risk management. J Hydrol (Amst) 394:118–133

Monkkonen P (2008) Using online satellite imagery as a research tool: mapping: mapping changing patterns of urbanization in Mexico. J Plan Educ Res 28:225–237

Nasrin T, Siddiqui MA, Sarif MN, Bera S, Ramana GV (2024) Assessing the physical vulnerability of the houses to cyclonic hazard and its relation with economic and environmental factors: a study from the Sundarban Biosphere Reserve of India. Nat Hazards. https://doi.org/10.1007/s11069-024-06808-6

Onyenechere EC (2010) Climate change and spatial planning concerns in Nigeria: remedial measures for more effective response. J Hum Ecol 32(3):137–148

Ouma YO, Tateishi R (2014) Urban flood vulnerability and risk mapping using integrated multi-parametric AHP and GIS: methodological overview and case study assessment. Water 6:1515–1545

Rimba AB, Setiawati MD, Sambah AB, Miura F (2017) Physical flood vulnerability mapping applying geospatial techniques in Okazaki City, Aichi Prefecture, Japan. Urban Sci 1(7):1–22

Saaty TL (1977) A scaling method for priorities in hierarchical structures. J Math Psychol 15(3):234–281. https://doi.org/10.1016/0022-2496(77)90033-5

Saini SS, Kaushik SP (2012) Risk and vulnerability assessment of flood hazard in part of Ghaggar Basin: a case study of Guhla block, Kaithal, Haryana, India. Int J Geomat Geosci 3(1):42–54

Samela C, Manfreda S, De Paola F, Giugni M, Sole A, Fiorentino M (2016) DEM-based approaches for the delineation of flood-prone areas in an ungauged basin in Africa. J Hydrol Eng 21:06015010

Samela C, Troy TJ, Manfreda S (2017) Geomorphic classifiers for flood-prone areas delineation for data-scarce environments. Adv Water Resour 102:13–28

Samela CR, Sole A, Manferda S (2018) A GIS tool for cost-effective delination of flood-prone areas. Comput Environ Urban Syst 70:43–52

Santato S, Bender S, Schaller M (2013) The European Floods directive and opportunities offered by land use planning. CSC Report 12. Climate Service Center, Germany

Schmidt-Thome P (2006) Integration of natural hazards, risks and climate change into spatial planning practices. Geological Survey of Finland, Espoo

Steinberg F, Lindfield M (2012) Spatial development and technologies for Green cities. In: Lindfield M, Steinberg F (eds) Green cities. Asian Development Bank, Mandaluyong City

Suriya S, Mudgal BV (2011) Impact of urbanization on flooding: the Thirusoolam sub watershed—a case study. J Hydrol (Amst) 412–413:210–219

Sutanta H (2012) Spatial planning support system for an integrated approach to disaster risk reduction. Doctoral Thesis. The University of Melbourne, Meelbourne, VIC

Uddin K, Gurung DR, Giriraj A, Shrestha B (2013) Application of remote sensing and gis for flood hazard management: a case study from Sindh Province, Pakistan. Am J Geogr Inf Syst 2(1):1–5

UNDP (2015) Strengthening disaster risk governance: UNDP support during the HFA implementation period 2005–2015. UNDP, New York

Wapwera SD, Egbu CO (2013) Master planning system: constraints for planning authorities in Jos Metropolis, Nigeria. Built Human Environ Rev 6:61–81

Watson V (2009) Seeing from the south: refocusing urban planning on the globe's central urban issues. Urban Stud 46(11):2259–2275

Wondim YK (2016) Flood hazard and risk assessment using GIS and remote sensing in lower Awash Sub-basin, Ethiopia. J Environ Earth Sci 6(9):69–86

Yahaya S, Ahmad N, Abdalla RF (2010) Multicriteria analysis for flood vulnerable areas in Hadejia-Jama'are River Basin, Nigeria. Eur J Sci Res 42(1):71–83

Zheng D, Ge Q, Zhang X et al (2005) Regionalization in China: retrospect and prospect. Geogr Res 20:330–344. https://doi.org/10.3321/j.issn:1000-0585.2005.03.002

Spatiotemporal Flood Impact Assessment Using Google Earth Engine—Case of Darrang District, Assam

11

Ranadeep Sarma, Subhas Bera, and P. K. Ryngnga

Abstract

Assam is subject to recurring floods due to its location on the Brahmaputra floodplain and the rivers flowing from nearby hilly regions to the valley. A study was conducted in Darrang district, Assam, using Google Earth Engine (GEE) and Sentinel data to observe the spatial and temporal variation of flood dynamics in 2022. The K-means classification results show that 41% of the district was submerged under water after extracting the permanent water body with the help of Modified Normalized Difference Water Index (MNDWI). Pre-flood, peak flood, and post-flood behavior were observed with a floodwater depth of about 7.2 m at peak duration. The main flooding event lasted almost 12 days, mainly affecting the southwestern and northwestern parts of the district. The intersection of the LULC map and the flood extent map helps identify the regions affected by flooding in and around the study area. The bar morphology of the Brahmaputra River in the study area shows a large amount of erosion and deposition during the peak- and post-flood event, respectively. Several large sand bar depositions took place during the post-flood event. Two rivers from each side of the district were considered to observe the sinuosity index of their river channels, to know the nature of these rivers and their impact on the surrounding environment. It was found that these rivers were meandering in nature with shallow alluvial plains, which is responsible for high flood events in the district.

Keywords

Flooding · Flood depth · GEE · Sentinel-1 · MNDWI · Bar morphodynamics

11.1 Introduction

Floods rank among the most destructive and frequent natural disasters globally, affecting millions annually. The increasing occurrence and intensity of floods in recent decades are closely associated with climate change and its influence on extreme weather events (IPCC 2013). As global temperatures rise, extreme precipitation events become more frequent, leading to higher

R. Sarma · P. K. Ryngnga
Department of Geography, North-Eastern Hill University, Shillong 793022, India

S. Bera (✉)
Department of Remote Sensing and GIS Applications, Hemvati Nandan Bahuguna Garhwal University, Srinagar Garhwal 246174, India
e-mail: sbera17@gmail.com

S. Talukdar et al. (eds.), *Communication, Science, Technology and Innovation in Disaster Risk Management*, GIScience and Geo-environmental Modelling, https://doi.org/10.1007/978-3-031-77189-7_11

flood risks in various regions (CRED 2019; Ramachandran et al. 2019). Developing nations, such as India and Bangladesh, are particularly vulnerable to the devastating consequences of floods (Ali et al. 2019; Bera et al. 2022). The changing climate has disrupted the monsoon season's rainfall pattern in the Indian sub-continent, causing more intense and concentrated rainfall over shorter periods, exacerbating flood hazards (Kulp and Strauss 2019). The impacts of these floods extend beyond immediate damages to infrastructure and properties, which cause disruptions in food and water supplies, displacement of communities, and negative consequences for public health and the environment. Understanding the drivers of flood occurrence and implementing effective mitigation and adaptation strategies are crucial for minimizing the impacts of floods and building resilient communities. Developing nations, with their limited resources and infrastructure, are particularly vulnerable to the devastating impacts of floods. India and Bangladesh, in particular, experience recurrent flooding because of factors such as their geographical location, monsoon climate, and increasing population density in flood-prone areas. Studies have highlighted the growing risks faced by these countries. For instance, research by Ali et al. (2019) utilized satellite remote sensing data and GIS to assess flood hazards in parts of India. Understanding the specific vulnerabilities and risk factors in developing countries is essential for implementing effective flood management strategies that can save lives and protect livelihoods.

Assam is considered one of the most flood-impacted regions in the northeastern part of India (IFRC 2003). Every year, a large part of the region is impacted by flood, which causes great losses and damage to lives and property (Hariprasanna & Lodrick, 2022). In the study, flood assessment was done in the district of Darrang, which lies in the central part of Assam. Every year, during monsoon season, most part of the district is affected by flood due to high precipitation and by its braided and meandering rivers that are highly prone to channel migration and avulsion (Pal et al., 2020). Brahmaputra that flows from east to western part of the district also has a great impact on the flood events. Most of the small rivers flow from Indo-Bhutan border, flow through the district, and merge into the Brahmaputra. High precipitation in the Indo-Bhutan region also causes an increase in the water levels of the rivers and highly sinuous, braiding, avulsion, and alluvial nature of the rivers are the main reasons of the flood events (Xie et al. 2018).

In the previous study, it was found that Synthetic Aperture RADAR data is highly useful for detection of inundated areas during peak flood behavior. Optical data cannot be used during the monsoon season due to cloud cover and extreme climatic conditions. The ability of the SAR data to operate during both day and night time and its ability to operate in extreme weather condition like rainfall and cloud cover have made it more applicable for flood studies and its dominance over optical data (Conde and De Mata Muñoz 2019; Rahman and Thakur 2018). In the study, Sentinel-1 SAR data was used, which can openly be accessed from Copernicus Open Hub. Sentinel data has a temporal resolution of 6–12 days (including 1A and 1B at the equator), a spatial resolution of 20×20 m, and a wide swath (Muro et al. 2016). It consists of dual polarization with VV and VH and from the recent studies, it was concluded that VV polarization data is highly significant for mapping of water extent areas (Martinis et al. 2018; Twele et al. 2016). Water surface acts as a specular reflector, reflects back low energy to the sensor, and enables us to differentiate it from another feature (Khan et al. 2022; Schlaffer et al. 2017). Floodwater depth was estimated for both large and small flood events and compared to model-stimulated water depth predictions (Cohen et al. 2018). Floodwater depth estimation was done using Google Earth Engine (GEE) to identify the flood depth of the inundated areas. Flood extent layer and Digital Elevation Model (DEM) were used in which Sentinel-1 data for flood extent layer and STRM DEM were used to calculate the flood depth (Tazmul Islam and Meng 2022). The study also focuses on bar morphology of the Brahmaputra River, which includes changes in the area of bar during peak flood and post-flood compared to pre-flood. Sentinel-1 and Sentinel-2 data were used to digitize the Sand Bar to identify the change detection. Two rivers were considered (one from eastern part of the district flowing from north to south and the other from western part of the district flowing from north to south and getting submerged into Brahmaputra River) to determine the nature of the rivers and their effects on the surroundings areas. In words, the study focuses on threefold objectives: (i) identify flood-inundated areas; (ii) estimate flood depth; (iii) identify changes in the Bar Morphology during and after the flood.

11.2 Study Area

Assam is the largest state in northeast India (covers about 2.4% of India's total land area), situated at the eastern Himalayan foothills and along the middle course of the Brahmaputra and Barak Rivers (Government of India 2011). The Brahmaputra basin, encompassing an area of 5,80,000 sq. km, includes 70,634 sq. km within Assam (Government of India 2001). The state has a diverse topography with hills and plains. Assam has a varied climate with temperatures ranging from 6 to 38 °C. The southwest monsoons bring high humidity and heavy rainfall to the region, averaging 120 inches or more per year in the Brahmaputra valley and surrounding areas. The monsoons support the rich biodiversity of Assam, which is comparable to that of equatorial rainforests (Govt of Assam 2023). Darrang district is located in the central part of Assam on the north bank of the Brahmaputra River (Fig. 11.1). The Brahmaputra River flows to the south of the district, which is bordered by Sonitpur district to the east and Kamrup district to the west.

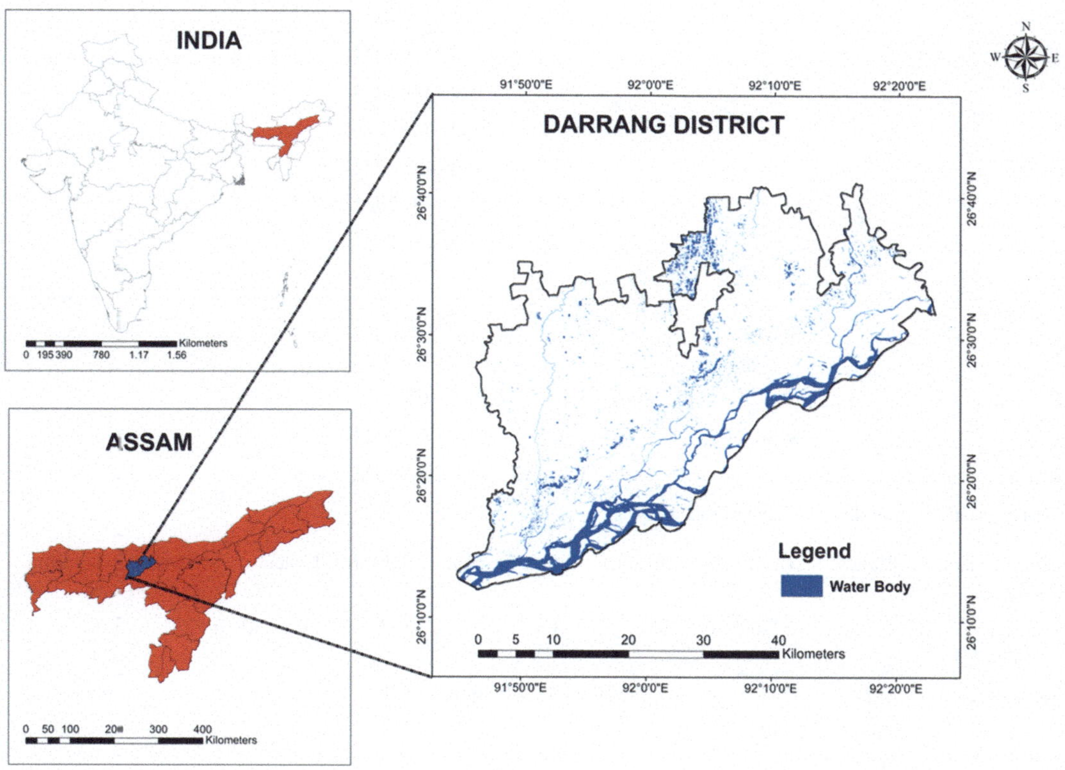

Fig. 11.1 Location and water body of Darrang district of Assam, India

11.3 Materials and Methods

The study uses Sentinel-1, SAR GRD data to analyze the flood behavior. Sentinel-1 images (Table 11.1) were used to capture the pre-, post-, and peak flood conditions of the district. Modified Normalized Difference Water Index (MNDWI) was applied to identify the permanent water bodies using Sentinel-2 data. All the data were obtained from Copernicus Open Access Hub. The rainfall data was collected from Indian Meteorological Department. Sentinel-2 data was also used to create Land Use Land Cover (LULC) map to assess the impact of flood on settlement and agricultural land. A graphical representation of the proposed methodology is shown in Fig. 11.2.

11.3.1 MNDWI Mapping

The study applies the Modified Normalized Difference Water Index (MNDWI) to extract water bodies using the green and short-wave infrared (SWIR) bands (Xu 2006). MNDWI is suitable for this purpose because water has a strong absorption in the SWIR band, resulting in positive values when calculated as MNDWI and expressed as (Eq. 11.1)

$$\text{MNDWI} = \frac{\text{GREEN} - \text{MIR}}{\text{GREEN} + \text{MIR}} \qquad (11.1)$$

The method employs two different approaches for unsupervised classification using SNAP software: one cluster with 10,000 random iterations (Xu 2006) or two clusters with 30,000 random

Table 11.1 Properties of the Sentinel-1 and Sentinel-2 datasets utilized in the research

Scene	Datasets	Date
1	Sentinel-1	01-05-2022
2		11-05-2022
3		23-05-2022
4		06-06-2022
5		16-06-2022
6		18-06-2022
7		28-06-2022
8		30-06-2022
9		12-07-2022
10		24-07-2022
11		03-08-2022
12		15-08-2022
1	Sentinel-2	17-04-2022
2		28-10-2022

Characteristics of the datasets used in the study

Characteristics/ Data used	Band	λ (nm)	Product type	Sensor mode	Platform	Order	Swath (km)	Resolution
Sentinel-1 (VH and VV)	C-band	55,465,760	SLC	IW	S1A	Descending	250	5 (m) × 20 (m)
Sentinel-2	Blue	492.1	S2MSI1C	MSI Imaging	S2B		290	10 m
	Green	559.0						
	Red	664.9						
	NIR	832.9						

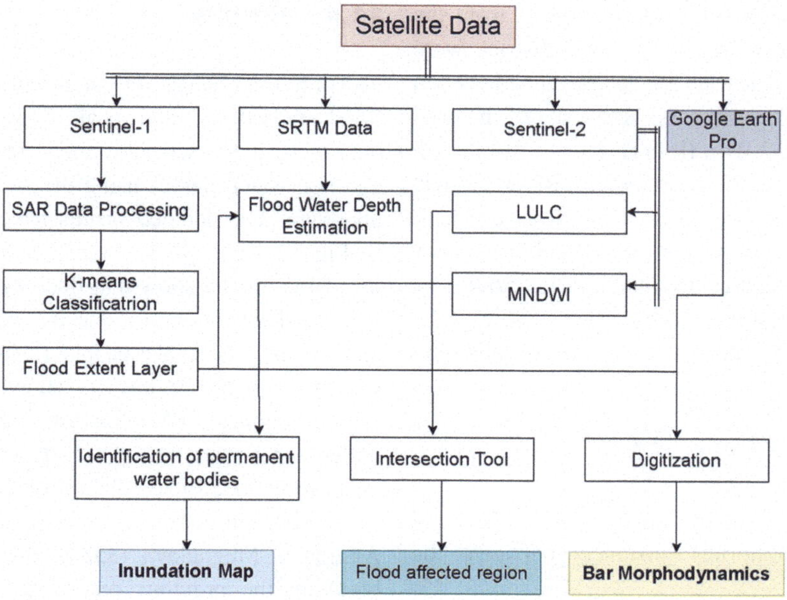

Fig. 11.2 Schematic diagram of methodology

iterations (Ruzza et al. 2019). Consequently, both soil and vegetation exhibit negative values in the results. The method enhances the contrast between built-up areas and water bodies, but there is a noticeable increase in water features and a decrease in built-up areas, transitioning from positive to negative values (Xu 2006).

To identify permanent water bodies, March 2022 data was used, as this time was considered optimal for detection before the monsoon season. The data utilized for this purpose involved Band 3 of Sentinel-2. Band 11 was resampled to 10-m resolution using ArcGIS 10.4 and incorporated into the equation through raster calculator.

11.3.2 LULC Mapping

The LULC map was generated using Sentinel-2 data. Bands 3, 4, and 8 were combined, and the classification process was performed using Maximum Likelihood Classification (MLC) algorithms in ERDAS software (Ha et al. 2020). The resulting LULC map consisted of five classified classes: settlement, fallow land, water body, sand, and agricultural land. To assess the accuracy of the map, 100 random points were selected and the overall accuracy achieved was 84%.

11.3.3 Estimation of Flood Depth

Flood depth was estimated using Google Earth Engine (GEE). Flood extent layer and a DEM layer were used as an input. It follows the following steps (Cohen et al. 2018):

i. Detecting boundary cells of the flood extent layer;
ii. Extracting elevation of the boundary cells;
iii. Determining the elevation of floodwater surface for each cell;
iv. To measure the depth of flood.

11.3.4 Bar Morphodynamics

To examine the changes in bar morphology, Sentinel-1 and Sentinel-2 were used for periods before, during, and after flooding. The mid-channel and side bars were digitized and their areas were computed. Two rivers, one from the east and one from the west of the district, were chosen to demonstrate the connection between hydrological parameters and river material (Sapkale et al. 2016). Both rivers run from north to south and join the Brahmaputra River. Google Earth Pro was used to digitize the river channels. The rivers were split into smaller sections to measure the sinuosity index, since cloud-free images from Sentinel-2 were not available.

The sinuosity index was derived by following the equation:

$$\mathrm{SI} = \frac{\mathrm{CL}}{\mathrm{VL}} \quad (11.2)$$

where SI = Sinuosity index, CL = Channel length, VL = shortest distance between the channels.

11.4 Results

To study the changes in bar morphology of pre-flood, post-flood, and peak floods, Sentinel-1 data were processed and categorized into water and non-water classes using the K-means cluster approach. The detailed results have been shown in Fig. 11.3 and Table 11.2. The flood dynamics of the district show that flooding rose in June compared to May and reduced gradually in July and August. The total area under water at the start of May was 230.93 km^2, as shown by the processed Sentinel-1 VV polarization data. The area under water increased slowly from 337.71 km^2 on 11 May 2022 to 363.40 km^2 on 23 May 2022.

The current period had the highest rainfall in Assam & Meghalaya (858.1 mm) since 1901, breaking the previous record of 789.5 mm in

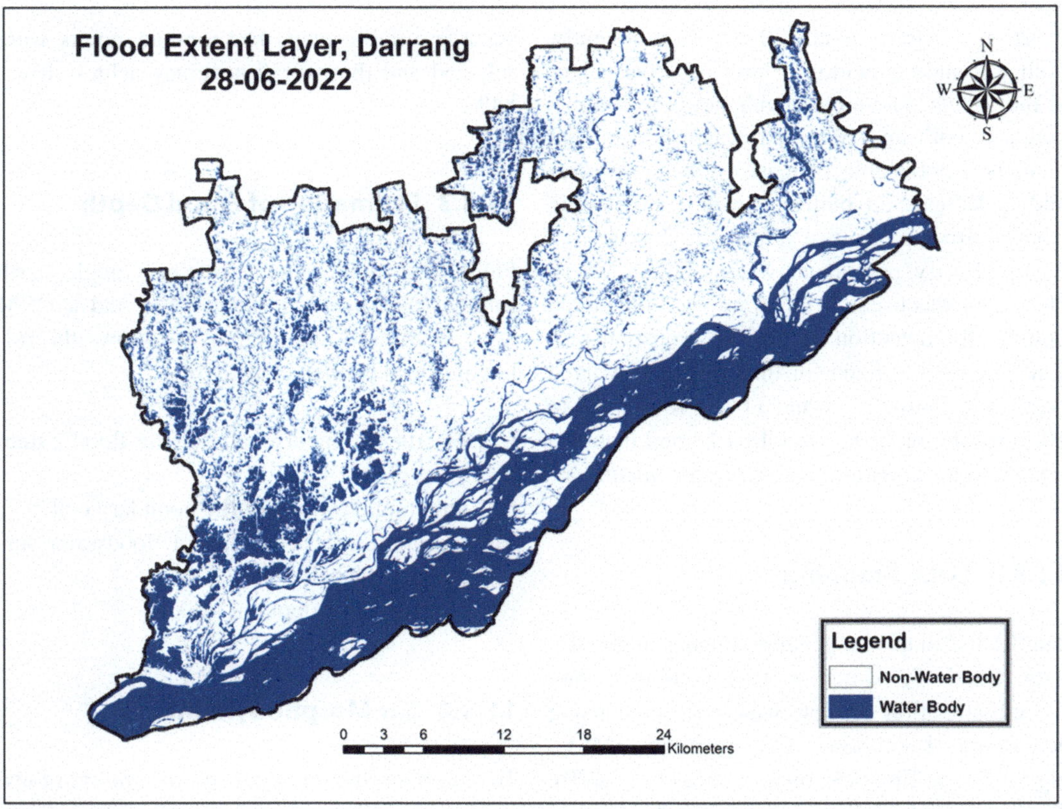

Fig. 11.3 Flood extent layer obtained from Sentinel-1 SAR image after processing

Table 11.2 Changes in flood-inundated area

Time	Area (km^2)	Net expansion (km^2)	Water occupied area (%)
1 May 2022	230.93	79.29	14.07681
11 May 2022	337.71	186.07	20.5858
23 May 2022	363.40	211.76	22.15178
6 June 2022	326.75	175.11	19.91771
16 June 2022	610.89	459.25	37.23804
18 June 2022	677.43	525.79	41.29412
28 June 2022	514.61	362.97	31.36909
30 June 2022	548.68	397.04	33.4459
12 July 2022	336.13	184.49	20.48948
24 July 2022	422.81	271.17	25.77324
3 August 2022	368.08	216.44	22.43706
15 August 2022	368.08	216.44	22.43706
Total area	1640.5 km^2		
MNDWI-based pre-monsoon occupied area	151.64 km^2		

1966. The results show that there was a small drop in the area covered by water (326.75 km^2) at the start of June (6 June 2022), but on the 16th and 18th of June, the area under water rose from 610.89 to 677.43 km^2 (Fig. 11.4). It was because of excessive heavy rainfall in June. The current period had the highest rainfall in Assam and Meghalaya (858.1 mm) since 1901, breaking the previous record of 789.5 mm in 1966 (IMD, Climate Summary report for the month of June 2022). There was a small drop in the area covered by water at the end of June, which continued to drop gradually in July and August.

The first peak of flooding occurred on 18 June 2022 and continued for about 10 to 12 days (Fig. 11.5). The second peak occurred on 24 July but was smaller than the first peak. During the first peak of flooding, water affected most of the northwestern and southwestern parts of the district. As the flooding decreased, there was a reduction in the area under water in the northwestern part of the district. Again, during the second peak of flooding, water covered some regions in the north and northwestern parts of the district.

11.4.1 MNDWI-Based Mapping

To distinguish between permanent bodies of water and areas affected by flooding, MNDWI mapping was applied to the pre-flood Sentinel-2 data. MNDWI mapping uses SWIR band, which is more absorbed by water than NIR band, and green band, which is more reflected by crop, settlement, and vegetation. The MNDWI map was overlaid with the Sentinel-1 processed water extent layer to analyze the flood dynamics and the percentage change in flooding behavior in May, June, July, and August. The MNDWI map showed that the permanent water body covered 151.6 km^2.

The percentage of flooded areas relative to the permanent water body increased from 14 to 22% by the end of June (Fig. 11.6). There was a sharp rise in the flooded area up to 37 and 41% on the 16th and 18th June, respectively. By the end of June, the flooded area decreased to 33%. In early July, there was a drop of up to 20%, and then it increased to 25% by the end of the month, indicating the second flood peak (Fig. 11.5). After that, the flood situation gradually improved in August.

11.4.2 Floodwater Depth Estimation

Flood depth ranges from 0 to 7 m and is grouped into five classes: 0–0.22 (Very Low), 0.22–0.66 (Low), 0.66–1.24 (Medium), 1.24–2.18 (High), and 2.18–7.06 (Very High) (Fig. 11.7). The maximum depth was observed in the northern and northwestern parts of the district. The northeastern part of the district had very low to low flood depth, while the eastern part had low to medium flood depth.

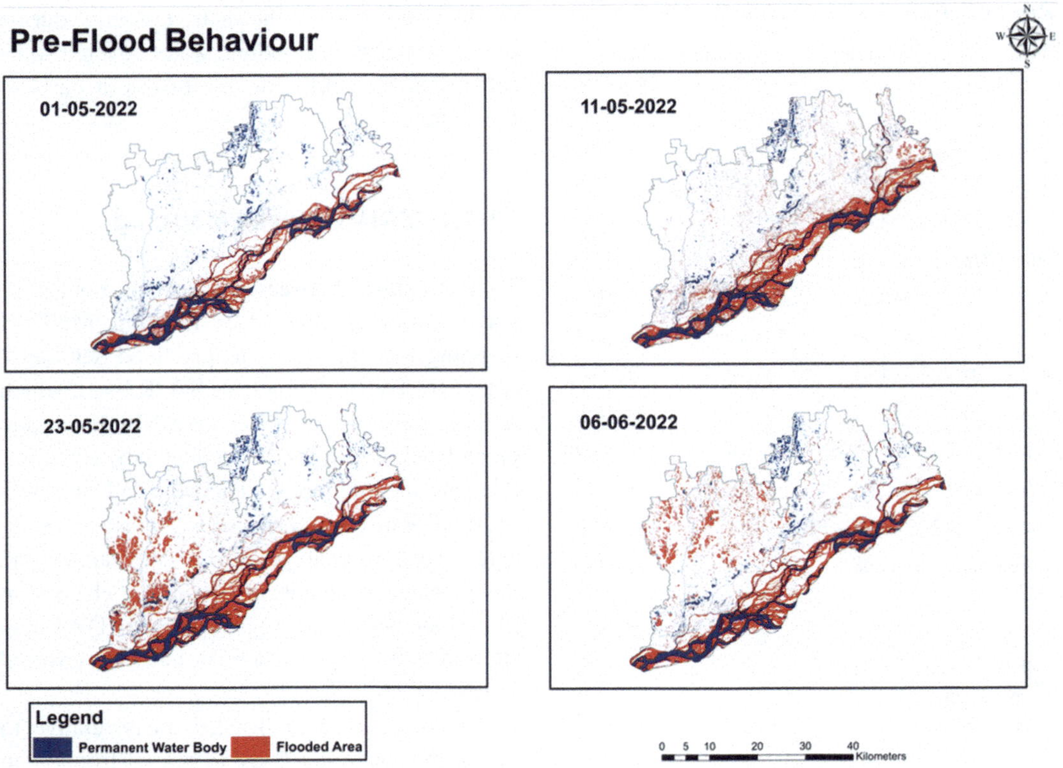

Fig. 11.4 Pre-flood behavior of the study area showing permanent water body and flooded area

11.4.3 LULC-Based Mapping

The LULC map (Fig. 11.8) was created using Sentinel-2 data with 84% accuracy and divided into seven classes: Agricultural land, Vegetation, Barren Land, Settlement, Water body, Fallow land, and Sand. Each class was reclassified separately and intersected with the flood extent layer using the ArcGIS Intersect tool to identify the flood-affected areas of each class.

The study found that out of the total area of 412.29 km^2 covered by agricultural land, 95.12 km^2 was submerged under water. Out of the total area of 87.25 km^2 covered by settlement, 20.10 km^2 was under water. Out of the total areas of vegetation, fallow land, and barren land, 35 km^2, 20.10 km^2, and 90.55 km^2 were flooded, respectively. Detailed distribution of area statistics is mentioned in Table 11.3. In terms of percentage, it was observed that about 23.08% of agricultural land, 10.84% of vegetation, 23.03% of settlement area, 52.26% of fallow land, and 43.48% of barren land were under water during the peak flood period.

11.4.4 Bar Morphodynamics

To study the bar morphological changes, the variation in the bar before, during, and after the flood was observed using optical and SAR data. Sentinel-2 data was used for pre- and post-flood bar morphology, and Sentinel-1 data was used for flood time as optical data is limited by extreme weather conditions. The results showed that before the flood, the sand bar covered 143.84 km^2 and the river covered 104.11 km^2. During the flood, the sand bar area reduced to 130.82 km^2 and the water body area expanded to 320.40 km^2 (Fig. 11.9). After the flood, the sand bar area increased to 152.94 km^2 and the water body area decreased to 270 km^2.

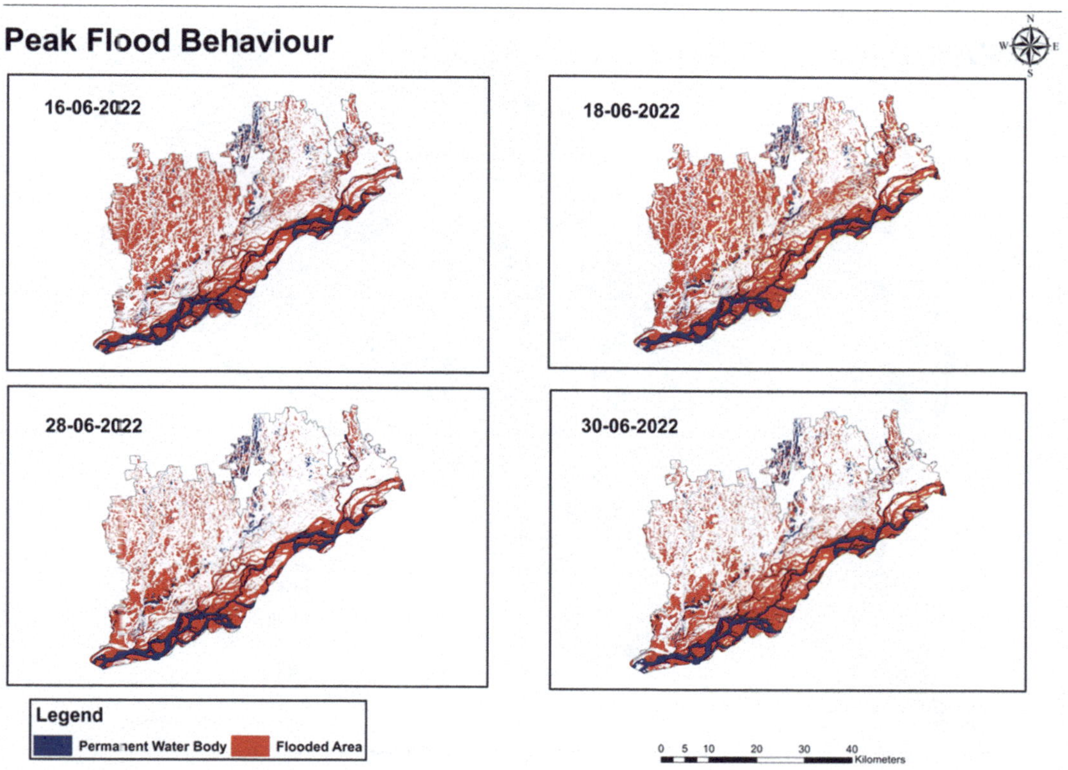

Fig. 11.5 Peak flood behavior of the study area showing permanent water body and flooded area

The number of small and large bars before the flood was 21 and 12, respectively, while after the flood, it was 11 and 36, respectively (Table 11.4). The peak flood period had only six large bars and two small bars. It can be seen that erosion occurred on the right side of the river and deposition on the left side during the peak flood period (Fig. 11.10).

11.4.5 Sinuosity Index

Two rivers (L1 and L2) were examined to determine the sinuosity index. River L1 was divided into three segments and river L2 into two segments (Fig. 11.11). The sinuosity index of L1 segments was 1.43 (L1-G1), 1.29 (L1-G2), and 1.50 (L1-G3) with an average sinuosity index of 1.40 (Table 11.5). The middle segment of the river had a lower sinuosity index than the upper and lower segments (Brice and U.S. Government Printing Office 1964). Based on the average sinuosity index and alluvial channels, river L1 can be classified as meandering. River L2 segments had a sinuosity index of 1.28 (L2-G1) and 1.66 (L2-G2), with an average value of 1.44, indicating its meandering nature as well.

11.5 Discussion

This study used SAR data to determine the maximum flooding area in the district, as it is very effective for surface water mapping. The SAR data can differentiate water bodies from other land cover objects by the low backscattering of radio waves. Sentinel-1 data, which is freely available, was especially useful for identifying flooded areas during extreme weather conditions, as it can penetrate through clouds and rainfall. Optical data, on the other hand, cannot provide information about flood behavior during heavy cloud cover and precipitation, making SAR data the preferred choice. However,

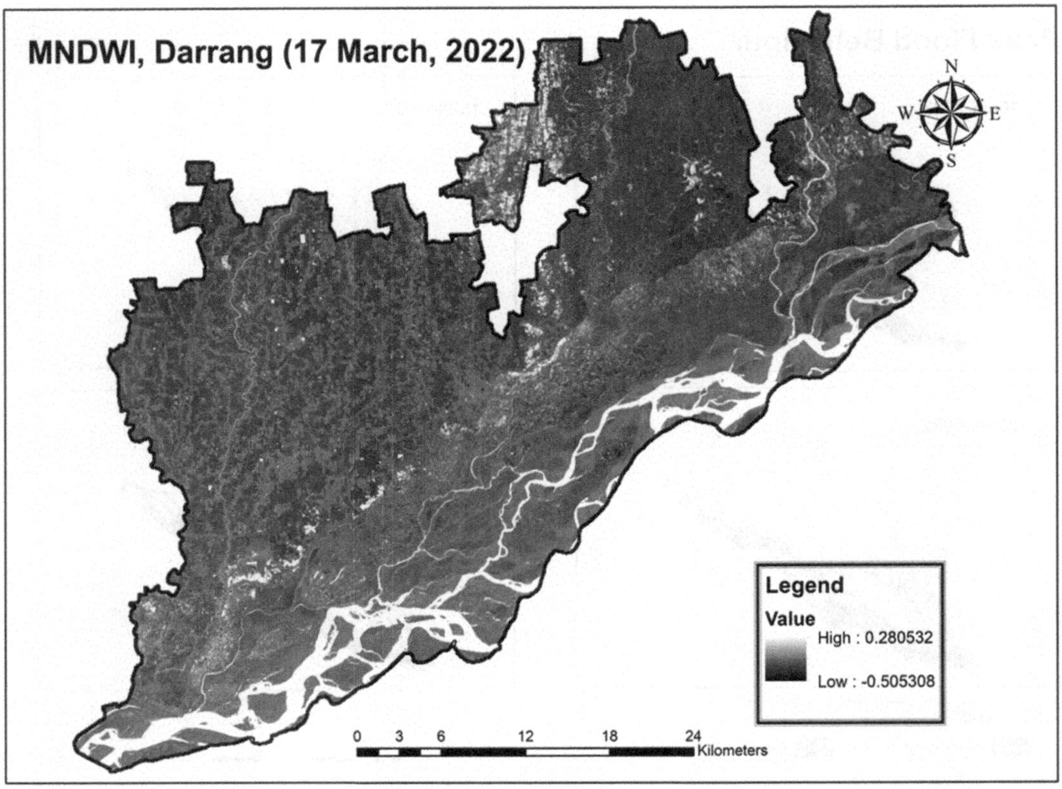

Fig. 11.6 MNDWI Map from Sentinel-2 for identification of permanent water body

the flooded area in the built environment is underestimated by the SAR data because of the side-looking geometry of the SAR system. This is due to the obstruction caused by buildings, which leads to layover and shadow effects that reduce backscatter (Twele et al. 2016). Moreover, urban flood mapping faces significant challenges due to double bounce scattering from buildings, adjacent ground, and roads, which have relatively low backscatter compared to water (Mason et al. 2021).

The results of the study area show a gradual increase in the flood-inundated area with higher rainfall (Bhatt et al. 2021). Notably, on the 16th and 18th June 2022, there was a sudden rise in the inundated area due to excessive rain in June. On 18th June 2022, about 41% of the district was submerged, with the most severe flooding occurring in the north-southern and northwestern parts. Other areas of the district were also inundated, but less severely than the northwestern and north-southern parts. The flooding behavior lasted for about one week to 10 days, as shown by the little change in flooding dynamics on June 28th and 30th. The MNDWI map was used to subtract permanent water bodies from the Sentinel-1 flood extent layer to identify the actual flooded regions.

Analyzing the LULC map of the study area helped identify different regions affected by flooding. The majority of the flood-affected regions were fallow land (52%), indicating that agricultural land was relatively less affected by the flooding. However, around 23% of agricultural land in the northeastern and southeastern parts of the district experienced flooding. Additionally, 23% of settlement areas, mostly in rural regions and villages, were also affected, with some parts of Mangaldoi, the district headquarters, being inundated. Furthermore, the eastern part of the district, including Orang National Park, witnessed extensive flooding.

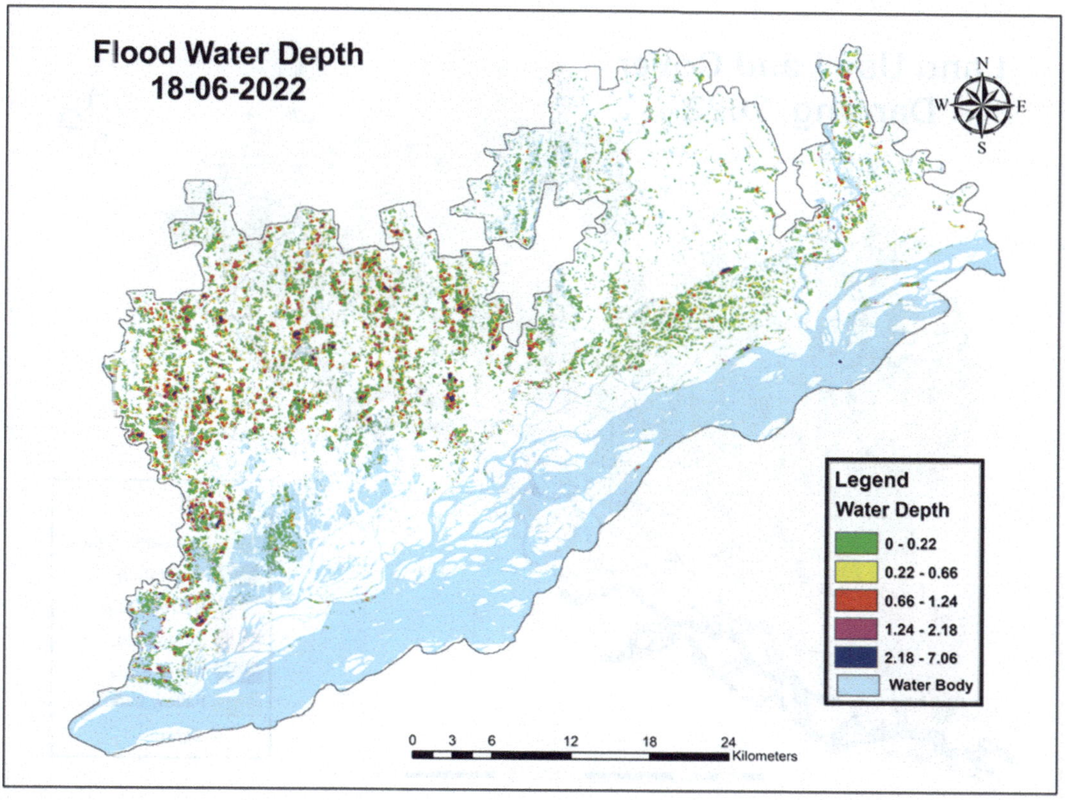

Fig. 11.7 Floodwater depth estimation on 18th June 2022 of the study area

Floodwater depth estimations using SRTM DEM revealed that the highest flood depths ranged from 2 to 7 m, mainly affecting areas in the northwestern part of the district. In most regions, floodwater depth ranged from very low to medium, with an average depth of around 1 to 2 m or less.

The sinuosity index of the two rivers played a crucial role in understanding the devastating flood events in the study area (Khan et al. 2022). The river with alluvial channels and meandering nature worsened the flooding behavior, leading to property damage in various parts of the district The morphodynamics of bars are a crucial indicator of whether the channel is eroding or depositing sediment and its geological impact (Wang et al. 2018). Specifically, the Brahmaputra River showed significant erosion and deposition in a short period, with areas that were previously covered with large sandbars being eroded during the flood and then experiencing deposition post-flood. This behavior demonstrated the high erosion and depositional tendencies of the Brahmaputra River in the study area.

11.6 Conclusion

The study used Sentinel-1 data and the K-means cluster method to separate water and non-water classes, which helped to detect the changes in flood patterns in a district. The results indicated that the flood extent increased in June compared to May, and then decreased gradually in July and August. There were two peaks of flooding, one on 18th June that lasted for about 10 to 12 days, and another on 24th July that was less severe. The main cause of the flooding was the heavy rainfall in Assam & Meghalaya in June, which reached the highest level recorded since 1901. The study also used pre-flood Sentinel-2

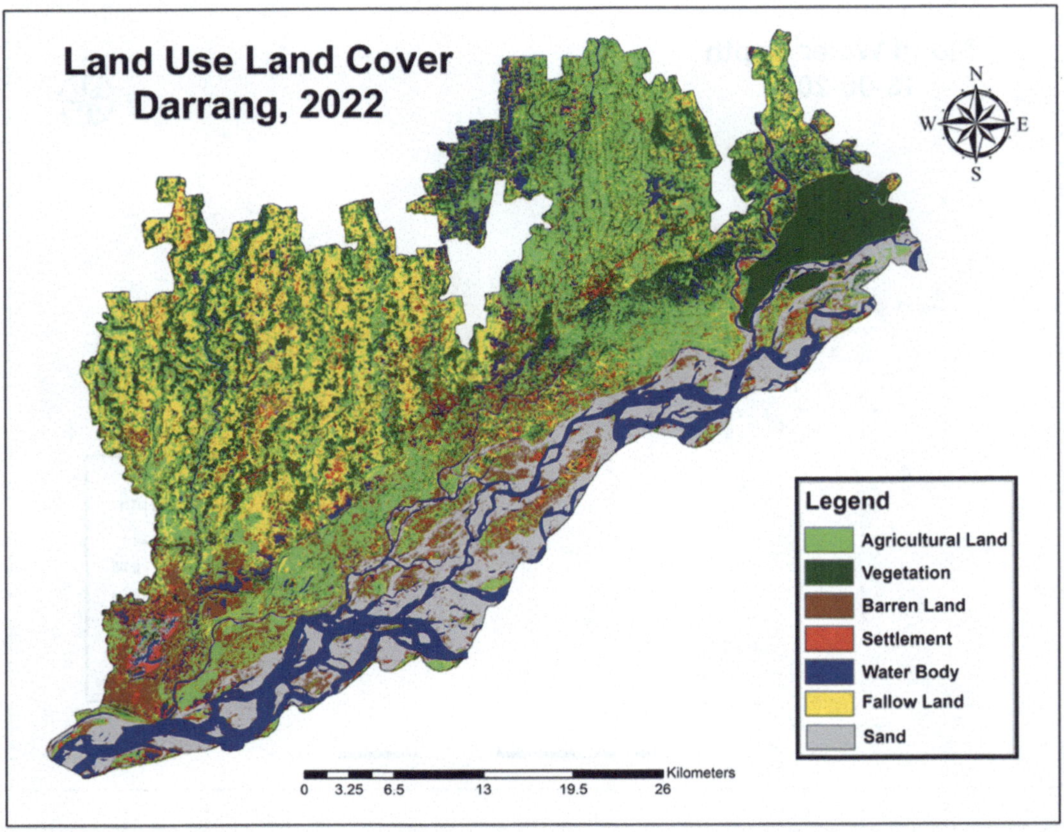

Fig. 11.8 LULC of Darrang

Table 11.3 LULC affected by flood

LULC			
Classes	Total area covered (km^2)	Area affected by flood (km^2)	Water occupied area (%)
Agricultural land	412.29	95.12	23.07
Vegetation	325.76	35.34	10.84
Settlement	87.25	20.10	23.03
Fallow land	243.32	127.22	52.26
Barren land	208.25	90.55	43.48

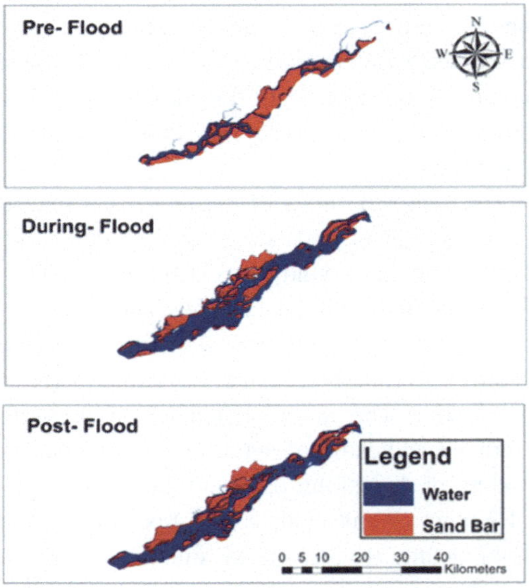

Fig. 11.9 Bar deposition in the different phases of flood

Table 11.4 Bar morphodynamics

	Bars (km^2)	River (km^2)	Total number of bars	Large bars	Small bars
Pre-flood area	143.84	104.11	33	12	21
Peak flooding area	130.82	320.40	28	6	22
Post-flood	152.94	270.83	47	11	36

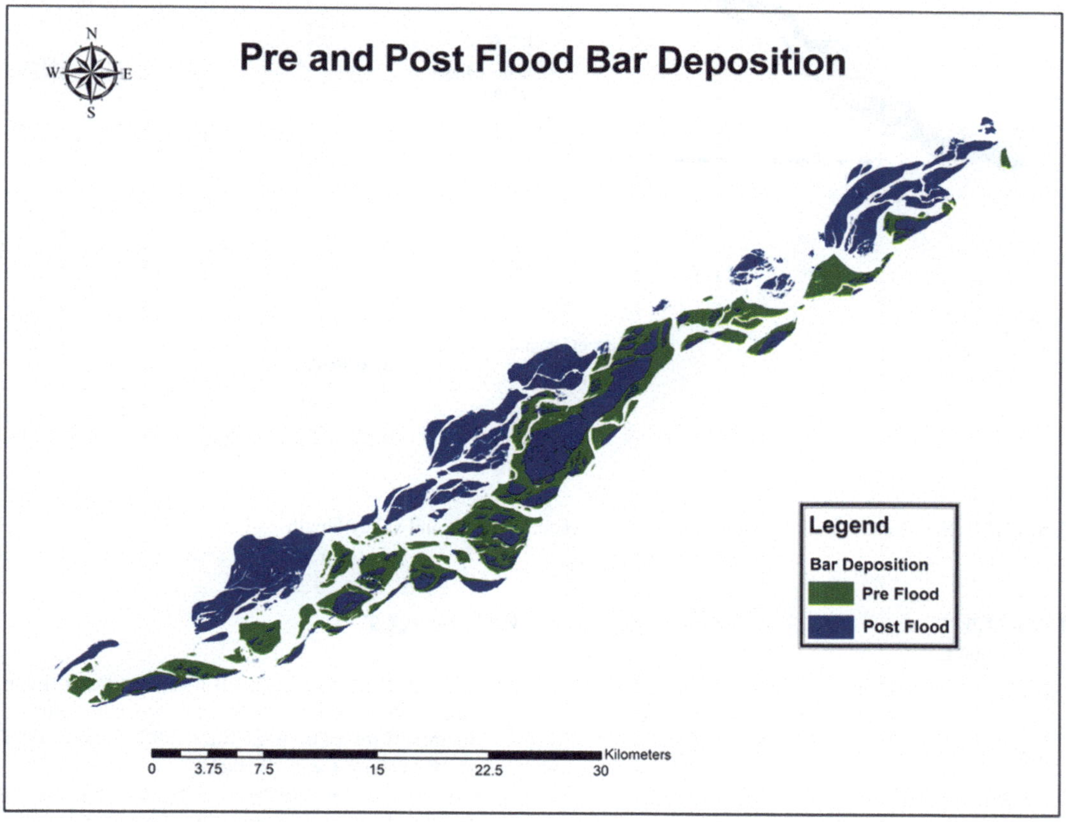

Fig. 11.10 Pre- and post-flood bar deposition

data to apply MNDWI mapping, which distinguished permanent water bodies from temporary flooded areas. This allowed for the observation of flood patterns in May, June, July, and August. The MNDWI mapping and the Sentinel-1 water extent layer were overlaid to show the differences in flood patterns during these months. Moreover, a LULC map was created using Sentinel-2 data with an accuracy of 84%. The analysis revealed that a large proportion of agricultural land, settlement areas, vegetation, fallow land, and barren land were affected by the flood, with varying percentages during the peak flood period. The sinuosity index of the two rivers indicated the nature and behavior of the flooding events in and around the district. It was important to study the morphological behavior of a meandering river with a shallow alluvial plain because it could cause significant damage during flooding events. The morphological analysis showed erosion on the right side of the river during the peak flood period, followed

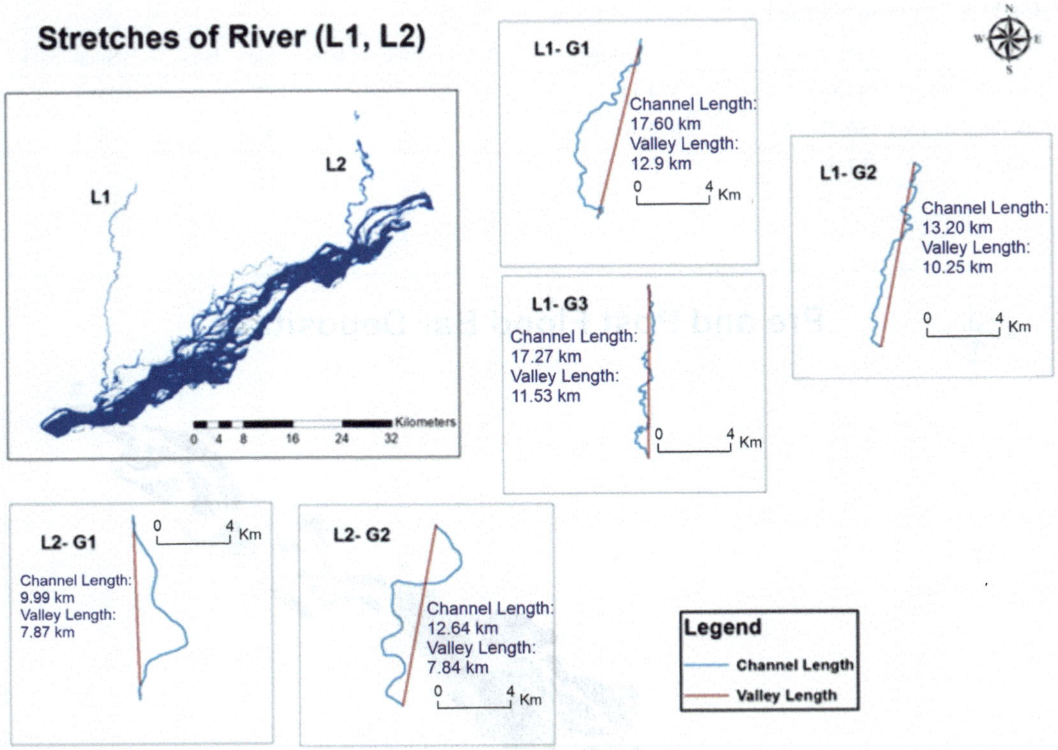

Fig. 11.11 Segments of river L1 and L2 to show the channel length and valley length

Table 11.5 Sinuosity indices of different segments of river L1 and L2

Segments	Channel length (km)	Valley length (km)	Sinuosity index
L1-G1	17.6	12.29	1.43
L1-G1	13.2	10.25	1.29
L1-G1	17.27	11.53	1.5
		Average	1.4
L1-G1	9.99	7.78	1.28
L1-G1	12.64	7.84	1.61
		Average	1.44

by deposition on the left side during the post-flood period. The combination of optical and SAR data was effective in monitoring and understanding the changes in the bar morphology of the Brahmaputra River during different flood stages, which provided useful information for river management and flood risk assessment of the region.

References

Ali H, Modi P, Mishra V (2019) Increased flood risk in Indian sub-continent under the warming climate. Weather Clim Extrem 25(May):100212. https://doi.org/10.1016/j.wace.2019.100212

Bera S, Das A, Mazumder T (2022) Evaluation of machine learning, information theory and multi-criteria decision analysis methods for flood susceptibility mapping under varying spatial scale of analyses. Remote Sensing Appl Soc Environ 25(January):100686. https://doi.org/10.1016/j.rsase.2021.100686

Bhatt CM, Gupta A, Roy A, Dalal P, Chauhan P (2021) Geospatial analysis of September, 2019 floods in the lower gangetic plains of Bihar using multi-temporal satellites and river gauge data. Geomat Nat Haz Risk 12(1):84–102. https://doi.org/10.1080/19475705.2020.1861113

Brice JC, U.S. Govt. Print. Office (1964) Channel patterns and terraces of the Loup Rivers in Nebraska. Professional Paper. https://doi.org/10.3133/pp422D

Cohen S, Brakenridge GR, Kettner A, Bates B, Nelson J, McDonald R, Huang Y-F, Munasinghe D, Zhang J (2018) Estimating floodwater depths from flood inundation maps and topography. JAWRA J Am

Water Resour Assoc 54(4):847–858. https://doi.org/10.1111/1752-1688.12609
Conde FC, De Mata Muñoz M (2019) Flood monitoring based on the study of Sentinel-1 SAR images: the Ebro River case study. Water (Switzerland) 11(12):2454. https://doi.org/10.3390/w11122454
CRED (2019) EM-DAT: the emergency events database. https://www.emdat.be/emdat_db/
Government of India (2001) Census of India 2001. Office of the Registrar General & Census Commissioner, India. https://www.censusindia.gov.in/2011-common/census_data_2001.html
Government of India (2011) Census of India 2011. Office of the Registrar General & Census Commissioner, India. http://censusindia.gov.in/2011census/dchb/WB.html
Govt of Assam (2023) Assam at a Glance
Ha NT, Manley-Harris M, Pham TD, Hawes I (2020) A comparative assessment of ensemble-based machine learning and maximum likelihood methods for mapping seagrass using sentinel-2 imagery in Tauranga Harbor, New Zealand. Remote Sensing 12(3):355. https://doi.org/10.3390/rs12030355
Hariprasanna D, Lodrick DO (2022) Assam (The Editors of Encyclopaedia Britannica) (Ed). Encyclopedia Britannica
IFRC (2003) India floods: Assam and Bihar. www.ifrc.org
IPCC (2013) Climate Change 2013: the physical science basis—contribution of Working Group I to the Fifth Assessment Report of the Intergovernmental Panel on Climate Change. https://www.ipcc.ch/report/ar5/wg1/
Khan A, Govil H, Khan HH, Kumar Thakur P, Yunus AP, Pani P (2022) Channel responses to flooding of Ganga River, Bihar India, 2019 using SAR and optical remote sensing. Adv Space Res 69(4):1930–1947. https://doi.org/10.1016/j.asr.2021.08.039
Kulp SA, Strauss BH (2019) New elevation data triple estimates of global vulnerability to sea-level rise and coastal flooding. Nat Commun. https://doi.org/10.1038/s41467-019-12808-z
Martinis S, Plank S, Ćwik K (2018) The use of Sentinel-1 time-series data to improve flood monitoring in arid areas. Remote Sensing 10(4):583. https://doi.org/10.3390/rs10040583
Mason DC, Bevington J, Dance SL, Revilla-Romero B, Smith R, Vetra-Carvalho S, Cloke HL (2021) Improving urban flood mapping by merging synthetic aperture radar-derived flood footprints with flood hazard maps. Water (Switzerland) 13(11):1577. https://doi.org/10.3390/w13111577
Muro J, Canty M, Conradsen K, Hüttich C, Menz G, Nielsen AA, Skriver H, Strauch A, Thonfeld F (2016) Identification of dynamic cover types in wetlands by using multitemporal cross-polarized Sentinel-1 images. https://orbit.dtu.dk/en/publications/identification-of-dynamic-cover-types-in-wetlands-by-using-multit
Pal S, Lee TR, Clark NE (2020) The 2019 Mississippi and Missouri river flooding and its impact on atmospheric boundary layer dynamics. Geophys Res Lett. https://doi.org/10.1029/2019GL086933
Rahman MR, Thakur PK (2018) Detecting, mapping and analysing of flood water propagation using synthetic aperture radar (SAR) satellite data and GIS: a case study from the Kendrapara District of Orissa State of India. Egypt J Remote Sens Space Sci 21:S37–S41. https://doi.org/10.1016/J.EJRS.2017.10.002
Ramachandran A, Palanivelu K, Mudgal BV, Jeganathan A, Guganesh S, Abinaya B, Elangovan A (2019) Climate change impact on fluvial flooding in the Indian sub-basin: a case study on the Adyar sub-basin. PLoS ONE 14(5):e0216461. https://doi.org/10.1371/journal.pone.0216461
Ruzza G, Guerriero L, Grelle G, Guadagno FM, Revellino P (2019) Multi-method tracking of monsoon floods using Sentinel-1 imagery. Water (Switzerland) 11(11):2289. https://doi.org/10.3390/w11112289
Sapkale JB, Kadam YU, Jadhav IA, Kamble SS (2016) River in planform and variation in Sinuosity Index: a study of Dhamni River, Kolhapur (Maharashtra), India. Int J Sci Eng Res 7(2):863–867
Schlaffer S, Chini M, Giustarini L, Matgen P (2017) Probabilistic mapping of flood-induced backscatter changes in SAR time series. Int J Appl Earth Observ Geoinform 56:77–87. https://doi.org/10.1016/j.jag.2016.12.003
Tazmul Islam M, Meng Q (2022) An exploratory study of Sentinel-1 SAR for rapid urban flood mapping on Google Earth Engine. Int J Appl Earth Observ Geoinform 113:103002. https://doi.org/10.1016/j.jag.2022.103002
Twele A, Cao W, Plank S, Martinis S (2016) Sentinel-1-based flood mapping: a fully automated processing chain. Int J Remote Sens 37(13):2990–3004. https://doi.org/10.1080/01431161.2016.1192304
Wang Z, Li H, Cai X (2018) Remotely sensed analysis of channel bar morphodynamics in the middle Yangtze River in response to a major monsoon flood in 2002. Remote Sens 10(8):1165. https://doi.org/10.3390/rs10081165
Xie Z, Huang HQ, Yu G, Zhang M (2018) Quantifying the effects of dramatic changes in runoff and sediment on the channel morphology of a large, wandering river using remote sensing images. Water (Switzerland) 10(12):1767. https://doi.org/10.3390/w10121767
Xu H (2006) Modification of normalised difference water index (NDWI) to enhance open water features in remotely sensed imagery. Int J Remote Sens 27(14):3025–3033. https://doi.org/10.1080/01431160600589179

Part IV
Innovation in Disaster Risk Management

Application of Unmanned Aerial Vehicle (UAVs) for Disaster Risk Reduction and Management in Nepal

12

Suraj Gautam, Basanta Raj Adhikari, Uttam Pudasaini, and Pravin Lamsal

Abstract

Nepal is highly prone to natural hazards where the frequency and magnitude of the occurrence of such events have been found to increasing in recent decades. Disaster risk reduction and management (DRRM) is of utmost necessity in this country to save lives and properties. While doing so, different forms of frontier technologies are being practiced for risk reduction. The Unmanned Aerial Vehicle (UAV) has rapidly emerged as a promising and highly effective solution for facilitating data collection and enhancing service delivery across all stages of the disaster cycle, including prevention/mitigation, preparedness, response, and recovery. The complexities of the rugged topography, steep slopes, and difficult terrains in the Nepal Himalayas can be overcome through the versatility of drones as they serve as an easy solution for high-resolution aerial surveys that contribute to enhancing situational awareness. Moreover, drones are quickly deployable technology that can help capture the ground situation in detail over the monsoonal hotspots, and planning processes. The use cases regarding the use of drones after the Gorkha Earthquake 2015, the Drone Optimized Therapy System (DrOTS) project, high-resolution aerial mapping during the COVID-19 lockdown, and drones-based rapid damage assessment during Melamchi Debris Flow 2021 suggest how the drone technologies have been evolving for a transformation and paradigm shift toward the adoption of Science, Engineering, Technology, and Innovation (SETI) for DRRM. The adoption of SETI does take ample time, but with the availability of science and evidence-based approaches, the acceptance and willingness toward the change becomes evident. However, it is also important to understand the underlying constraints and barriers while deploying drones for effective DRRM.

Keywords

UAV · Frontier technology · SETI · Paradigm shift · DRRM

S. Gautam (✉) · B. R. Adhikari · U. Pudasaini · P. Lamsal
U-Inspire Nepal, Kathmandu, Nepal
e-mail: ersurajgautam@gmail.com

S. Gautam · B. R. Adhikari
Institute of Himalayan Risk Reduction, Lalitpur, Nepal

B. R. Adhikari
Centre for Disaster Studies, Institute of Engineering, Tribhuvan University, Kathmandu, Nepal

S. Gautam · U. Pudasaini
Nepal Flying Labs, Kathmandu, Nepal

P. Lamsal
Geovation Nepal, Kathmandu, Nepal

S. Talukdar et al. (eds.), *Communication, Science, Technology and Innovation in Disaster Risk Management*, GIScience and Geo-environmental Modelling, https://doi.org/10.1007/978-3-031-77189-7_12

12.1 Introduction

12.1.1 Disaster Scenario in Nepal

The Nepal Himalayas are strongly affected by natural hazards, with regular occurrences attributed to a combination of active seismo-tectonic activities, steep and rugged topography, and complex geology. The persistent collision between the Indian and Eurasian plates has resulted in the formation of numerous tectonic faults, making earthquakes a common occurrence. Significantly, nearly 80% of the country's total annual rainfall occurs during the months of June to September, giving rise to hydro-meteorological disasters (Shrestha 2000; Yihui and Chan 2005). Further, the impacts of climate change are increasingly becoming apparent through the emergence of unprecedented weather conditions, intense rainfall, and so on. As a result of which, the frequency and severity of the disasters are also found to be increasing. Nepal is among the top ten countries that were most affected in the period 2000–2019, indicating that there were almost 0.82 fatalities per 100,000 inhabitants contributing to about 0.39% losses per unit GDP (Eckstein et al. 2021). The number of consecutive wet days and intense and heavy precipitation days are increasing while the warm days and nights are also getting observed more frequently (Adhikari and Tian 2021; Chapagain et al. 2021; DHM 2015). Disasters result in disruption of the infrastructure services and service delivery, thereby leading to the need of external sources for the response and recovery.

Nepal is exposed to a multitude of hazards including geophysical, hydrological, meteorological, and climatological hazards. The exposure of underlying elements at risk such as the population, infrastructures, and so on to the hazards and existing vulnerability in the region makes it highly susceptible to disaster risks. It has been observed that strong earthquakes are observed every 60–80 years in Nepal, which has resulted in significant losses of lives and infrastructure (Gautam and Chaulagain 2016). The complexity of the region has also led to the interaction of hazards, thereby resulting in cascading or domino impacts (Pandey et al. 2021). Some of the major disasters in the region include the Melamchi Debris Flow of 2021 with more than 5 fatalities and 20 missings, Gorkha Earthquake of 2015 with more than approx. 9000 fatalities, Jure Landslides of 2014 with more than 156 fatalities, and the cloudburst event in central Nepal of July 1993 with more than 62 fatalities (Gautam 2017; Gautam et al. 2022; Panthi 2021; Upreti and Dhital 1996). The distribution of landslides in Nepal indicates that fatal landslides resulted in an average of 113 fatalities annually between 1971 and 2021 (Adhikari and Gautam 2022). Likewise, the overall countrywide lightning fatalities from the year 1971 to 2019 are about 1.77 per million per year (Adhikari 2021). The recent debris flow events of 2023 in Kagbeni, Mustang also demonstrated the geo-hazard risk even in the rain-shadow zone of the Himalayas. These events destroyed more than 29 houses and displaced more than 150 people.

12.1.2 Evolution of UAVs

An unmanned aerial vehicle gets its name from the fact that the vehicle can be operated without the help of a pilot or a crew. The demonstration of a hot-air balloon in the year 1783 in France is reported to be the first public demonstration of an unmanned aircraft taking flight (Scott 1984). Afterward, the use of UAV, also known as drones, was found to be observed for military purposes when the Austrian soldiers attacked the city of Venice by attaching balloons equipped with bombs in 1849 (Chamola et al. 2021). A French photographer named Gaspard Felix Tournachon, also known as Nadar, is reported to have clicked the first known aerial photograph from the hot-air balloon in Paris, France in 1858 where he carried his bulky cameras in the hot-air balloon (Cohen 2000). With the help of aerial photography, the British forces prepared a map of the German front during the battle of

Neuve Chapelle. This is considered to be one of the earliest examples of an orthomosaic, which was prepared by laying out the photographs on top of one another (Chasseaud 2002). The first unmanned aircraft aerial target controlled by radio technology was tested in 1917 and since then, the evolution of drones continued. Further, it kept on evolving to deploy the first modern drone named DH. 82B Queen Bee aircraft in 1935 which was low-cost radio-controlled target aircraft. Since then, it has been evolving for multiple sectors not limited to the purposes of photography, surveying and mapping, agriculture, environmental conservation and management, journalism, disaster risk management, and so on (Bendig et al. 2014; Chamberlain 2017; Giordan et al. 2018; Rahman et al. 2017; Smith 2015).

In 2000, researchers from the Japanese government employed a Yamaha RMAX helicopter drone equipped with advanced sensors to gather crucial data during the volcanic eruption of Mt. Usu. This specific instance showcases the effective utilization of drones in disaster risk management, highlighting their significant role in acquiring valuable information during such critical events (Sato 2003). Likewise, the application of fixed-wing drones for search and rescue operations and the evaluation of river levels in the aftermath of Hurricane Katrina in 2005 exemplifies another significant instance of utilizing drones in disaster risk management (Murphy et al. 2008). In the aftermath of the 2008 Wenchaun Earthquake, Chinese responders employed both manned and unmanned micro aircraft equipped with remote sensors to gather crucial information on the extent of damage (Lei et al. 2017). One of the most popular uses of drones in a post-disaster humanitarian context was in the year 2010 when the United Nations Satellite Imagery Analysis Organization (UNOSAT) used a drone to capture high-resolution imagery of Haitian camps for persons displaced by the 2010 Haiti earthquake for the International Organization for Migration (IOM) to support camp planning and management. Since then, drone imageries have been used on a periodic basis for camp management planning activities (Martini et al. 2016). The widespread media coverage of drone footage captured in the aftermath of Typhoon Haiyan in the Philippines in November 2013 marked a turning point in recognizing the substantial potential of drone technology for post-disaster scenarios on an international scale. Since then, the possibilities of using drones in disaster response and recovery efforts have seen a remarkable surge in interest and exploration (Greenwood et al. 2020).

12.1.3 Evolution of UAVs in Nepal

The first recorded use of drones in Nepal was reported on its applications toward the wildlife conservation related program. The World Wildlife Fund (WWF) Nepal conducted a successful test on 12 June 2012, deploying two drones specifically designed for monitoring wildlife and detecting poachers within Chitwan National Park (Poppleton 2012). Besides the use of UAV in anti-poaching efforts, the use of UAV was also found in the study and investigation of the impact of climate change on glaciers by ICIMOD Nepal (Immerzeel et al. 2014). In this study, the drone was deployed before and after the melt and monsoon season over the Lirung Glacier. The first-ever training on "Professional UAV training for Disaster Preparedness and Recovery" was organized at Kathmandu University in September 2015 by a group of organizations. Further, during the Gorkha Earthquake, a number of organizations used UAVs for the photography and videography. With the use of aerial images and videos for news content and coverage by the international media, the significance of UAVs was then highlighted.

Similarly, Kathmandu University used UAVs for post-disaster rapid assessment, crop monitoring, and biomass estimation. For the geo-hazard assessment, a consortium of humanitarian organization, Medair and Nepal Flying Labs, carried out an aerial mapping of a Gyappche landslide in Ramechhap to assess the debris flow (Pangeni et al. 2020). Similarly, the use of drones by multiple user groups such as academicians, researchers, consultants, and students started to increase since the 2015 Nepal Earthquake. A local entity, called Nepal Flying Labs, to carry forward the development of drone ecosystem got started after the training which worked to localize UAVs and robotics solutions (Pangeni et al. 2020). The Institute of Engineering, Tribhuvan University, Nepal has started a Bachelor's program in Aerospace engineering where students have been doing research on UAV synthesis. Gorkha Earthquake 2015, Gyappche Landslide Mapping 2017, Drone Optimized Therapy System to improve healthcare access are some of the key notable works that got started in Nepal, which ultimately contributed to the development of drone ecosystem in Nepal.

12.2 Drones in Disaster Risk Management

12.2.1 Drones in Prevention/Mitigation

The images captured through drone survey can be used in the preparation of the Orthophoto and Digital Surface and Terrain Model (Ahmed et al. 2021; Ruzgienė et al. 2015). By using the point clouds, the Digital Terrain Model and Digital Surface Model can be developed. The high-resolution datasets can thus be used for different studies such as geomorphic mapping, time series analysis of landslide, high-resolution monitoring of Himalayan glacier dynamics, and so on, which will ultimately contribute toward the prevention and mitigation of risk of disasters (Eker et al. 2018; Immerzeel et al. 2014; Turner et al. 2015). The datasets/imageries obtained from the drone flights can be used to assess the locations for the construction of structural measures such as dams, retaining structures, damage inspection and assessment of bridge infrastructure, estimation of quantities for cut and fill operation (Jain and Jagtap 2023; Mandirola et al. 2022). While for nonstructural measures such as early warning systems, drones can support the assessment of the instrumentation location, used with existing early warning systems to support in predicting the outbreak time and scope, reduce the cost, and so on (Erdelj and Natalizio 2016; Luo et al. 2019).

12.2.2 Drones in Preparedness

Drones can be used smartly to identify the large open spaces that can serve the management of displaced populations during emergencies. Further, drones are being used for urban Green Space Mapping, a tool for the promotion and management of urban space (Erdelj and Natalizio 2016; Lahoti et al. 2020; Mohd Daud et al. 2022). It is very evident to explore, identify, and monitor the inaccessible areas in underground mines, hazardous areas, and forest areas to have pre-information on forest fires, and drones provide effective and promising solutions (Shahmoradi et al. 2020; Suh and Choi 2017). Likewise, drones have significant applications in the surveillance of traffic management, forest ecosystems, river systems, and so on (Allauddin et al. 2019).

12.2.3 Drones in Response

During any emergency, there are multiple stakeholders working on the ground. In order to assess the existing situation and scenario, drones can be quickly deployed to capture high-resolution images and videos close to the ground to give multiple perspectives on the affected areas (Furutani and Minami 2021; Velev et al. 2019). The captured high-resolution datasets of the ground can then be analyzed to identify the disaster-hit areas with greater damage, to guide the Search and Rescue mission based on a need-based approach, and ultimately create a proper data-driven response plan to aid emergency responders to effectively and efficiently coordinate response activities (Claessens et al. 2005; Erdelj and Natalizio 2016; Greenwood et al. 2020; Velev et al. 2019). Likewise, drones can be equipped with multiple types of sensors like thermal, infrared, etc. to aid the search and rescue. In certain instances, drones, when outfitted with sensors like noise detection, binary detection, vibration sensors, and thermal imaging, can also be employed to locate survivors trapped beneath debris (Abrahamsen 2015). The drones can be flown very close to buildings or rivers, capturing data from multiple angles to capture the image and video datasets of the elements-at-risk, which can be overlaid to assess the exposure and damages. The captured images can then be used to characterize the severity of damages and hence for the impact assessment (Chou et al. 2010). It is very obvious that critical infrastructures such as transportation, telecommunication, drinking water, power utilities, etc. can be interrupted by disasters. This could lead to the limited use of mobile phones, electricity, and internet and thus affect the emergency response. Under such circumstances, drones can be a savior by providing internet access using aerial communication infrastructures, delivery of wireless networks, and low-tech radio coverage (Antoine et al. 2020; de Albuquerque et al. 2016). The likelihood of problems of partial or full obstruction of access roads during emergencies, such as flooding, landslides, earthquakes, etc., is very much obvious. In such situations, drones can transport medical and emergency supplies considering the types of drones and payload they can carry (Yakushiji et al. 2020).

12.2.4 Drones in Recovery

Drones have emerged as indispensable tools in the aftermath of disasters, facilitating recovery efforts through their functionalities. With the help of drones captured high-resolution aerial imageries, high-resolution Digital Elevation Models, three-dimensional models, and orthophotos can be derived which is crucial for modeling and damage assessment aftermath of disasters (Balcik 2017; Cabello et al. 2021). The use of detailed 3D models and high-resolution maps of disaster-affected regions can aid in the planning and coordination of efforts. These maps support the urban planners and responders in understanding the terrain and planning infrastructure repair and reconstruction. Similarly, the rapid and detailed visual documentation of damage through drones can expedite insurance claims, simplifying the process for affected individuals and businesses.

12.3 Application of Frontier Technologies in Disaster Risk Management in Nepal

The procedures for Remotely Piloted Aircraft (RPA)/Drone Flight implemented from 15 Feb 2019 have classified the drones under four categories in Nepal, i.e. (i) Category A for Very Low-Risk operations weighing below and up to 250 g (ii) Category B for Low-Risk Operations weighing between 250gm and 2 kg, (iii) Category C for Regulated, Low-Risk Operations weighing between 2 and 25 kg, and (iv) Category D for Regulated, High-Risk Operations weighing above 25 kg (MOHA 2019). These drones can be applied in photogrammetry, videography, search and rescue missions, mailing and delivering, environmental protection, space missions, marine missions, and so on. The application of drones has been evolving rapidly in the context of disaster risk management. Starting from the prevention/mitigation to the recovery, drones have provided different innovative solutions. Some of the exemplary initiatives of drone use in Nepal are described below.

12.3.1 Use Cases of Drone After Gorkha Earthquake 2015

Nepal stands as one of the early adopters of drone utilization in response to a major disaster event. In the aftermath of the 2015 Gorkha Earthquake, numerous drones were employed for crisis mapping and humanitarian efforts, capturing images of the devastation. Despite various challenges like drone congestion, flights by inexperienced operators, and an emphasis on experimentation rather than practical applications, drone usage during the earthquake prompted a wider acceptance of drones in the international humanitarian and disaster management field (Giordan et al. 2018). The aerial photography and content by the international media agencies induced a widespread message regarding the potential usage of drones in immediate disaster response scenarios. The Gorkha Earthquake 2015 not only brought devastation but also induced the opportunity for a training activity on drones for disaster response and crisis mapping, a first-of-its-kind capacity-building training in Nepal. The Flying Labs initiatives by WeRobotics started in Nepal (Fig. 12.1) and are now across 30+countries most of them from the global south with disaster management challenges across the globe (Pangeni et al. 2020). More than 10 DJI Phantom drones were donated by DJI and altogether 30 Kathmandu University students were capacitated on using drones while PIX4D also shared its application and provided assistance on the usage of its software to stitch images to prepare orthomosaic and 3D models (Meier 2015). This initiative also contributed to the aerial surveys of Panga, which is one of the hardest-hit towns by the 2015 Gorkha Earthquake in the Kathmandu valley. Further, these projects also supported the capacity enhancement of the local partners to develop the drone ecosystem in Nepal.

Fig. 12.1 Kathmandu University's students working with Pix4Dcapture app and DJI Phantom 3 Advanced drones

12.3.2 Drone Optimized Therapy System

The majority of the advanced and sophisticated healthcare services of Nepal are located in the urban cities. The complex terrain of the country, average infrastructure, and limited healthcare facilities in rural settings make a significant portion of the population compelled to travel on foot for an average of 6–8 h to ensure enhanced healthcare accessibility (Pudasaini 2019). On the 2nd of June 2019, DrOTS was successfully inaugurated in Pyuthan district. The program was attended by stakeholders such as the Ministry of Health and Population, National TB Center, Health Office Pyuthan, municipalities, and so on. The aerial mapping of 12 health post sites and 2 central health centers was made for the planning of the project. The flights were planned and made operational for 8 health centers out of 12 health posts, and drone flights were carried out from Pyuthan Hospital and Bhingri Primary Health Center. For the operationalization of this project, a drone lab was set up at the hospital. The community healthcare worker from health facilities in remote areas makes a drone flight request after collecting the sputum sample of probable Tuberculosis patients. Next, the drone pilot located at the central drone hub would send the drone to the health facilities based on that request. The healthcare staff in the remote healthcare facility can take manual control of drones in case if required, land the drones, load the sputum samples for Tuberculosis diagnosis, and send them back to the hospital for testing using a highly accurate Gene expert machine. The hospital then receives the test samples to perform diagnostic tests and the same process would be repeated for all other flights.

The DrOTS (Drone Optimized Therapy System) project implemented in two municipalities of Pyuthan district, Nepal aimed to support the accessibility of healthcare facilities in rural villages of Nepal. With the help of drones, the Tuberculosis diagnostic tests were carried out through the linkages of Community Health Workers (CHWs) and state-of-the-art diagnostic tools (GeneXpert). The drones, took off from central hospitals, flew to remote healthcare centers, collected the sputum samples, and returned back with sputum samples for diagnostic purposes. The project was piloted to assess the feasibility of the drones-based service in the health sector and provide a basis for the Ministry of Health and Population (MoHP) and National Tuberculosis Center (NTC) for the expansion of such service in other remote locations of Nepal.

The mobility through the road infrastructure was quite tough and time-consuming. The travel time through the hilly roads from Majhkot to Pyuthan Hospital took about 4–5 h. However, the aerial distance between the remote health center and the district/central hospital is significantly lesser and the travel time through the drone flight was just 5 min in a single flight (Fig. 12.2). Hence, the use of drones significantly reduced the travel time.

Regarding the type of the hardware, the DJI M600 drone was used. The maximum take-off weight of the drone is 15.5 kg with a maximum flight time of 35 min. The drone was customized by the technical team of the project to add a cargo box that was capable of collecting 16 sputum samples (up to 1 kg). This was the first time the Government of Nepal officially gave permission to fly a multicopter drone such as DJI M600 for medical cargo purposes and Beyond Visual Line of Sight (BVLOS) flight in particular. Flight routes between hospital and health centers in rural areas were carefully created using QGround control software, for each drone path. Two different measures were taken for safety considerations. First, the flight path was planned in such a way that it falls above areas where buildings are not location. Second, intermediate stations were also planned at certain locations for landing on drones in case of any emergencies (Fig. 12.3).

The project team consisted of experts spanning various sectors: Public Health partner represented by Birat Nepal Medical Trust (BNMT), Drones and Technology partner represented by a group of the technical team from WeRobotics, Nepal Flying Labs and DroNepal, and the Research partner led by The Liverpool School of Tropical Medicine. Collaborating alongside were governmental stakeholders such as the Ministry of Health and Population (MoHP Nepal), the National TB Center, and the District Public Health Office (DPHO Pyuthan). This initiative received support from Stony Brook University and the Simons Foundations. Over a span of 6 months, the project successfully conducted locally led cargo drone deliveries, facilitating the transportation of more than 700 patient samples from 8 clinics to 2 regional hospitals. The purpose of these deliveries was to expedite

Fig. 12.2 Flight plan adopted during DrOTS project

Fig. 12.3 Ground control software showing the flight plans

the process of TB testing, aiding in swift diagnoses. The impact of this effort was profound, significantly contributing to the diagnosis of Tuberculosis cases in the Pyuthan region.

The effectiveness of this endeavor garnered widespread recognition, notably receiving the AUVSI XCELLENCE Awards in 2020 under the Humanitarian category during the AUVSI XPONENTIAL 2020 event. Furthermore, the project's contributions were acknowledged by the International Air Cargo Association, which bestowed upon it the TIACA Air Cargo Sustainability Awards in 2020.

12.4 Use Case of Drones during COVID-19 Lockdown

Considering the COVID-19 outbreak, the Government of Nepal enforced a decision of lockdown in the country after detecting a second COVID-19 case on 24 March 2020. The lockdown resulted in the shutting down of vehicular movements and it also contributed to a significant drop in air traffic (around one to three international and domestic flights per day). These limitations on road travel and physical movement resulted in the gradual improvement of air quality and visibility providing a unique opportunity for drone-based high-resolution mapping and spatial data gathering over urban clusters of Nepal.

As such, a group of organizations came together and conducted a drone-based aerial survey of Changunarayan, an urbanizing municipality from the Kathmandu Valley. For the high-resolution aerial survey of the Changunarayan municipality during the COVID-19 lockdown, Wingtra—a hybrid Vertical Takeoff Landing (VTOL) drone equipped with a high-resolution camera—was used. The KML file of the site was imported to the Wingtra Hub, which is auto-piloting software for drone flights. The flight plan was set on the software with a front overlap of 70%, side overlap of 70%, and flying height of 300 m for the whole site (Fig. 12.4). The local government took the leadership and helped the technical team to secure all necessary flight permissions including field-based monitoring support by involving a human resource from the air traffic control office from the International airport, which is located next to the municipality. The project team assessed the area if it felt under a no-fly zone or restricted area for the uninterrupted survey. The geo-tagged images were captured from the field and were processed to obtain different types of outputs. The initial results of data processing were the point clouds with point density. The dense stereo matching algorithm was used to generate millions of 3D

Fig. 12.4 Flight planning for the Changunarayan Municipality aerial survey

textured point clouds that looked exactly like the ground itself (Joglekar et al. 2014). These points were used to obtain the 3D mesh of the project area. Considering the digital solutions, the orthomosaic was shared with Changunarayan municipality. The orthomosaic generated was later on used to digitize the building and road inventory, finally updating the base map of the municipality. Furthermore, the municipality utilized a high-resolution orthophoto map to implement a metric addressing system in Changunarayan Municipality.

12.4.1 Melamchi Debris Flow 2021

The 11th of June marked the onset of monsoon in Nepal for the year 2021. The devastating flood of 15 June was a result of multiple factors including the processes in the high-altitude glacial environment, weather conditions, complexities in the Bhremathang old landslide region, river damming, bank erosion, and debris deposition. The flood resulted in the displacement of more than 525 families and caused damages to more than 337 households alongside the critical infrastructures such as a hydropower project, the headworks of Melamchi Water supply project, 13-foot trail bridges, and 7 Reinforced concrete bridges (Gautam et al. 2022). The complex flow resulted in the riverbed aggradation and the infrastructures near the riverbank were significantly damaged. Disaster response during these tough situations is a challenging task considering the air visibility, weather conditions, and need of surveillance throughout the whole stretch of the affected areas. There have been several advances in flood response activities with the evolution of various frontier technologies. Among such, the use of drones has established a solid benchmark of supporting evidence-based planning during emergencies for effective humanitarian response. Likewise, immediately aftermath the Melamchi Flood event on the 18th of June, a consortium of U-InspireNepal, Institute of Himalayan Risk Reduction, Nepal Flying Labs, and Geovation Nepal conducted an aerial mission over some

Fig. 12.5 Images from the Melamchi Bazar: satellite image with cloud cover on the left and a portion of the Orthophoto map (2021/06/20) from the Drone survey on the right

Fig. 12.6 A drone photo showing the road section near the headworks of National Hydro Power Company Limited in the Indrawati River Stretch

of the sections of the Melamchi Bazar. The team was able to collect images over a small region only as there were frequent helicopter flights operating in the area which kept the airspace busy. For the rapid damage assessment of Melamchi Debris Flow 2021, the permission document was secured from the Sindhupalchowk District Administration Document and coordinated with the local municipality and Nepal Engineers' Association for the study and detailed data collection..

The limitations of the satellite images were clearly observed during this event. Satellite images of the upstream region of the Melamchi River were extensively obscured by clouds, rendering it unfeasible to evaluate ground conditions. In contrast, the team captured the clear aerial images of the ground surface by flying the drones flown below the cloud (Fig. 12.5). This capability also makes the drone an effective tool than satellite for close range and high-resolution imaging of ground surface during cloudy weather during an immediate disaster situation (Fig. 12.6).

Accessibility is a crucial factor to plan immediate rescue and relief activities during a disaster situation. The high-resolution images from drones provide a clear picture of the accessibility conditions at the disaster-affected area. The extent of damage to the roads was also clearly visible on the oblique images from drones as shown in Fig. 12.7. Likewise, the interpretation of the orthophoto map from the images showed the sweeping away of two concrete motorable bridges and around half a dozen suspension bridges from Helambu to Melamchi Bazar. Furthermore, the authorities in the region utilized both orthophoto maps and topographical data derived from elevation models to devise alternative access routes. The orthophoto map processed with drone images was also compared to the satellite images of different temporal references to better understand the before and after disaster scenarios. The before and after image clearly shows the area suffering higher aggradation, deposition of sand/debris, and washing away of the agricultural area (Fig. 12.7).

The images obtained from drone survey show the majority of the area including structures along the river corridor and other infrastructures in the floodplains being inundated and even washed away. It also helped to further identify that the flood has entered into the

Fig. 12.7 Orthophoto comparison of the river extents in Melamchi Bazar: Google Earth Satellite image (2021/01/02) before the flood on the left and drone captured image (2021/06/20) after the flood on the right

Fig. 12.8 Oblique image from drone survey at Indrawati and Melamchi confluence

settlements displacing several families, depositing debris over two and half floors of houses. The debris flow was noted to be significant along the Melamchi River stretch and only about 1.5 km upstream on the Indrawati River from the Melamchi-Indrawati confluence (Fig. 12.8).

12.5 Barriers to Adopting Frontier Technologies

Amid the opportunities, there exist noteworthy obstacles in embracing cutting-edge technologies such as drones. There are limited skilled human resources who are capable of properly flying the drones and then also process the datasets to generate multiple outputs. Furthermore, there is no certification program in the country to certify the drone pilots. In the context of DRRM, there are even more challenges as using the drones in the emergency settings and also in the geographically complex region like Nepal is not an easy job. Knowledge of operating drones in a normal context alone wouldn't work as a disaster situation is complex. Besides, there are no organizations that offer drone pilot certification in Nepal. Hence, the absence of well-defined policies and guidelines for licensing has resulted in a detrimental effect on the quality of drone services in the market. This circumstance has also added complexity to the process for professionals seeking to use UAS on a frequent basis. To address this constraint, it is essential to develop comprehensive guidelines for pilot licensing standards and establish appropriate training programs. The other major barrier is the hardware issue as most of the drones used in Nepal are imported. There are only a few service centers, and even those are uncertain of making a successful recovery in case of any serious damages. Drones with serious crash need to be sent out of the country for recovery which makes it time-consuming and expensive. The absence of insurance providers with proper UAS guidelines has made the situation even worse. Further, drone data processing requires the procurement of a license for expensive software. Hence, it involves the investment in the software, selection of appropriate hardware for processing data, and engaging skilled personnel capable of effectively utilizing the software (Aiyetan and Das 2022). In addition to it, there are also the shortcomings of the hardware which may not function well during the bad weather in monsoon. The heavy wind or strong intense rainfall can often result in the complexity during the operation. For the sustainability of drone ecosystems in the country, the country requires the expansion and development of drone networks. However, this needs to be rational and shall strictly comply with the compliances related to the flying of drones and country aviation protocols. The unregulated use of the frontier technology, without considering the privacy and data protection concerns, is a significant misconduct that raises its credibility and significance. Hence, a disciplined and regulated approach toward humanitarian responses is required. For this, the relevant agencies should come up with a national UAS code of conduct that includes data capture, data protection, data sharing, and community engagement guidelines.

12.6 Conclusions and Ways Forward

Frontier technologies have made notable progress in the assessment, analysis, and reduction of disaster impacts. Nevertheless, the constraints of limited time and resources frequently demand swift and adaptable decision-making in response to the dynamic environment. The importance of any cutting-edge technology is gauged by its practicality, adaptability, user-friendly attributes, economic viability, and tangible real-world effects. The ability to quickly deploy in the field and acquire high-resolution datasets and information for analytics and visualization makes the drone popular and very relevant. Despite the barriers and challenges, the significance of drones for social good and humanitarian action has immensely increased the demand for drones. The advances in technology have opened up the horizons to generate high-resolution outputs which in turn will contribute toward the precise assessment. High-resolution aerial images from drones are very useful for developing baseline information to understand the scale and impact of any disaster; quantify the amount of damage; assess the accessibility conditions; provide more accurate ground situation details; and also support the emergency responders in effective search and rescue missions.

Nepal has good prospects to apply such frontier technology considering the existing complex topography, need for economic solutions, and data-based decision-making in disaster risk reduction and management. Considering the exposure to multiple disasters, technologies such as drones could evolve to bring an economic and effective solution to assess the existing disaster risk and explore the potential solutions and implementation. The use of such cutting-edge technology is rapidly growing in the country and could certainly be adopted for the rational and effective problem-solving approaches. However, it is very crucial to have a centralized integrated system capable of tracking the drones, drone pilots, and entire drone missions that are happening across the country. The drone technology should be coordinated complying with the standard operating procedures for the controlled and regulated operations. Further, there should be simplified protocols for flight operation, including fly/no-fly weather criteria, validated training modules, robust and durable hardware, and customized software. The usage of UAV in different context such as deploying drones during the emergencies of Gorkha Earthquake 2015 and Melamchi Debris Flow 2019 has put forward some low-cost tangible outputs in the approach toward the risk understanding. Further, the use of drones in municipal mapping and its application toward the adoption of metric addressing system are the positive signs of rational use of data into action. Similarly, the deployment of drones for healthcare services in the complex topographical environment of the rural settings has provided an alternative solution to enhance better accessibility to life-saving medical services in hard-to-reach areas.

These are some of the representative solutions embraced by the youth and young professionals toward the disaster risk reduction. Given the opportunity to access and adapt with these technologies, the youth and young professionals can certainly bring the paradigm shift in the approaches and aim for the sustainable efforts in disaster resilience. This could however take some time for the progression, but with Science, Engineering, Technology, and Innovation the acceptance and willingness toward adapting to such technology will certainly be evident.

References

Abrahamsen HB (2015) A remotely piloted aircraft system in major incident management: concept and pilot, feasibility study. BMC Emerg Med 15(1):12. https://doi.org/10.1186/s12873-015-0036-3

Adhikari BR (2021) Lightning Fatalities and Injuries in Nepal. Weather Clim Soc 13(3):449–458. https://doi.org/10.1175/WCAS-D-20-0106.1

Adhikari BR Gautam S (2022) A review of policies and institutions for landslide risk management in Nepal. Nepal Public Policy Rev. https://doi.org/10.3126/nppr.v2i1.48397

Adhikari BR, Tian B (2021) Spatiotemporal distribution of landslides in Nepal. In: Eslamian S, Eslamian F (eds) Handbook of disaster risk reduction for resilience. Springer International Publishing, Cham, pp 453–471. https://doi.org/10.1007/978-3-030-61278-8_20

Ahmed R, Mahmud KH, Tuya JH (2021) A GIS-based mathematical approach for generating 3D terrain model from high-resolution UAV imageries. J Geovis Spatial Anal 5(2):24. https://doi.org/10.1007/s41651-021-00094-7

Aiyetan AO, Das DK (2022) Factors and strategies for improving construction management on sites in mega-projects in South Africa: an explorative survey. Infrastructures 7(2):19. https://doi.org/10.3390/infrastructures7020019

Allauddin MdS, Kiran GS, Kiran GSS, Srinivas G, Mouli GUR, Prasad PV (2019) Development of a surveillance system for forest fire detection and monitoring using drones. In: IGARSS 2019—2019 IEEE International Geoscience and Remote Sensing Symposium, pp 9361–9363. https://doi.org/10.1109/IGARSS.2019.8900436

Antoine R, Lopez T, Tanguy M, Lissak C, Gailler L, Labazuy P, Fauchard C (2020) Geoscientists in the sky: unmanned aerial vehicles responding to geohazards. Surv Geophys 41(6):1285–1321. https://doi.org/10.1007/s10712-020-09611-7

Balcik B (2017) Site selection and vehicle routing for post-disaster rapid needs assessment. Transp Res Part E Log Transp Rev 101:30–58. https://doi.org/10.1016/j.tre.2017.01.002

Bendig J, Bolten A, Bennertz S, Broscheit J, Eichfuss S, Bareth G (2014) Estimating biomass of barley using crop surface models (CSMs) derived from UAV-based RGB imaging. Remote Sens 6(11):10395–10412. https://doi.org/10.3390/rs61110395

Cabello KE, Germentil MQ, Blanco AC, Macatulad EG, Salmo III SG (2021) Post-disaster assessment of mangrove forest recovery in Lawaan-Balangiga, Eastern Samar using NDVI time series analysis. In: ISPRS Annals of the Photogrammetry, Remote Sensing and Spatial Information Sciences, V-3-2021, pp 243–250. https://doi.org/10.5194/isprs-annals-V-3-2021-243-2021

Chamberlain P (2017) Drones and Journalism. Routledge. https://doi.org/10.4324/9781315618470

Chamola V, Kotesh P, Agarwal A, Naren, Gupta N, Guizani M (2021) A comprehensive review of unmanned aerial vehicle attacks and neutralization techniques. Ad Hoc Netw 111:102324. https://doi.org/10.1016/j.adhoc.2020.102324

Chapagain D, Dhaubanjar S, Bharati L (2021) Unpacking future climate extremes and their sectoral implications in western Nepal. Clim Change 168(1–2):8. https://doi.org/10.1007/s10584-021-03216-8

Chasseaud P (2002) British, French and German mapping and survey on the western front in the First World War, pp 171–204. https://doi.org/10.1007/978-94-017-1550-8_12

Chou T-Y, Yeh M-L, Chen Y-C, Chen Y-H (2010) Disaster monitoring and management by the unmanned aerial vehicle technology. 38:137–142

Claessens M, Lewyckyj N, Biesemans J, Everaerts J (2005) Pegasus, A UAV project for disaster management

Cohen CJ (2000) Early history of remote sensing. In: Proceedings 29th Applied Imagery Pattern Recognition Workshop, pp 3–9. https://doi.org/10.1109/AIPRW.2000.953595

de Albuquerque JC, de Lucena SC, Campos CAV (2016) Evaluating data communications in natural disaster scenarios using opportunistic networks with Unmanned Aerial Vehicles. In: 2016 IEEE 19th International Conference on Intelligent Transportation Systems (ITSC), pp 1452–1457. https://doi.org/10.1109/ITSC.2016.7795748

DHM (2015) Study of climate and climatic variation over Nepal. https://www.dhm.gov.np/uploads/climatic/47171194ClimateandClimaticvariabilityofNepal-2015.pdf

Eckstein D, Künzel V, Schäfer L (2021) Global Climate Risk Index 2021. www.germanwatch.org/en/cri

Eker R, Aydın A, Hübl J (2018) Unmanned aerial vehicle (UAV)-based monitoring of a landslide:

Gallenzerkogel landslide (Ybbs-Lower Austria) case study. Environ Monit Assess 190(1):28. https://doi.org/10.1007/s10661-017-6402-8

Erdelj M, Natalizio E (2016) UAV-assisted disaster management: applications and open issues. In: 2016 International Conference on Computing, Networking and Communications (ICNC), pp 1–5. https://doi.org/10.1109/ICCNC.2016.7440563

Furutani T, Minami M (2021) Drones for disaster risk reduction and crisis response, pp 51–62. https://doi.org/10.1007/978-981-16-0360-0_4

Gautam D (2017) Unearthed lessons of 25 April 2015 Gorkha earthquake (M_W 7.8): geotechnical earthquake engineering perspectives. Geom Nat Hazards Risk 8(2):1358–1382. https://doi.org/10.1080/19475705.2017.1337653

Gautam D, Chaulagain H (2016) Structural performance and associated lessons to be learned from world earthquakes in Nepal after 25 April 2015 (M_W 7.8) Gorkha earthquake. Eng Fail Anal 68:222–243. https://doi.org/10.1016/j.engfailanal.2016.06.002

Gautam D, Adhikari R, Gautam S, Pandey VP, Thapa BR, Lamichhane S, Talchabhadel R, Thapa S, Niraula S, Aryal KR, Lamsal P, Bastola S, Sah SK, Subedi SK, Puri B, Kandel B, Sapkota P, Rupakhety R (2022) Unzipping flood vulnerability and functionality loss: tale of struggle for existence of riparian buildings. Nat Hazards. https://doi.org/10.1007/s11069-022-05433-5

Giordan D, Hayakawa Y, Nex F, Remondino F, Tarolli P (2018) Review article: the use of remotely piloted aircraft systems (RPASs) for natural hazards monitoring and management. Nat Hazard 18(4):1079–1096. https://doi.org/10.5194/nhess-18-1079-2018

Greenwood F, Nelson EL, Greenough PG (2020) Flying into the hurricane: a case study of UAV use in damage assessment during the 2017 hurricanes in Texas and Florida. PLoS ONE 15(2):e0227808. https://doi.org/10.1371/journal.pone.0227808

Immerzeel WW, Kraaijenbrink PDA, Shea JM, Shrestha AB, Pellicciotti F, Bierkens MFP, de Jong SM (2014) High-resolution monitoring of Himalayan glacier dynamics using unmanned aerial vehicles. Remote Sens Environ 150:93–103. https://doi.org/10.1016/j.rse.2014.04.025

Jain S, Jagtap D (2023) Improving cut and fill operations in construction using drone technology and aerial analytics. E3S Web Conf 405:02008. https://doi.org/10.1051/e3sconf/202340502008

Joglekar J, Gedam SS, Mohan BK (2014) Image matching using SIFT features and relaxation labeling technique—a constraint initializing method for dense stereo matching. IEEE Trans Geosci Remote Sens 52(9):5643–5652. https://doi.org/10.1109/TGRS.2013.2291685

Lahoti S, Lahoti A, Saito O (2020) Application of unmanned aerial vehicle (UAV) for urban green space mapping in urbanizing indian cities. In: Unmanned Aerial Vehicle: Applications in Agriculture and Environment. Springer International Publishing, pp 177–188. https://doi.org/10.1007/978-3-030-27157-2_13

Lei T, Zhang Y, Wang X, Fu J, Li L, Pang Z, Zhang X, Kan G (2017) The application of unmanned aerial vehicle remote sensing for monitoring secondary geological disasters after earthquakes. In: Falco CM, Jiang X (eds), p 104203H. https://doi.org/10.1117/12.2281909

Luo C, Miao W, Ullah H, McClean S, Parr G, Min G (2019) Unmanned aerial vehicles for disaster management. Springer, Singapore, pp 83–107. https://doi.org/10.1007/978-981-13-0992-2_7

Mandirola M, Casarotti C, Peloso S, Lanese I, Brunesi E, Senaldi I (2022) Use of UAS for damage inspection and assessment of bridge infrastructures. Int J Disaster Risk Reduc 72:102824. https://doi.org/10.1016/j.ijdrr.2022.102824

Martini T, Lynch M, Weaver A, van Vuuren T (2016) The humanitarian use of drones as an emerging technology for emerging needs, pp 133–152. https://doi.org/10.1007/978-94-6265-132-6_7

Meier P (2015) UAVs and humanitarian. Drones and aerial observation: new technologies for property rights, human rights, and global development a primer, p 57. http://www.iapad.org/wp-content/uploads/2015/07/DronesAndAerialObservation.pdf#page=57

MOHA (2019) Remotely piloted aircraft (RPA) popularly known as "drone" procedural guidelines. MOHA. https://caanepal.gov.np/storage/app/media/drone/sep%2024%202022/droneb.pdf. https://caanepal.gov.np/storage/app/media/drone/sep%2024%202022/droneb.pdf

Mohd Daud SMS, Mohd Yusof MYP, Heo CC, Khoo LS, Chainchel Singh MK, Mahmood MS, Nawawi H (2022) Applications of drone in disaster management: a scoping review. Sci Justice 62(1):30–42. https://doi.org/10.1016/j.scijus.2021.11.002

Murphy RR, Steimle E, Griffin C, Cullins C, Hall M, Pratt K (2008) Cooperative use of unmanned sea surface and micro aerial vehicles at Hurricane Wilma. J Field Robot 25(3):164–180. https://doi.org/10.1002/rob.20235

Pandey V, Gautam D, Gautam S, Adhikari R, Lamsal P, Talchabhadel R, Puri B, Niraula S, Karki S, Thapa B, Subedi S, Adhikari T, Lamichhane S, Shah S, Bastola S, Bhattarai P, Dahal B, Acharya I, Kandel B, et al (2021) Multi-perspective field reconnaissance after the melamchi debris flow of June 15, 2021 in central Nepal. Lalitpur

Pangeni B, Pudasaini U, Pudasaini D, Pradhan S (2020) Evolution of unmanned aerial vehicles in Nepal. J Geoinform Nepal. https://doi.org/10.3126/njg.v19i1.50963

Panthi KK (2021) Assessment on the 2014 Jure Landslide in Nepal—a disaster of extreme tragedy. IOP Conf Ser Earth Environ Sci 833(1):012179. https://doi.org/10.1088/1755-1315/833/1/012179

Poppleton L (2012) WWF Nepal launches conservation drones. https://www.nathab.com/blog/wwf-nepal-launches-conservation-drones/

Pudasaini U (2019) Drones optimized therapy system (DrOTS): use of drones for tuberculosis diagnosis in Nepal. In: J Human Health Sci (IJHHS) 14. https://doi.org/10.31344/ijhhs.v0i0.132

Rahman AAA, Maulud KNA, Mohd FA, Jaafar O, Tahar KN (2017) Volumetric calculation using low cost unmanned aerial vehicle (UAV) approach. IOP Conf Ser Mater Sci Eng 270:012032. https://doi.org/10.1088/1757-899X/270/1/012032

Ruzgienė B, Berteška T, Gečyte S, Jakubauskienė E, Aksamitauskas VČ (2015) The surface modelling based on UAV photogrammetry and qualitative estimation. Measurement 73:619–627. https://doi.org/10.1016/j.measurement.2015.04.018

Sato A (2003) The RMAX helicopter UAV. https://api.semanticscholar.org/CorpusID:26842120

Scott AF (1984) The invention of the balloon and the birth of modern chemistry. Sci Am 250(1):126–137

Shahmoradi J, Mirzaeinia A, Roghanchi P, Hassanalian M (2020) Monitoring of inaccessible areas in GPS-denied underground mines using a fully autonomous encased safety inspection drone. In: AIAA Scitech 2020 Forum. https://doi.org/10.2514/6.2020-1961

Shrestha ML (2000) Interannual variation of summer monsoon rainfall over Nepal and its relation to Southern Oscillation Index. Meteorol Atmos Phys. https://doi.org/10.1007/s007030070012

Smith KW (2015) The use of drones in environmental management. World Environ Water Resour Congress 2015:1352–1361. https://doi.org/10.1061/9780784479162.133

Suh J, Choi Y (2017) Mapping hazardous mining-induced sinkhole subsidence using unmanned aerial vehicle (drone) photogrammetry. Environ Earth Sci 76(4):144. https://doi.org/10.1007/s12665-017-6458-3

Turner D, Lucieer A, de Jong S (2015) Time series analysis of landslide dynamics using an unmanned aerial vehicle (UAV). Remote Sens 7(2):1736–1757. https://doi.org/10.3390/rs70201736

Upreti BN, Dhital MR (1996) Landslide studies and management in Nepal

Velev D, Zlateva P, Steshina L, Petukhov I (2019) Challenges of using drones and virtual/augmented reality for disaster risk management. Int Arch Photogramm Remote Sens Spatial Inf Sci XLII-3/W8:437–440. https://doi.org/10.5194/isprs-archives-XLII-3-W8-437-2019

Yakushiji K, Fujita H, Murata M, Hiroi N, Hamabe Y, Yakushiji F (2020) Short-range transportation using unmanned aerial vehicles (UAVs) during disasters in Japan. Drones 4(4):68. https://doi.org/10.3390/drones4040068

Yihui D, Chan JCL (2005) The East Asian summer monsoon: an overview. Meteorol Atmos Phys 89(1–4):117–142. https://doi.org/10.1007/s00703-005-0125-z

13 Collectivization and Local Leadership for Resilience to Disasters

Sanchita Das, Peehu Pardeshi, Niti Mishra, and K. P. Soma

Abstract

The struggle for access to power and rights competes with the struggle for access to services and income opportunities. Most individuals, communities, and local grassroots organizations are not able to balance both these struggles and can prioritize only one of them. This chapter presents an approach for building community resilience by NGOs/CSOs, which intervene with a rights-based approach and foster empowerment and local leadership. We argue in particular that resilience relies on access to resources including natural, social, economic, and political, and that a significant way to build resilience is to ensure rights to these essential resources. This chapter is based on insights gathered from qualitative researches conducted among three distinct grassroots organizations: an SHG group and an informal collective in Uttarakhand, West Bengal, and Maharashtra. The ethnographic method has been accepted as an appropriate approach in social design, as it enables to gather an in-depth contextual understanding of people's everyday life in a comprehensive way via the use of participatory observation and interviews. The study depicts that CSOs play a significant role through collective action act for adaptive capacity against drought, managing impact of compound disaster such as COVID-19 and cloudburst by building trust in health system and collectivizing for continuing education, and empowering through rights-based dignified livelihood. The study concludes that collectivization is an innovative way to build resilience in the neoliberal world.

Keywords

Social innovation · Community response · Collectives · SHGs · Indigenous practices · Coping strategies

S. Das (✉)
Research and Development, Tata Institute of Social Sciences, Mumbai, India
e-mail: sanchitatiss@gmail.com

P. Pardeshi
Tata Centre for Technology and Design, Indian Institute of Technology Bombay, Powai, Mumbai 400076, India

N. Mishra
Centre for Disaster Management, Jamsetji Tata School of Disaster Studies, Tata Institute of Social Sciences, Mumbai, India

K. P. Soma
Makaam (Mahila Kisan Adhikaar Manch), Secunderabad, India

S. Talukdar et al. (eds.), *Communication, Science, Technology and Innovation in Disaster Risk Management*, GIScience and Geo-environmental Modelling, https://doi.org/10.1007/978-3-031-77189-7_13

13.1 Introduction

Disasters disrupt the social construction and distribution of resources in the community, sometimes exacerbating the vulnerabilities of the marginalized groups. Government and civil society intervention is needed for recovery, reconstruction, and restructuring of society and the environment. It is well recognized that the first responders to any disasters are the locally affected communities themselves and the neighborhood.[1,2,3] It is thus tempting to conceptualize that training and capacity building of local people in disaster management will lead to disaster-resilient communities. Various attempts on these lines have been made by government and non-government agencies.[4] However, more often than not incomplete and misguided community resilience has led to the exploitation of the communities or disaster capitalism.[5,6] The perception of people toward the government may also change during the disasters. If a disaster is mishandled, the frustration and dissatisfaction among the people may lead to disillusionment and polarization.[7]

There has been a shift in policy and praxis in viewing the community as victims and recognizing them as actors possessing capacities to adapt to dysfunctions. This has led to emphasis by decision-makers and researchers on community resilience in every phase of disaster[8] Each community has its traditions and cultural practices which are intricately linked with their livelihood practices and provide a moral code of conduct to preserve the societal hierarchy as well as ecology. However, power dynamics are inherent in such a community and the vulnerable populations like scheduled caste/tribes, women, disabled, and children may experience exclusion and marginalization. Community resilience needs to be conceptualized and practiced in innovative ways such that people are empowered to become self-reliant and at the same time hold the government accountable for service delivery as well as civic rights.

The critical literature on resilience problematizes the thrust upon individuals and communities in the pretext of resilience building. The responsibility of the state gets shifted to the individuals/communities' agency, openness to learn and accept change as inevitable, and how to improve their inherent capacity to remain cohesive despite poverty and other external pressures. Government institutions are primarily the key players in disaster responses. However, there is often a technocratic organizational perspective that guides such disaster responses. Civic society's engagement is also getting constricted. CSOs engaged in work related to advocacy and human rights are finding it difficult to operate. This results in the replacement of NGOs by state-controlled/headed enterprises addressing socio-technical systems rather than focusing on socio-structural changes (Joseph et al. 2023). The struggle for access to power and rights competes with the struggle for access to services and income opportunities. Most individuals, communities, and local grassroots organizations have to balance and prioritize between the two.

This chapter proposes an approach for building community resilience by NGOs/CSOs with a focus on rights-based approaches that foster empowerment and local leadership. We argue in particular that resilience relies on access to resources including natural, social, economic, and political, and that a significant way to build resilience is to ensure rights to these essential resources. To operationalize resilience, we use the DRIFT framework and define it as the "ability" or "capacity" of a community exposed to a destabilizing event "to do something" positive before, during, and after the destabilizing event in order to reduce its impact and to transform to a new state of existence. This aligns with the "bouncing forward" rather than bouncing back

---

[1] Ellebrecht and Joval (2022).
[2] Ashkenazi and Hunt (2019).
[3] O'Brien et al. (2020).
[4] Harris et al. (2018).
[5] Imperiale and Vanclay (2021a).
[6] Imperiale and Vanclay (2021b).
[7] Schwaller et al. (2022).
[8] Gil-Rivas and Kilmer (2016).

to the status quo. The collectives, civil society organizations, activist groups, and local leaders can be the catalyst for this kind of innovative resilience building. They can create a cohesive society that learns from traditional practices while integrating new knowledge and progressive human rights framework along with a modern understanding of the creation of disasters and risk.

In this backdrop, this chapter aims to understand the role of local groups/CSOs/NGOs/collectives in building community resilience in the context of everyday risks and recurring disasters. It seeks to critically review the community resilience paradigm for disaster risk reduction and identify the gaps, challenges, and innovative approaches for disaster resilience. We analyze the various sectors of development as well as community life from this lens and present the changes brought about by the CSOs/collectives to build resilience and reduce vulnerability. Our findings are situated around our interactions and field visits to various grassroots organizations such as Yusuf Meherally Centre, MAKAAM and informal community-based collective, and SHGs.

13.2 Methodology

This chapter is based on insights gathered from qualitative research conducted among two distinct grassroots organizations and informal collectives, in Uttarakhand, West Bengal, and Maharashtra in the context of environmental hazards such as cloudburst and drought. The study has adopted an ethnographic method that enabled us to gather in-depth contextual understanding of people's everyday life using participatory observation and interviews (Kuure and Miettinen 2017; Margolin and Margolin 2002). This method enables us to contextualize through the use of local narratives and understand people's perspectives (ibid).

Insights into collectivization and leadership building in resilience that are narrated here are based on the work of an SHG and a NGO working in Uttarakhand Yusuf Meherally Centre, a national alliance Mahila Kisan Adhikar Manch (MAKAAM) working nation-wide, and a doctoral research work carried out between 2014 and 2021 (Das 2021) in a tribal hamlet in Purulia, West Bengal. The Yusuf Meherally Centre (YMC) was established in 1962 as a voluntary non-profit organization to honor the legacy and principles of the late Shri Yusuf Meherally, a prominent figure in India's independence movement. Since its inception, YMC has undertaken various social initiatives spanning education, healthcare, women's empowerment, rural development, youth mobilization, employment creation, environmental conservation, and the promotion of freedom movement ideals. Additionally, it has provided aid and rehabilitation during and after natural disasters. MAKAAM, founded in 2014, is a national alliance comprising women farmers alongside various organizations and networks. Its objective is to advocate for the acknowledgment and representation of women as farmers, as outlined in the National Policy on Farmers 2007, which encompasses all individuals involved in agriculture and related activities, including laborers, forest workers, and livestock handlers. MAKAAM aims to incorporate a nuanced understanding of patriarchy and other socio-economic disparities such as class, caste, and religion into discussions surrounding women's roles in managing natural resources (Sangam and Hope 2021). The work of these independent organizations and collectives focused on addressing people's everyday risks, understanding challenges to collective action, promoting solidarity groups, understanding people's vulnerabilities and threats to livelihood practices among farming and forest dwelling communities. The authors came together to critically look at resilience from the interdisciplinary lens and shared conversations that emerged from the interventions taken up by the respective organizations.

This chapter is based on the belief that innovation in building community resilience would require us to understand the complexities of everyday risks. The civic societies engagement narrated here is related to (a) addressing the drivers of risk—vulnerability, hazard, and exposure; (b) managing the immediate impacts of COVID-19 lockdown and cloudburst—a double whammy to an already slow economy; (c) solidarity architecture of collective action for adaptive capacity against drought; and (d) fostering youth empowerment and leadership post-recovery phase of disaster for social transformation.

Twenty-one interviews were conducted, comprising in-depth interviews, oral history interviews, and key informant interviews, involving individuals from various backgrounds within the farming community and representatives from organizations. Field notes were utilized to document the daily routines and discussions prevalent in the lives of participants. Additionally, digital recorders were employed to capture interviews conducted in the local language. Subsequently, these interviews were transcribed into English and subjected to thematic analysis using Atlas.Ti software.

The framework discussed in Table 13.1 draws from the DRIFT framework of resilience (Manyena et al. 2019) and the discourse on resilience as transformative by community-led network actions (Cretney 2018). The DRIFT framework deliberates on the resilience capacities as Preventive, Anticipative, Absorptive, Adaptive, and Transformative indicators. The deliberation by Cretney (2018) is referred to provide a more comprehensive understanding of the role of community-led networks on restructuring resilience through reconstruction and resilience programs. The idea of resilience calls for radical action that challenges existing structures of dominance through empowering communities. They recognize that alternatives exist which can lead to creation of new possibilities of mechanism such that "strengthens community economies, facilitates new systems for food production and distribution, and influences processes of local governance".

13.3 Addressing the Drivers of Risk—Vulnerability, Hazard, and Exposure

13.3.1 Vulnerability of Subsistence Farmers and Distress Migration

In most households in Uttarakhand region, the young and able-bodied persons, mostly men, migrate to urban areas for a job to support their family and farms. Subsistence farming is carried out by the remaining members, which produces food mostly for household consumption. The land parcels received from the government are, however, unsuitable for growing fruit trees or even grass, as the land is either barren *(bajri zameen)* or rocky. About 90% of the people have received such land. In the hilly region, there is very little land for agriculture. Lands that are suited for farming are occupied by powerful and upper caste people. Most of the Dalit households got the land that is far away from the villages. The following narrative by Mr. Singh (in-charge of the Uttarakhand center) of Yusuf Meherally Centre illustrates this.

> Land that was never noticed by people or they never bothered to look at it is what people have got. A place where there are monkeys and other wild animals. They didn't get a farm but a piece of forest land that was rocky or muddy. There are high slopes and you can't grow trees. The pieces of land in those nearby villages they got are still as they were. No use. They could just use that for grazing their cows. That's it and it has no use at all.

Farming activities are routinely disrupted by animals leading to loss in productivity and increase in frustration among the farmers. As a means to support livelihood activities, women's self-help groups have been formed in this region. Each group contains 8–10 members and each member contributes between 100 and 500/- per month to the savings pool. A member can take a loan from that pool and start an enterprise of their own—garments, handicrafts, candles,

Table 13.1 Indicators of different types of capacities required to build resilience

Prevention	Anticipation	Absorption	Adaptation	Transformation
Creation and enforcement of land-use policies Disease prevention activities such as immunization coverage Investment in multi-hazard resilient design for critical infrastructure Addressing sustainable livelihood assets and root causes of vulnerability These resources encompass various aspects such as healthcare and welfare (including access to clean water, healthcare services, and sanitation), environmental stewardship (addressing issues like land degradation, landslides, droughts, fires, and managing catchment areas), educational opportunities, social safety nets, risk management strategies, financial access, and the safeguarding of physical infrastructure	Understanding risk knowledge with various stakeholders The scope of early warning systems, disaster readiness, and response strategies Availability of risk and early warning data, methods for managing risk, and strategies for transferring risk	Enhancing absorption capacities will contribute to robust, resourceful, resistant, responsive systems and structures Creation of robust systems for humanitarian assistance, support for victims and survivors, rehabilitation, and recovery Absorptive capacity aims to reduce disaster mortality, people affected by disasters, losses, and the humanitarian aid required following a disaster. These are outcomes of SFDRR indicators that are able to provide values of absorptive capacity of the system	Preparedness and response, diversification livelihoods, coping, and self-organization in order to moderate future damages Linkage and the capacity of diverse assets including the human, natural, physical, social, financial, and political of the affected community or system Sustainable solution to cope with changing use of natural resources such as land, forests, water, and ecosystems. These include activities that incentivize sustainable management practices to manage natural resources taking into change climatic conditions Boosting socio-economic resilience involves government investment in social safety nets, expanding income sources and livelihood options, and facilitating access to financial services. It also entails reinforcing infrastructure such as electricity provision and ensuring the availability of medical professionals to address prevalent and emerging health challenges	Shift toward values and norms away from individualized ideology to a more local community-based network that is responsive to encouraging social justice, support, and connection toward a post-capitalist politics of recovery and reconstruction Nurtured values of social support and community that stood in opposition to the individualist and self-responsibilization discourses of hegemonic neoliberal ideology Ideas that question structural elements of constraints for resilience of people; actions that align with people's everyday struggle and the extent to which communities organize and subvert structures of domination to asset ones right

Source Adapted from Rejecting and recreating resilience after disaster. In *The Resilience Machine* (pp. 80–93). Routledge; Manyena et al. (2019)

packaged food items, homemade herbal remedies, etc. which are sold in local shops. Women in SHGs had taken out some loans before the lockdown which they could use for household expenditures.

People's vulnerability has further deepened due to the hill Development Projects in the form of roads, highways, and large dams that have over time destroyed forests, mountains, and displaced people. The recent Char Dham project and the Joshimath incident are a few to name. This has led to increasing incidences of floods, where villages are washed away. Trees are uprooted in order to create tunnels and highways. Mr. Singh from YMC narrates the scale of destruction that has made way:

> They build highways but where do they throw the debris? Down in the river. It all goes through sewers to rivers which have destroyed farm lands. During rain all the sand and mud that is down there in the river turns into a big torrent and it all takes it away with a flow. All the cement debris gets mixed with mud and collects on people's farm

Vulnerability is accentuated by the lack of a support structure, the lack of space to discover one's own strengths, the lack of recognition of those strengths, as well as the lack of socio-economic opportunities and material assets and opportunities to be able to improve one's own ability to cope with risk, to be able to seek opportunities, and claim a life of dignity and well-being.

13.4 Managing the Immediate Impacts of COVID-19 Lockdown and Cloudburst—A Double Whammy to an Already Slow Economy

The COVID-19 lockdown had disrupted urban employment such that all those who had left their villages to find jobs in cities had to travel back with the hope that at least they would be able to work on their farms and sustain their families during the lockdown period. However, soon after that, northern regions of India were faced with a cloudburst. In Uttarakhand, many people ended up losing their lands altogether, due to flooding. Those having petty shops and small businesses also suffered due to lack of demand and market linkages during lockdown and loss of equipment and transport after cloudburst. Few narratives from the field illustrate the loss people faced,

> All throughout the lockdown the farms were closed, somehow, we had started cultivating after things opened up and now this rain has ruined everything…NGOs giving jobs to artisans stopped getting orders from their buyers during the lockdown…70% of the members of the self-help group have taken loan… Loan money used for household expenses instead of opening business… Earlier covid was killing, now continuous rains for 3 days are killing…We had just tilled the land to prepare it for the crop ... now we don't have the energy in the bodies to do it all over again…

To respond to the flooding, local NGOs started organizing and collecting donations for immediate relief activities like food and shelter provision. They also coordinated short-term livelihood projects so as to provide a dignified means of income to the inhabitants, such as stitching cotton masks,

> So we are talking to donors and philanthropists if they can help some of the women as well… A lot of work orders for masks came so women stitched masks too.

The government response to the floods was inadequate and short term. People relied more on their networks, i.e., their relationships with the community as well as with the CSOs/NGOs to start the recovery process. There is an inherent agency and ability to organize within the community, if need arises. They stood up to help each other and coordinated the donations coming in the form of cash or kind, as stated here,

> The Government has given 2–3 thousand for house and crop/farmland damage. But that is not enough—not even a wall can be built in that money. A few people had come to our office to provide flood relief, but now nobody is there. Even our office was washed away… The local municipality arranged for the stay in the banquet hall and arranged food and water. There were some businesspersons as well…like restaurant owners who had got food prepared and distributed.

13.4.1 Allaying Fear and Building Trust on Health System

For common ailments, many people in Uttarakhand don't go to hospitals and manage with herbal traditional medicines at home. Public Hospitals may not be liked by all, so those who can afford it would choose to go to private hospitals. During COVID-19, those who had heart attacks or breathing trouble got treatment in hospitals but many people who could be at home treated themselves at home with help of steam and medicines prescribed by doctors. They were not sure of the facilities and quality of food and care at the hospitals. In fact, there was fear and mistrust among the local people which made them stay away from all health facilities.

> Most people who went to hospitals died. At home, at least they would get food from time to time…-hospitals were full. One good thing that happened is that due to corona, private hospitals were not keeping the patients for undue long periods. Those who had surgeries had to stay a bit longer, but all others were quickly discharged…Eating giloy, chyawanprash and other herbal concoctions helped prevent infections... we took these from the very beginning…as a routine….black pepper, garam masala powders were available in the market for making concoctions (kaadha).

To help people get the appropriate medical treatment and execute their right to healthcare, Yusuf Meherally Centre worked with the communities in Uttarakhand to build their trust in the hospitals and allay their fear of government authorities. This was narrated by the member of the organization here,

> …villages were sanitized, medical camps were organized, medicines and treatment were provided. Those who were tested covid positive, they were linked to government hospitals. The organization took care of all their expenses. People were scared to go to the hospital, they thought they'd be beaten up but when they realized this organization would back them up, will monitor, people's trust developed. All patients and people with some symptoms were closely monitored. Medical camps were set up frequently in order to take preventive measures. This obviously instils faith in the villagers and gives them some confidence to trust the organization and the people.

13.4.2 Adapting by Collectivizing for Continuing Education

Many people in the hills in Uttarakhand have also migrated due to lack of education facilities. The government schools lack infrastructure and adequate teachers. Those who have money due to tourism and want to educate their kids in good schools would leave so that their kids go to good private schools.

When schools were closed due to lockdown, all education shifted to online mode. Children got distracted due to phones, since both education and entertainment were on phones. Parents had to spend extra time sitting with the children to monitor their online studies and progress. The school teachers in Uttarakhand and UP organized the parents of students to coordinate the schoolwork and homework activities. Those families, who could not pay the school fees, were assigned duties by the school teachers to monitor the online schoolwork and homework of students in lieu of reduced school fees. This way, they found innovative ways to rise to the occasion even in the midst of crisis.

> Online mode—those who could afford a laptop, used them, but those who were poor had a lot of trouble—no phone or just one phone for 2–3 kids. In intercollege, even for a short while when the schools opened, there was low attendance, so classes were clubbed together…In Nainital, most of the students have phones—both boys and girls—for online education, many people got wifi connection in their houses. But girls missed out on education because of the lack of phones and internet in poorer families... Most girls were shifted from private to public schools as the household incomes were reduced…The school teachers were not paid their full salaries during the lockdown which also impacted their household situation…

13.5 Solidarity Architecture of Collective Action for Adaptive Capacity Against Drought

Su et al. (2018) conducted a study examining the intricate connection between livelihood risks and livelihood capitals within farming communities in China's Shiyang River Basin. Their research underscores how certain risks, such as health and social risks, detrimentally impact rural farmers' ability to withstand livelihood risks and improve their well-being (Su et al. 2018). The study further notes that it is not solely individual capital but rather a combination of different capitals that determines an individual or household's resilience. Human and social capital emerge as significant factors in managing livelihood risks. To address uncertainties arising from hazardous natural environments and limited information, households rely on their human capital, particularly in terms of health (health investments and disease prevention) and education. Additionally, the study reveals an intriguing and unexpected finding: the relationship between physical capital (including natural capital) and social risk is directly proportional. Farmers in the Shiyang River Basin with weaker social connections tend to accumulate more natural and physical capitals, such as livestock, as a buffer against livelihood risks.

The Santhals residing in the Sirkaband gram panchayat of Purulia, West Bengal, are a traditional community deeply reliant on farming and forest resources, frequently confronting severe droughts. Their longstanding connections with forests and land form a vital aspect of their life, traditions, culture, and indigenous knowledge, which are pivotal in sustaining their livelihood practices and coping with extreme conditions such as drought. To mitigate the risks associated with their livelihoods, farmers often opt for cultivating a diverse range of crops and strive to maintain self-sufficiency in seeds. However, not all farmers can achieve this self-reliance, particularly single women farmers who often encounter difficulties accessing seeds during the peak agriculture season.

In response to this challenge, a customary practice has emerged among Santali households, wherein seeds are exchanged and used as a form of payment for labor to villagers who rely on neighbors and family for seed access. This practice aids resource-poor individuals in accessing seeds during challenging times. It is recognized that vulnerability can be mitigated to some extent by ensuring secure access to valuable resources through investments in social institutions such as kinship and social relations.

Farmers' activities involving seed production, storage, and exchange within their local environment constitute vital elements of a dynamic system, serving as the primary seed source for food crops within the community. Through planting, harvesting, storing, and exchanging seeds, farmers not only sustain themselves but also contribute to the food security of others. In carrying out these activities, they accumulate extensive knowledge about various crop varieties, their attributes, potential applications, and their interactions with the surrounding ecosystem. An elderly female farmer elaborates on this process,

> In the seeds produced locally, there is local reproduction, using local seed selection, production and careful observation of the environment. The process involves saving some for the next planting season and informal distribution of seed to other farming households often within the community or extended family in the uplands.

The utilization of local seeds and their exchange among farmers not only fosters ecological stability but also facilitates the interaction of different cultivars with each other and with the agro-ecological environment. These exchanges are rooted in cooperation and reciprocity, where seeds are often exchanged on a mutual basis. Consequently, this process involves the sharing of knowledge and practices to optimize seed performance. Farmers acquire insights by observing the growth of seeds in other fields, forming an open-access system that transcends market functions. Both seed storage and exchange practices possess the capacity to adapt positively during uncertainties, such as crop loss or restricted seed access. Furthermore, households experience varying levels of risk exposure during drought years due to differences in resource endowments and capabilities among households. In such instances where individual access and capacity are insufficient, collective arrangements prove beneficial. The expertise of these women in utilizing a diverse range of seeds and maintaining significant in situ reserves of genetic resources guides the adaptive measures undertaken by the group.

The significance of solidarity groups also surfaced during the COVID times. During this time, the water and food needs of the family escalated, increasing the burden on women, as in India, cooking and caring work mostly falls on women due to gendered roles. Long-term association with self-help groups had made women aware about their rights. Solidarity with other women helped them gain confidence and take up income-generating activities and also gave them voice Even if they could not change their fates, they ensured that their daughters had better opportunities.

> Children are not married off early unlike in the past. Women take up work together and provide solidarity with each other. We go together wherever we need to go. Earlier we were scared of men—men would stop them from going somewhere, now we answer back and go wherever we need to go, wherever the good work takes us ... what is the problem in going out for work, I learnt so many things through this work…women have become very smart. Men are a bit scared to misbehave now…

Informal collectives and solidarity groups are also the cornerstone of Mahila Kisaan Adhikaar Manch (MAKAAM). MAKAAM as a collective organizes people's voice and nurtures women's leadership so that the collective agency may be asserted to be able to claim spaces, resources, and representative forums. So that women can negotiate their rights as farmers, producers, as well as civil society citizens and agents for change. So, from the perspective of forest dwelling communities, the whole point of living with ecological resources in a way that is harmonious is a principle of resilience itself. It's an underlying way of being, of living. And if that is compromised by destructive development, then their ability to be resilient in their own definition shifts to becoming dependent on state-provided benefits, such as the ration system wherein the food security basket gets disrupted, and the processes of understanding the rhythms of resource availability are threatened.

In this context, as resources become scarce or captured, the resilience becomes dependent much more on their agency to be able to negotiate with others. So, it is a composite and a dynamic process of resilience. As a collective, MAKAAM has focused upon building resilience through recognition. Recognition is a pillar of their work and it starts with making women aware of their contribution to the economy, which is about "Atma Samman" and their contribution to the economy in terms of GDP. That if the economy is being counted in terms of GDP, women's work remains invisible and undercounted and therefore not represented in their economic contribution. Recognition is needed at multiple levels, at the household level because women are invisible labor and can be manipulated for productive care and reproductive work and controlled and managed by the men. So, the patriarchal issues come into that framework. The class aspect comes into the framework by talking about the marginalization of farmers, especially small and landless farmers and how that marginalization is perpetuated by the neoliberal pattern of development and the patriarchal footprint of resourcefulness for women. What this leads to and enables women to do is to form their collectives, to decide what it is they want to do together, emphasizing that individual women can only fight their battles in their households and in their societies through collective action. Building the solidarity architecture of collective action becomes another cornerstone.

So along with institution building and collectivization, leadership development is much needed. MAKAAM is instrumental in building agency through assertion as collectives and building leadership for claiming that space to assert and negotiate as it is believed that building solidarity among women is the key to building agency.

13.6 Fostering Youth Empowerment and Leadership Post-recovery Phase of Disaster for Social Transformation

An ideology-driven socialist organization, Yusuf Meherally Centre, has made it its priority to work in disaster-prone regions during the relief phase and continue post-recovery. They believe that through constructive activities, the credibility of the organization improves. They envisioned working toward empowering people, especially the oppressed. In Uttarakhand, a rehabilitation center was established for people who were homeless or were threatened or vulnerable in any form. As in the words of the organization's founding member,

> …through this, wherever there were disasters we worked there and 90% of our branches were built because of the work we did after the disaster. It increased outreach. Our capability to put forward our thoughts has increased…

Another strategic move by the organization was to work in the periphery, as there is a greater need to work with people who remain unreached. This strategy arose from the observation that relief is immediately provided where there is maximum damage, but not to those who are located at the margins but are equally vulnerable. Disasters served as a "window of opportunity" to enter the community and understand it better while carrying out recovery activities.

YMC's center in Uttarakhand has played a crucial role in supporting the people who are marginalized and have faced some form of caste or gender-based violence. Such that the perpetrators of crime have now become aware of this and remain vigilant of their malpractices. The center provides legal, financial, and logistics support to people who have been oppressed.

The center also organizes youth camps 4–5 times a year. Young boys and girls are selected from the villages and are trained in legal matters such as POSCO and SC ST Acts. They are also made aware of different government schemes that can benefit the villagers. These trained youth are sent to different parts of the country to attend seminars and workshops and conduct training programs in their respective villages as trainers. Such trained cadres of youth are presently working in 250 villages. Through them, the organization spreads awareness in their respective villages through street plays or organizing meetings. There are books distributed to the schools or children and a few libraries are also set up in the villages.

13.6.1 Empowering Through Rights-Based Dignified Livelihood

People are provided training in diverse fields as a livelihood promotion activity depending on their interests and skills. They are provided training in woodwork, handloom weaving, basic stitching, etc. They are also handheld to establish their own production unit. Nonetheless, they can work in shops or on contract once they have received the training.

The first 10 years, Yusuf Meherally Centre worked at the ideological level, people were made aware of Gandhian values, Ambedkar's principles, and the socialist movement. Over the years, there was a realization that there is a deep-rooted historical injustice in this region, which is Jaunsar-Bawar tribal belt of Uttarakhand region. The irony is that the upper castes of this region are tribals and so are the Brahmin Rajputs of this region. The tribal upper castes have reservations. 50% are upper caste and the rest are Dalit or Namashudra. So, when there is some form of violence on the Dalits or SC people by the upper caste tribes, such offenses are not recorded under the ST ACT. The upper caste tribal people are responsible for the violence and promote untouchability. The Dalits are not allowed to enter the temple's yard. If they want to offer money in the temple, they keep it on the floor on the side and the priest would then pick it up. Regarding the mark on the forehead (tilak/teeka), they are given on a leaf. The Dalits cannot have a procession during their wedding ceremony. They can only get married in a building provided by the government. If they ever protest or voice against the injustice, they are boycotted.

In this context, the organization intervened and started working in 150 villages, under the POCSO. In the last 1.5 years, six people were imprisoned for life. There are similar cases still pending with the court. Such acts of violence against women were never reported in the police files. The sarpanch would never want such cases to be recorded. On the contrary, they would bribe or threaten the people from reporting. They would bribe or threaten the people. Since the organization started to work, official complaints have been lodged against the gram panchayats. The organization has now become the point of contact for three zilla (administrative districts), Uttarkashi that borders China, Dehradun the capital area, and Tehri Garhwal. The narrative below illustrates this.

> Any crime against a Dalit, girl child or women they come to us. We have a whole system and we file cases on behalf of those people... According to my record in 2-3 police stations there are no cases under POCSO or SC ST Act because people come to us and ask what they must do. With whatever expertise and ideas, we have we file reports and do other things as fast as we can. If we wouldn't have been there such cases of violence and atrocities against the Dalits would have continued. In the last 1.5 years we filed so many cases in the district court and high court. We have so many cases. In fact, we have declared some areas as sensitive zones under POCSO and SC ST Act. We worked so much in 2-3 years and that's how it happened.—Jabar Singh, Uttarakhand.

13.7 Conclusion

Livelihood generation, ecological awareness and resilience building, negotiation—these are all activities that need to be conducted in rural and urban areas. CSOs are best suited to do this. NGO, SHGs, collectives, and such groups increase the participation in community level activities and increase awareness of rights. In many cases, they even deliver services and ensure outreach and networking. There are organizations that have approached resilience from the right perspective. Experiences of Yusuf Meherally Centre in Uttarakhand show that it's not just disaster, but it's also sexual harassment, caste-based discrimination, and other forms of everyday risks that need attention. Resilience can be understood as emotional, political, and social. Resilience begins from within, but within is not necessarily the individual but when one comes in solidarity with multiple people, their self-esteem grows. This process nurtures and builds confidence in the collective and individually to grow and withstand all hardships as well as oppression within the households and outside. Reflected in MAKAAM's work is to be able to go beyond the struggle to survive and establish one's ability to live with a certain dignity and adequacy of material and non-material needs, to be able to withstand hardship, rise above it, and be able to secure oneself for the future is resilience.

Self-reliance could be a part of resilience building. But it has often become a gimmick. Promotion of self-employment or self-help for people has failed when the state is unable to provide for the basic social and economic opportunities. It is important to use that as an opportunity to build on communities' abilities to demand it, so that they have opportunities economically, socially, and politically. So that it becomes the first point for assertion of agency, building of leadership, and building of capacity to negotiate. These are processes that are necessary and really non-negotiable in the process of self-reliance and are important as part of the architecture for resilience also. Because if people's voice is missing, collective solidarity is absent, then one's issues can very easily be pushed under the carpet.

Resilience is not just about reconstruction or rehabilitation post-disaster, but it is also about bringing transformative change, by addressing the power negotiation and access to services and resources to reduce or manage everyday risks and conflicts. As described by the DRFIT framework, resilience in the true sense must help vulnerable populations bounce forward from shocks and disasters. This bouncing forward through social transformation and empowerment is a work in progress rather than a tangible goal or milestone. Different communities may be unique in their levels of empowerment and may define it differently. At the core of the resilience building and empowerment paradigm is the recognition that people and communities are partners and stakeholders in the development, not just beneficiaries or victims. Yet, there are tensions between government and communities over resources and power such that their interests are more often than not juxtaposed and conflicted. A bridge can be formed by CSOs and community leaders to increase the power of the vulnerable populations and to increase the accountability of government institutions such that human rights and development processes go hand in hand.

In the neoliberal world where individualism has fragmented and broken down the social capitals and increased the costs of development, collectivization is an equitable and innovative way to build resilience. With migration, floating populations, and elderly populations, this may be an emerging need. While communities continue to adapt to the existing livelihood situations, CSOs are positioned to catalyze a social transformation to bounce forward. Nevertheless, preventive, anticipatory, and absorptive capacity building requires state interventions and regulations for land rights, protection, and risk transfer. In the absence of these, the communities remain vulnerable while at the same time being resilient owing to their own agency and adaptive capacities. For a truly resilient community, government and CSOs both need to partner with the vulnerable people as equal stakeholders with the power to influence the course of development of the region and the country in a democratic manner.

References

Ashkenazi I, Hunt RC (2019) You're it—you've got to save someone: immediate responders, not bystanders. Front Public Health

Cretney R (2018). Rejecting and recreating resilience after disaster. In: The resilience machine. Routledge, New York, NY, pp 80–93

Das S (2021) State-community interfaces in drought risk reduction: a study on the vulnerability contexts of a Santal community in Purulia, West Bengal. Doctoral dissertation, Tata Institute of Social Sciences, Mumbai

Ellebrecht N, Joval A (2022) Active violent incident: Medical first responders' perception of people's willingness to help. Méd Catastr Urgences Collect 6:241–246

Gil-Rivas V, Kilmer RP (2016) Building community capacity and fostering disaster resilience. J Clin Psychol 72(12):1318–1332. https://doi.org/10.1002/jclp.22281

Harris C, McCarthy K, Liu EL, Klein K, Swienton R, Prins P, Waltz T (2018) Expanding understanding of response roles: an examination of immediate and first responders in the United States. Int J Environ Res Public Health 15(3):534. https://doi.org/10.3390/ijerph15030534

Imperiale AJ, Vanclay F (2021a) Conceptualizing community resilience and the social dimensions of risk to overcome barriers to disaster risk reduction and sustainable development. Sustain Dev 29(5):891–905. https://doi.org/10.1002/sd.2182

Imperiale AJ, Vanclay F (2021b) The mechanism of disaster capitalism and the failure to build community resilience: learning from the 2009 earthquake in L'Aquila, Italy. Disasters 45(3):555–576. https://doi.org/10.1111/disa.12431

Joseph J, Awasthi S, Mulla ZR (2023) Leadership for disaster resilience: lessons from India. Routledge, New York, NY

Kuure E, Miettinen S (2017) Social design for service: building a framework for designers working in the development context. Des J 20(sup1):S3464–S3474

Manyena B, Machingura F, O'Keefe P (2019) Disaster resilience integrated framework for transformation (DRIFT): a new approach to theorising and operationalising resilience. World Dev 123:104587

Margolin V, Margolin S (2002) A 'social model' of design: issues of practice and research. Des Issues 18(4):24–30

O'Brien A et al (2020) Situation awareness in multi-agency emergency response: models, methods and applications. Int J Disas Risk Reduc 48:101634

Sangam V, Hope S (2021) Women's collectives create pathways for change—Volume 4 of the extraordinary work of 'ordinary people—beyond pandemics and lockdown. Vikalp Sangam Core Group, Pune

Schwaller NL, Campbell L, Nguyen MT et al (2022) (Mis)trusting the process: how post-disaster home buyout processes can degrade public trust. Nat Hazards 111:2681–2702. https://doi.org/10.1007/s11069-021-05153-2

Su F, Saikia U, Hay I (2018) Relationships between livelihood risks and livelihood capitals: a case study in Shiyang River Basin, China. Sustainability 10(2):1–20

14 A Study of Biophilic Design and Resilience in Malabar Region: An Innovation Towards Organizational Sustainability

Sandra Suresh, Dharmendra Kumar Singh, Chetna Jaiswal, and Unnikannan P. Santosh Kumar

Abstract

Based on people's innate love for nature, biophilic design aims to incorporate natural elements into built environments to promote effectiveness and well-being. The incorporation of biophilic principles stands out as an innovative development in the context of organizational sustainability and GHRM fostering a sense of connection with nature that facilitates ecological symbiosis such as bionomics, person–environment fit towards the readiness to disaster and risk management. This study investigates how elements of biophilic design, such as sunlight, gravitational features, curves, colour, water, life, representations of nature, fractals, and organized complexity, support improved psychological restoration and foster an HSR (healthy, sustainable and resilient) workplace. Following the purposive sampling method, the study used a sample of 150 employees working in corporate organizations based in the Malabar region because of its socio-geographic significance and ecological interface. The study followed a correlational research design, and statistical methods including Pearson's correlation and simple regression analysis were used to see the relationship between the study variables, biophilic designs and psychological restoration. Results have shown a significant positive correlation between biophilic designs (measured by biophilic index) and psychological restoration and its dimensions fascination, being away and compatibility at ($p \leq 0.01$) level of significance. Also, biophilic design has shown prediction with psychological restoration and its dimension. In conclusion, organizations can become more functional and adaptable by implementing biophilic measures, promoting a healthier environment, more resilient urban structures, and better resilient outcomes for employees.

Keywords

Biophilic designs · Psychological resilience (restoration) · Innovative architectural design · Organizational sustainability (HSR)

14.1 Introduction

Biophilia, a fundamental human inclination towards nature (Wilson 1986), is harnessed through biophilic design, which incorporates

S. Suresh (✉) · D. K. Singh · C. Jaiswal · U. P. S. Kumar
Department of Psychological Sciences, Central University of South Bihar, Gaya, Bihar 824236, India
e-mail: sandrasuresh@cusb.ac.in

S. Talukdar et al. (eds.), *Communication, Science, Technology and Innovation in Disaster Risk Management*, GIScience and Geo-environmental Modelling, https://doi.org/10.1007/978-3-031-77189-7_14

elements of the natural environment and structural landscape characteristics into the built environment to establish connections between humans and nature (Joye 2007; Hes and Du Plessis 2014). Rooted in humanity's intrinsic connection to the natural world, biophilic design integrates natural elements, configurations, and processes into constructed spaces, fostering a sense of harmony and affinity. This approach embraces principles such as optimizing natural light and views of the outdoors to reduce stress and enhance well-being (Bano et al. 2023), incorporating indoor plants to improve air quality and aesthetics, introducing water features for a calming effect, utilizing natural materials like wood and stone to create a sensory link with the environment, and integrating nature-inspired patterns, colours, and forms into architectural and interior design (Ryan and Browning 2020).

14.1.1 Biophilic Designs and Connection with Nature and Sustainability

Biophilic designs seamlessly integrate natural components into diverse spaces, fostering sustainability through a harmonious fusion of environmental, social, and economic attributes. These designs transcend aesthetics, nurturing a profound bond with the natural world and providing a plethora of advantages that align with sustainable principles.

Biophilic designs contribute to environmental, social and economic attributes of sustainability. It embraces ecological stewardship by curtailing resource consumption, optimizing energy efficiency, and bolstering ecological robustness. Approaches such as harnessing natural airflow, utilizing daylight, and implementing rainwater collection systems contribute to mitigating environmental impact (Beatley 2011). Biophilic designs help in gaining social sustainability by designing spaces that help cultivate more healthy and supportive environments for occupants (Gür and Kaprol 2022). Access to nature, verdant spaces, and natural illumination has been correlated with heightened mental well-being (Phillips et al. 2023), decreased stress (Gaekwad et al. 2023), and improved productivity (Scott et al. 2020). These designs foster social cohesion and interpersonal interactions, nurturing a sense of community and well-being (Kellert et al. 2008). Biophilic designs confer economic advantages by enhancing employee contentment, which can lead to enhanced retention rates, elevated productivity, and reduced absenteeism. Furthermore, energy-efficient strategies like harnessing natural light and passive heating/cooling systems can engender long-term cost savings (Browning et al. 2014).

Moreover, biophilic designs often draw inspiration from biomimicry, a concept that derives insights from nature's efficient designs and processes. This approach fosters inventive, sustainable solutions in design and technology, contributing to both environmental and economic sustainability (Benyus 1997).

Another concept called Green human resource management (GHRM) plays an active part in going green at the office, where they are engaged in managing the environment within an organization (Ahmad 2015). Green HRM is the use of HRM policies to encourage the sustainable use of resources within business enterprises and promote the cause of environmentalism which further boosts employee morale and satisfaction (Mampra 2013). These green initiatives within HRM form part of wider programs of corporate social responsibility with a major element called environment-friendly HR practices (Mandip 2012). Green HRM and green architecture together provide a holistic approach to sustainability with a common goal of reducing environmental impact and promoting sustainable practices within the organization creating an eco-conscious workforce.

14.1.2 Biophilic Designs and Innovation

Biophilic approach to design has been associated with fostering innovation through various means, promoting ingenuity, effectiveness, and overall cognitive capacity as it is involved in; amplified creativity and problem-solving, alleviation of stress and cognitive aptitude, and heightened productivity.

Biophilic settings have demonstrated the power to stimulate imaginative reasoning and proficiency in problem-solving. Exposure to natural elements and observations of nature can lead to enhanced cognitive performance (Collado and Manrique 2020; Mason et al. 2021) and the generation of innovative ideas, as evidenced by research by Lee and Shin. Embracing a biophilic framework can contribute to stress reduction and improved cognitive function, empowering individuals to think more clearly and creatively. Lower stress levels often correlate with better decision-making and inventive thinking (Hartig et al. 1991). Integrating biophilic elements into workspaces has been linked to heightened productivity and efficiency. Environments that incorporate features like natural light, greenery, and organic materials create a conducive atmosphere for focused effort and pioneering endeavours (Browning et al. 2014).

Biophilic design not only fosters social interactions but also encourages collaboration among individuals. Environments designed to be natural and inviting play a crucial role in facilitating communication and the exchange of diverse perspectives, ultimately leading to novel problem-solving approaches and creative idea formulation (Kellert 2008). Spaces adorned with biophilic elements have the potential to cultivate mindfulness and cognitive flexibility, both of which are essential attributes for innovative thinking. By practising mindfulness within natural surroundings, individuals can approach challenges with fresh viewpoints, as highlighted by Joye (2007). Moreover, biophilic design inherently provides innovative solutions for both ecologically sustainable (Mahmoud Ibrahim 2021) and aesthetically pleasing architecture. The integration of nature-inspired patterns and materials can yield inventive and groundbreaking architectural blueprints (Kellert et al. 2008).

This biophilic innovation in architecture has also incorporated technology in their designs to regulate illumination, temperature, and ventilation to emulate natural conditions (Makram 2019). For instance, dynamic illumination setups can emulate the colour, warmth and intensity of sunlight across the day, harmonizing with occupants' circadian rhythms (Vedvik 2019). Additionally, the utilization of virtual reality (VR) and augmented reality (AR) technologies can offer immersive encounters with nature within indoor settings, fostering tranquillity and stress reduction (Mollazadeh and Zhu 2021).

This innovation of biophilic designs is also helpful in urban planning as with an aim of establishing healthier and more habitable cities by infusing green areas, rooftop gardens, and vertical greenery into urban landscapes can ameliorate air quality, mitigate urban heat island effects, and elevate community welfare (Beatley 2011, 2016).

Incorporating biophilic design into diverse domains, including workplaces, educational institutions, and healthcare facilities, can nurture an atmosphere conducive to innovation and resourceful thinking. The provided references corroborate the correlations between biophilic design and innovation by underscoring nature's affirmative impacts on cognitive function, creativity, and collaborative efforts.

14.1.3 Biophilic Designs for Disaster Risk Management

In the realm of hazard and disaster situations, the primary focus revolves around ensuring personal safety and security. Remarkably, instances have emerged wherein a swift "green response" occurs during crises, such as the creation of community gardens among war veterans and widows in Bosnia (Brdanovic 2009), cultivating flowers within World War I trenches (Helphand 2006), and the utilization of horticulture therapy, involving healing gardens and gardening in rehabilitation efforts (Relf 2005). This therapeutic approach has been employed in diverse contexts, including aiding returning war veterans (Helphand 2012), assisting refugees (Moore 2012), and benefitting inmates

(Lindemuth 2012) to alleviate stress and anxiety, as well as promoting well-being among employees (Karthikeyan and Eswar 2018). Louv (2005) delved into the profound connection with living organisms in urgent circumstances in the journey to resilience.

Within disaster risk management, biophilia gains significance as it emphasizes nature's potential role in lessening disaster impacts and bolstering resilience. Exposure to natural elements like green spaces and landscapes can alleviate stress, anxiety, and trauma for those affected by disasters improving emotional well-being (Hartig et al. 2014; Sutton-Grier and Sandifer 2019). Moreover, interventions like biophilic designs such as communal gardens or natural parks, encourage social cohesion and mutual support encouraging unity in community (Tabb 2020), as natural ecosystem help enhance a community's ability to withstand and recuperate from disasters (Bratman et al. 2019).

Also, diversified and sustainable economic activities amplify adaptive capacity as a resilient economy with various sectors and income sources is better poised to endure and rebound from disasters. Developing resilient infrastructure and incorporating nature-based solutions heighten adaptive capacity as well-designed infrastructure, such as flood-resistant structures and green spaces, minimizes hazard impacts and accelerates recovery (Girardet 2020).

Discrete biophilic designs help to withstand and recover from disasters. Living infrastructure encompassing green roofs and living walls integrates vegetation onto building surfaces serves as natural insulation, diminishes stormwater runoff, and elevates air quality which thereby enhances disaster readiness and combat issues like heat and pollution (Arabi 2015). Crafting landscapes with rain gardens and bioswales proves effective in managing stormwater, which thwarts flooding and soil erosion while promoting biodiversity and water absorption (Dunnett and Kingsbury 2008). Versatile green spaces including multipurpose parks and green areas deliver both recreational value and flood mitigation capabilities (Childers et al. 2015). Furthermore, the use of fire-resistant native plants and landscaping techniques helps mitigate the risk of wildfire spread to structures (Wang et al. 2021).

Beatley and Newman (2013) gave the idea of biophilic routes to urban resilience, emphasizing the valuable role of biophilic urban planning in safeguarding and enhancing favourable climatic and micro-climate conditions within cities. Biophilic cities offer an array of favourable outcomes for individuals and families, contributing to their capacity to adeptly manage and adjust to forthcoming risks and unforeseen events (Beatley and Newman 2013). Biophilic urban planning holds the potential to establish a groundwork for elevated well-being and ways of life by incorporating an expansive network of pedestrian pathways and hiking trails, sizable green spaces seamlessly integrated into urban neighbourhoods, and urban land use arrangements that prioritize the proximity of parks. This configuration has been correlated with notably positive health outcomes (Schantz and Stigell 2008) like resilience of individuals and families to cope with future stresses. Furthermore, the direct way that biophilia leads to making things stronger involves putting resources into green projects, like bringing back wetlands or planting drought-resistant plants in cities which helps to make things more resilient and brings positive results, such as lowering high temperatures during summer and decreasing flooding from storms near the coast.

14.1.4 Biophilic Designs in Organizations

Biophilic design within corporate environments involves integrating natural elements and motifs into workspaces to enhance the well-being, efficiency, and overall contentment of employees. Organizations employ a variety of strategies to infuse biophilia into office environments, each contributing to a harmonious and invigorating atmosphere. Indoor vegetation and harnessing natural illumination are recognized practices that not only improve air quality but also have a positive impact on emotional well-being, stress reduction, and employee satisfaction (Browning et al. 2014). Biophilic architectural features, such as incorporating elements like wood and stone into designs, establish a seamless connection with the natural world, evoking feelings of tranquillity and comfort (Kellert et al. 2008). Workspaces inspired by nature, characterized by patterns, textures, and colours reminiscent of the outdoors, instill a sense of calm and invigoration, fostering heightened creativity and focused attention (Beatley 2011). Providing employees with views of natural landscapes or outdoor settings contributes to job satisfaction, stress alleviation, and overall enhanced well-being (Ulrich 1984; Yin et al. 2018). Artistic elements inspired by biophilia, such as murals and decorations, add visual allure and stimulation, elevating employee morale and innovative thinking (Joye 2007). Creating spaces dedicated to restoration, like relaxation zones, rooftop gardens, or outdoor seating areas, not only rejuvenates employees but also forges a deeper connection with nature, as emphasized by Ryan et al. (2014).

14.1.5 Psychological Restoration

"It is the set of processes leading to the renewal of physiological, psychological and social resources that become depleted when coping with the tasks and demands of everyday life" (Hartig 2004). The concept of psychological restoration draws attention to the pivotal role of nature and revitalizing surroundings in ameliorating stress, augmenting cognitive aptitude, and nurturing comprehensive psychological well-being.

Nature's influence on stress alleviation is undeniable, with engagement in natural settings like parks or lush landscapes demonstrating the ability to soothe and restore the psyche. Kaplan and Kaplan (1989) highlight the calming and immersive qualities of nature, aiding in cognitive recovery from fatigue. The Attention Restoration Theory (ART) further supports this notion, suggesting that exposure to nature allows focused attention to rest, resulting in enhanced cognitive efficiency and reduced cognitive fatigue (Kaplan 1995). Urban designs, when thoughtfully infused with restorative elements such as green havens, water features, and natural elements, offer urban dwellers opportunities for psychological restoration and relaxation, thus contributing to overall well-being (Herzog et al. 1997). Nature-centric views, whether through windows or within indoor spaces, wield positive effects on psychological restoration and stress alleviation, as emphasized by Ulrich et al. (1991). The impact of natural sounds on psychological welfare further reinforces this narrative, where sounds like cascading water and avian melodies impart a tranquillizing effect, fostering psychological restoration and reducing stress (Alvarsson et al. 2010).

According to Attention Restoration Theory (ART), being in nature may help us focus and be more concentrated in addition to being enjoyable (Ohly et al. 2016). A restorative environment should have four key features, according to ART: (1) Being Away which entails both mental and physical detachment from everyday stressors. Natural landscapes offer a respite from the demands of contemporary life, affording individuals an opportunity to temporarily disengage from routine pressures and partake in revitalizing experiences (Bratman et al. 2012). (2) Fascination: it refers to the captivating attributes of an environment that effortlessly seize one's attention (Berman et al. 2008). (3) Coherence: it encompasses the sensation of order and harmony emanating from an environment. Natural settings frequently showcase coherent patterns, such as fractals, that align with human cognitive processes, inducing a feeling of serenity and mental refreshment (Taylor et al. 2011). (4) Compatibility: it denotes the harmony between an individual's preferences and the features of an environment. A restorative setting that resonates with an individual's inclinations augments the overall revitalizing encounter (Joye and Van den Berg 2011).

14.1.6 Rationale of the Study

In the pursuit of sustainable and health-conscious office spaces, there is a growing emphasis on the integration of natural elements into the constructed work environment. This concept, referred to as biophilic design, aims to establish office settings that not only adhere to sustainability principles but also promote the well-being and resilience of employees. By seamlessly blending natural features like plants, water, sunlight, curves and organic materials, these designs aspire to alleviate stress, improve health, and cultivate a workforce that is both robust and productive. This exploration of the intricate relationship between nature, sustainability, and employee welfare emerges as a central focus, offering a promising avenue towards a more balanced and effective workspace (Fig. 14.1).

Biophilic design and strategic planning hold the potential to drive the progression of cities and spaces that are both resilient and sustainable. This is particularly noteworthy considering the comprehensive focus on human well-being, which aligns seamlessly with the increasingly popular lens of resilience. By intertwining the principles of biophilic design with the imperatives of health, sustainability, and resilience, a harmonious model emerges, fostering the creation of urban environments that comprehensively elevate the quality of life (Beatley 2009) while simultaneously supporting communities against various challenges. This integrated strategy underscores the intricate interconnections between human wellness, ecological sustainability, and the capacity to effectively navigate and recover from adversities (Cacique and Ou 2022; Fig. 14.2).

14.2 Method

This study explores how biophilic design enhances psychological restoration and its connection to sustainability by integrating nature-inspired elements into the built environment.

14.2.1 Hypotheses of the Study

H1: Biophilic designs will have a significant impact on psychological restoration of employees.

H1a: Biophilic designs will have positive impact on being away dimension of psychological restoration.

H1b: Biophilic designs will have positive impact on fascination dimension of psychological restoration.

H1c: Biophilic designs will have positive impact on coherence dimension of psychological restoration.

H1d: Biophilic designs will have positive impact on compatibility dimension of psychological restoration.

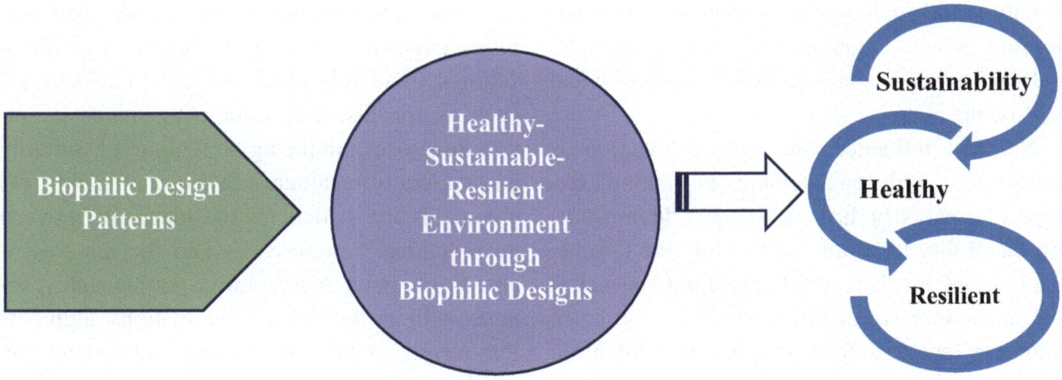

Fig. 14.1 Biophilic design for sustainable resilient environment

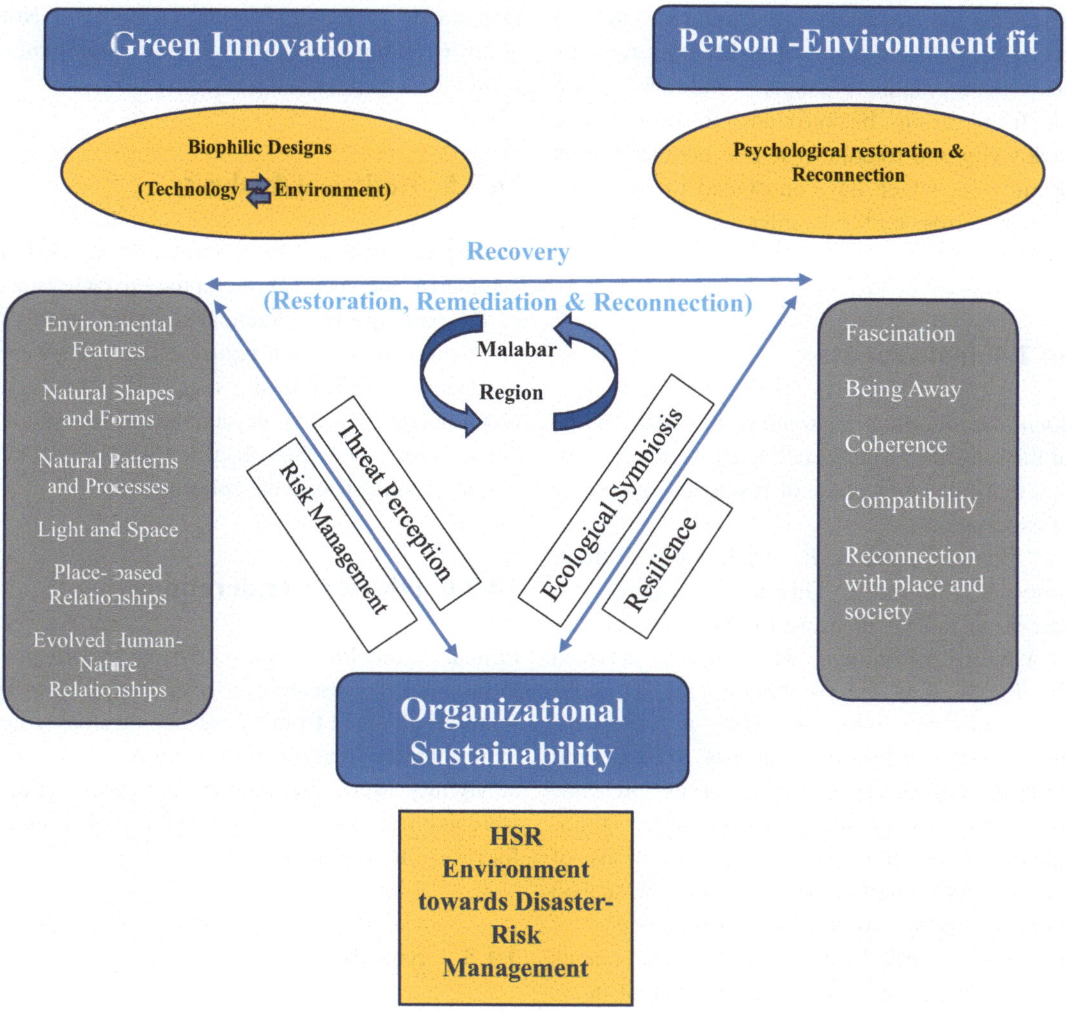

Fig. 14.2 A conceptual framework using biophilic and sustainable solutions for employee resilience

14.2.2 Sample and Design

The sample consisted of 150 corporate employees working in private sector organizations of Malabar region, following purposive sampling method. The study has used correlational research design.

The following inclusion criteria were taken for the selection of the sample:

1. Only private firm employees in managerial and educative positions are considered in this study.
2. Only employees in the fields of marketing, finance, operations, human resources, information technology, and engineering fields were taken into account for the sample.
3. Employees within the age range 20 to 50, both male and female employees, and those with at least one year of experience and with a minimum graduation degree were taken into account for the sample.

14.2.3 Study Area

Malabar region is rich in cultural diversity due to historical trade and migration, which is home to various ethnicities, religions and traditions.

They are hub for trade and commerce and are economically significant. The Malabar region's socio-demographic makeup embraces people from various backgrounds, fostering social inclusivity and tolerance. It is a popular tourist destination, which has contributed to economic growth, cross-cultural interactions and job creation.

14.2.4 Instruments

Socio-demographic personal data sheet—Some of the relevant variables collected were age, gender, marital status, place of residence, and years of experience.

"Biophilic index B"—Salingaros (2019) proposed this index. This index is used as a screening tool to measure the biophilic quality of a built environment, which ranges between $0 \leq B \leq 20$. It is the combination of 10 qualities, including light (sunlight), gravity (balance and equilibrium), details (rough and unfinished textured surfaces), curves (presence of arches), colour (earth tones), water (fountains), life (living plants), representations of nature (photographs), fractals (nested scales) and organized complexity (arrangement of details in an orderly manner). The respondents were instructed to evaluate the quality of items with an estimate of none = 0, some = 1, a large amount = 2.

"Perceived Restoration Scale"—developed by Korpela and Hartig (1996) which measures psychological restoration of the subject. This included 16 items intended to draw on the factors being away (e.g., "spending time here gives me a good break from my day-to-day routine"); fascination (e.g., "my attention is drawn to too many interesting things"); coherence (e.g., "there is a great deal of distraction"); and compatibility (e.g., "I have a sense that I belong here"). Answers were graded on a 7-point scale based on how well the statement represented the respondent's experience in the context (0 = Not at all; 6 = completely). The internal consistency shows the Cronbach's alphas as > 0.75.

14.2.5 Statistical Analysis

Descriptive statistics like frequencies were used to summarise the socio-demographic characteristics of the participants. Inferential statistics, such as correlation analyses and regression analysis were conducted to examine the relationship between biophilic designs and psychological restoration among employees. Statistical software SPSS-22 was utilized to analyse the collected data.

14.2.6 Ethical Considerations

Ethical guidelines were strictly followed throughout the research process. Informed consent was obtained from all participants, ensuring their voluntary participation. Confidentiality and anonymity were maintained, and participants were assured that their responses would be used for research purposes only.

14.3 Result

14.3.1 Statistics of Participants

Table 14.1 shows the socio-demographic details of participants among the samples. It shows that the majority of employees were males (66.2%) and 33.8% were females. 45.9% of samples were between the age of 20–30, 42.8% between the age of 31 to 40 and 11.3% between 41 to 50 years of age. 56.3% of the samples were married and 43.2% were unmarried. The majority of participants (75.2%) were from urban and 24.8% from rural. Out of the total 150 samples, 51.4% have experience of 1–3 years and rest 48.6% have more than 3 years of experience.

Table 14.1 Socio-demographic details of participants

Variable	%
Gender	
Male	66.2
Female	33.8
Other	0
Age	
20–30	45.9
31–40	42.8
41–50	11.3
Marital status	
Married	56.3
Unmarried	43.2
Other	0
Place of residence	
Rural	24.8
Urban	75.2
Years of experience	
1–3 years	51.4
More than 3 years	48.6

14.3.2 Relation Between Biophilic Index and Perceived Restoration

Table 14.2 has shown a significant positive correlation between biophilic designs (measured in biophilic index) and psychological restoration ($r=0.354$, $p \leq 0.01$). Also, biophilic index was found to be correlated with all the dimensions of perceived restoration, being away ($r=0.362$, $p<0.01$), fascination ($r=0.398$, $p<0.01$), compatibility ($r=0.289$, $p<0.01$), and coherence ($r=0.179$, $p<0.05$). This shows that as the person experience high biophilia in their setting, they most likely experiences better restorative capacity.

Table 14.3 showed that biophilic design was a significant predictor of perceived restoration with a maximum variance of 12.5% explained by biophilic index. The analysis reveals that the predictor variables collectively account for a modest amount of variance in Psychological Restoration, with R^2 values ranging from 0.032 to 0.159. The results suggest that fascination has the highest impact ($\beta=0.398$), followed by being away ($\beta=0.362$), compatibility ($\beta=0.289$), and coherence ($\beta=0.179$).

14.4 Discussion

The findings of the study offer valuable contributions to the expanding body of knowledge concerning biophilic design, psychological restoration, and sustainability in the context of constructed settings. The notable positive correlation between exposure to biophilic design components and mental rejuvenation promotes the significance of incorporating organic elements into architectural and urban planning endeavours.

The outcomes harmonize with prior research emphasizing nature's role in advancing human welfare. In 1989, Kaplan and Kaplan introduced the "Attention Restoration Theory," which

Table 14.2 The correlation between biophilic design (biophilic index) and perceived restoration (being away, fascination, coherence and compatibility)

0	*n*	*M*	SD	1	2	3	4	5	6
1	IND B	150	10.01	4.71					
2	PR	150	3.51	1.24	0.354**				
3	BA	150	3.49	1.56	0.362**	0.326**			
4	FA	150	3.49	1.24	0.398**	0.318**	0.831**		
5	CO	150	3.46	1.43	0.179*	0.312**	0.576**	0.659**	
6	COM	150	3.68	1.42	0.289**	0.544**	0.660**	0.702**	0.623**

Note PR Psychological restoration, *BA* Being away, *FA* Fascination, *CO* Coherence, *COM* compatibility, *IND B* biophilic index

** Correlation is significant at the 0.01 level (2-tailed)

* Correlation is significant at the 0.05 level (2-tailed)

Table 14.3 Regression analysis summary using biophilic designs (biophilic index) as predictor variables (IV) and psychological restoration and its dimensions as the criterion variables (DV)

Predictor variable: biophilic index					
Criterion	*R*	R^2	Beta	*F* change	Sig. of *F* change
PR	0.354	0.125	0.354	21.18	0.001
BA	0.362	0.131	0.362	22.34	0.001
FA	0.398	0.159	0.398	27.91	0.001
CO	0.179	0.032	0.179	4.89	0.029
COM	0.289	0.083	0.289	13.49	0.001

PRS Psychological restoration, *BA* Being away, *FA* Fascination, *CO* Coherence, *COM* Compatibility

proposes that exposure to natural environments can facilitate cognitive restoration and stress alleviation. This aligns with the current study's results, where respondents communicated heightened relaxation and diminished stress levels upon experiencing biophilic design elements. Elements such as natural light, indoor greenery, and glimpses of nature appear to evoke feelings of serenity and interconnectedness, thereby promoting mental restoration.

The Stress Reduction Theory (SRT) support the study findings which highlight that observing or visiting natural surroundings following stress-inducing situations has the potential to facilitate the recovery of an individual's emotional and physiological well-being (Ulrich 1984). Research has demonstrated that exposure to natural environments contributes to the alleviation of both physiological and psychological stress levels (Berto 2014; Shuda et al. 2020; Yao et al. 2021). Additionally, a study has indicated that, in contrast to built environments lacking natural elements, expansive green spaces featuring vegetation play a role in enhancing restorative experiences and diminishing negative emotions and perceived stress (Martínez-Soto et al. 2019).

When biophilic architectural features, such as abundant vegetation, flowing water, and natural illumination, are integrated into indoor spaces, they elicit a feeling of newness and captivation that resonates with the fascination component of psychological restoration. This phenomenon finds empirical support, as demonstrated in research by Herzog et al. (2003a, b) and Joye (2007), which indicated that exposure to nature-inspired surroundings within constructed environments triggered heightened levels of fascination and involvement among individuals, resulting in the restoration of cognitive faculties and the reduction of stress. The inclusion of elements rooted in biophilic design, which arouses sentiments of wonder and fascination, presents avenues for individuals to break away from ordinary stimuli and fully engage with settings that naturally command attention.

Designs that integrate natural-inspired elements into constructed spaces, provide individuals with an avenue to mentally disengage from the demands of modern life. When incorporating features like calming nature sounds, serene visuals of foliage, and tranquil water elements, these designs trigger a perceptual shift, transporting occupants to a simulated natural setting, creating a sense of "being away" where individuals experience a mental escape that aids in restoration (Staats et al. 2003; Roe and Aspinall 2011).

These designs artfully merge nature-inspired characteristics, such as organic motifs, natural textures, and rhythmic visual cues, into built spaces. This thoughtful integration creates an ambience that aligns with our innate appreciation for natural order and structure, contributing to an elevated sense of coherence (Tennessen and Cimprich 1995; Stigsdotter and Grahn 2003). The coherence dimension, as explained by Kaplan and Kaplan (1989), pertains to the perceived unity and comprehensibility of surroundings, influencing cognitive balance and renewal.

The thoughtful incorporation of nature-inspired elements within biophilic designs creates an environment that naturally exudes comfort and harmony. These designs skilfully blend elements like wood textures and plant life, evoking optimistic emotions and fostering a sense of connection, ultimately cultivating emotional well-being, which in turn, contributes to an enhanced perception of compatibility (Herzog et al. 2003a, b). Individuals occupying such spaces often experience a heightened sense of ease and emotional resonance, fostering a profound restoration of well-being (Kellert et al. 2008).

Furthermore, insights extracted from qualitative conversations with architects and urban planners provide a deeper comprehension of biophilic design's potential to enrich mental well-being and cultivate sustainability. Kellert et al. (2008) highlighted biophilic design's role in nurturing a sense of place and environmental rapport This sense of affiliation can foster heightened community involvement and an amplified reverence for nature, subsequently driving ecologically aware conduct. Also, this study could contribute to the discourse surrounding sustainable constructed environments. Biophilic design can be seen as an integral facet of green construction practices, furnishing a multi-pronged strategy that addresses energy efficiency, resource preservation, and human well-being concurrently. The infusion of biophilic elements into architectural ventures can contribute to the formation of healthier, more sustainable urban landscapes.

14.5 Conclusion

This study emphasizes the importance of biophilic designs in amplifying mental restoration and advancing sustainability within constructed environments. By weaving natural elements and principles into architectural and urban planning pursuits, we can craft spaces that not only enhance human welfare but also nurture a sense of obligation and alliance with the natural world. The association between biophilic design, psychological restoration, and sustainability embodies a comprehensive approach to urban development that prioritizes the requisites of individuals and the well-being of the planet.

Biophilic and smart cities may work together to improve urban resilience. Given these benefits of biophilic designs, decision-makers might favour them over traditional constructions in order to adapt to environmental changes, socio-economic needs, and community well-being. By incorporating these measures, cities can become more functional and adaptable, fostering a healthier environment, more resilient urban structures, and better resilient outcomes for the residents. Additionally, constructing offices with biophilic designs may entice occupants to adopt green behaviours and thereby support organizational sustainability.

References

Ahmad S (2015) Green human resource management: policies and practices. Cogent Bus Manag 2(1):1030817

Alvarsson JJ, Wiens S, Nilsson ME (2010) Stress recovery during exposure to nature sound and environmental noise. Int J Environ Res Public Health 7(3):1036–1046

Arabi R, Shahidan MF, Kamal MM, Jaafar MFZB, Rakhshandehroo M (2015) Mitigating urban heat island through green roofs. Curr World Environ 10(1):918–927

Bano F, Sehgal V, Singh R (2023) Healthy workplaces: the impact of building design on health, productivity, and social well-being of an occupant

Beatley T (2009) Biophilic urbanism: Inviting nature back to our communities and into our lives. Wm Mary Envtl L Pol'y Rev 34:209

Beatley T (2011) Biophilic cities: integrating nature into urban design and planning. Island Press

Beatley, T. and Newman P (2013) Biophilic cities are sustainable, resilient cities

Beatley T, Beatley T (2016) Singapore City, Singapore: city in a garden. Handbook of biophilic city planning and design, pp 51–66

Benyus JM (1997) Biomimicry: innovation inspired by nature. William Morrow Paperbacks

Berman MG, Jonides J, Kaplan S (2008) The cognitive benefits of interacting with nature. Psychol Sci 19(12):1207–1212

Berto R (2014) The role of nature in coping with psychophysiological stress: a literature review on restorativeness. Behav Sci 4(4):394–409
Bratman GN, Anderson CB, Berman MG, Cochran B, De Vries S, Flanders J, Folke C, Frumkin H, Gross JJ, Hartig T, Kahn Jr PH, Kuo M, Lawler JJ, Levin PS, Lindahl T, Meyer-Lindenberg A, Mitchell R, Ouyang Z, Roe J, Smith JR, van den Bosch M, Wheeler BW, White MP, Zheng H, Daily GC (2019) Nature and mental health: an ecosystem service perspective. Sci Adv 5(7):eaax0903
Bratman GN, Hamilton JP, Daily GC (2012) The impacts of nature experience on human cognitive function and mental health. Ann N Y Acad Sci 1249(1):118–136
Brdanovic D (2009) Gardens for peace and reconciliation. In: Campbell L, Wiesen A (eds) Restorative commons: creating health and well-being through urban landscapes. General Technical Report NRS-P-39, U.S. Forest Service, Newtown Square, PA, USA, pp 248–263
Browning WD, Ryan CO, Clancy JO (2014) 14 patterns of biophilic design: improving health & well-being in the built environment. Terrapin Bright Green
Cacique M, Ou SJ (2022) Biophilic design as a strategy for accomplishing the idea of healthy, sustainable, and resilient environments. Sustainability 14(9):5605
Childers DL, Cadenasso ML, Grove JM, Marshall V, McGrath B, Pickett ST (2015) An ecology for cities: a transformational nexus of design and ecology to advance climate change resilience and urban sustainability. Sustainability 7(4):3774–3791
Collado S, Manrique HM (2020) Exposure to awe-evoking natural and built scenes has positive effects on cognitive performance and affect. Environ Behav 52(10):1105–1132
Dunnett N, Kingsbury N (2008) Planting green roofs and living walls. Timber Press, Portland, OR
Gaekwad JS, Moslehian AS, Roös PB (2023) A meta-analysis of physiological stress responses to natural environments: biophilia and stress recovery theory perspectives. J Environ Psychol 102085
Girardet LH (2020) United Nations Office for Disaster Risk Reduction (UNDRR)
Gür M, Kaprol T (2022) The participation of biophilic design in the design of the post-pandemic living space. In Emerging approaches in design and new connections with nature. IGI Global, pp 75–106
Hartig T (2004) Restorative environments. In: Encyclopedia of applied psychology, vol 3. Academic, pp 273–279
Hartig T, Mang M, Evans GW (1991) Restorative effects of natural environment experiences. Environ Behav 23(1):3–26
Hartig T, Mitchell R, De Vries S, Frumkin H (2014) Nature and health. Annu Rev Public Health 35:207–228
Helphand KI (2006) Defiant gardens: making gardens in wartime. Trinity University Press, San Antonio, TX, p 21
Helphand KI (2012) Halprin in Israel. Landsc J 31(1–2):199–217
Herzog TR, Black AM, Fountaine KA, Knotts DJ (1997) Reflection and attentional recovery as distinctive benefits of restorative environments. J Environ Psychol 17(2):165–170
Herzog TR, Black AM, Fountaine KA, Knotts DJ (2003a) Reflection and attentional recovery as distinctive benefits of restorative environments. J Environ Psychol 23(3):201–218
Herzog TR, Maguire CP, Nebel MB (2003b) Assessing the restorative components of environments. J Environ Psychol 23(2):159–170
Hes D, Du Plessis C (2014) Designing for Hope: Pathways to Regenerative Sustainability. https://www.amazon.com/Designing-Hope-Pathways-Regenerative-Sustainability/dp/1138800627
J Sustain 5:3328–3345
Joye Y (2007) Architectural lessons from environmental psychology: The case of biophilic architecture. Rev Gen Psychol 11(4):305–328. https://doi.org/10.1037/1089-2680.11.4.305
Joye Y, Van den Berg AE (2011) Is love for green in our genes? A critical analysis of evolutionary assumptions in restorative environments research. Urban Forestry Urban Green 10(4):261–268
Kaplan R, Kaplan S (1989) The experience of nature: a psychological perspective. Cambridge University Press
Kaplan S (1995) The restorative benefits of nature: toward an integrative framework. J Environ Psychol 15(3):169–182
Karthikeyan V, Eswar R (2018) Horticultural therapy activities reduce stress levels: lessons from activities conducted for adults in Bangalore. Int J Sci Res (IJSR) 7(5):1042–1045
Kellert SR, Heerwagen JH, Mador ML (2008) Biophilic design: the theory, science and practice of bringing buildings to life. Wiley
Korpela K, Hartig T (1996) Restorative qualities of favorite places. J Environ Psychol 16(3):221–233
Lindemuth AL (2012) Beyond the bars: landscapes for health and healing in corrections. In: Tidball KG, Krasny ME (eds) Greening in the red zone: disaster, resilience, and community greening. Springer, New York, NY, USA
Louv R (2005) Nature deficit. Orion 70:71
Mahmoud Ibrahim M (2021) Beyond sustainability-towards restorative interior spaces through biophilic design. Int J Adv Res Plan Sustain Dev 4(2):8–44
Makram A (2019) Nature-based framework for sustainable architectural design-biomimetic design and biophilic design. Archit Res 9(3)
Mampra M (2013) Green HRM: does it help to build a competitive service sector? A study. In: Proceedings of tenth AIMS international conference on management, pp 1273–1281. Retrieved from: http://www.scribd.com/doc/126544005/green-HRM-competitive-service-sector-pdf

Mandip G (2012) Green HRM: people management commitment to environmental sustainability. Res J Recent Sci 1:244–252

Martínez-Soto J, de la Fuente Suárez LA, Gonzáles-Santos L, Barrios FA (2019) Observation of environments with different restorative potential results in differences in eye patron movements and pupillary size. IBRO Rep 7:52–58

Mason L, Ronconi A, Scrimin S, Pazzaglia F (2021) Short-term exposure to nature and benefits for students' cognitive performance: a review. Educ Psychol Rev 1–39

Mollazadeh M, Zhu Y (2021) Application of virtual environments for biophilic design: a critical review. Buildings 11(4):148

Moore R (2012) Refugees from Nazi Germany in the Netherlands 1933–1940, vol 9. Springer

New report links office design with staff health and productivity (n.d.) World Green Building Council. Retrieved 16 Aug 2023, from https://worldgbc.org/article/new-report-links-office-design-with-staff-health-and-productivity/

Ohly H, White MP, Wheeler BW, Bethel A, Ukoumunne OC, Nikolaou V, Garside R (2016) Attention Restoration Theory: a systematic review of the attention restoration potential of exposure to natural environments. J Toxicol Environ Health 19(7):305–343. https://doi org/10.1080/10937404.2016.1196155

Phillips TB, Wells NM, Brown AH, Tralins JR, Bonter DN (2023) Nature and well-being: the association of nature engagement and well-being during the SARS-CoV-2 pandemic. People Nat 5(2):607–620

Relf PD (2005) The therapeutic values of plants. Pediatr Rehabil 8(3):235–237

Roe J, Aspinall P (2011) The restorative benefits of walking in urban and rural settings in adults with good and poor mental health. Health Place 17(1):103–113

Ryan CO, Browning WD (2020) Biophilic design. Sustainable built environments, pp 43–85

Ryan CO, Browning WD, Clancy JO, Andrews SL (2014) Biophilic design patterns: emerging nature-based parameters for health and well-being in the built environment. Int J Archit Res: Archnet-IJAR 8(2):62–76

Salingaros NA (2019) The biophilic healing index predicts effects of the built environment on our wellbeing

Schantz P, Stigell E (2008) Are green elements principal pull factors for physical activity. Swedish School for Sport and Health Sciences: Stockholm, Sweden

Scott TL, Masser BM, Pachana NA (2020) Positive aging benefits of home and community gardening activities: older adults report enhanced self-esteem, productive endeavours, social engagement and exercise. SAGE Open Med 8:2050312120901732

Shuda Q, Bougoulias ME, Kass R (2020) Effect of nature exposure on perceived and physiologic stress: a systematic review. Complement Ther Med 53:102514

Staats H, Kieviet A, Hartig T (2003) Where to recover from attentional fatigue: an expectancy-value analysis of environmental preference. J Environ Psychol 23(2):147–157

Stigsdotter UA, Grahn P (2003) Experiencing a garden: a healing garden for people suffering from burnout diseases. J Ther Hortic 14:48–63

Sutton-Grier AE, Sandifer PA (2019) Conservation of wetlands and other coastal ecosystems: a commentary on their value to protect biodiversity, reduce disaster impacts, and promote human health and well-being. Wetlands 39(6):1295–1302

Tabb PJ (2020) Biophilic urbanism: designing resilient communities for the future. Routledge

Taylor RP, Spehar B, Van Donkelaar P, Hagerhall CM (2011) Perceptual and physiological responses to Jackson Pollock's fractals. Front Hum Neurosci 5:60

Tennessen CM, Cimprich B (1995) Views to nature: effects on attention. J Environ Psychol 15(1):77–85

Ulrich RS (1984) View through a window may influence recovery from surgery. Science 224(4647):420–421

Ulrich RS, Simons RF, Losito BD, Fiorito E, Miles MA, Zelson M (1991) Stress recovery during exposure to natural and urban environments. J Environ Psychol 11(3):201–230

Vedvik R (2019) Human-centric lighting explained: lighting designers can create natural-looking lighting options using various technologies and methods. Consult Specify Eng 56(7):26–31

Wang HH, Finney MA, Song ZL, Wang ZS, Li XC (2021) Ecological techniques for wildfire mitigation: two distinct fuelbreak approaches and their fusion. For Ecol Manag 495:119376

Wilson EO (1986) Biophilia. Harvard University Press

Yao W, Zhang X, Gong Q (2021) The effect of exposure to the natural environment on stress reduction: a meta-analysis. Urban Forestry Urban Green 57:126932

Yin J, Zhu S, MacNaughton P, Allen JG, Spengler JD (2018) Physiological and cognitive performance of exposure to biophilic indoor environment. Build Environ 132:255–262

15 Co-creating Public Spaces with Adolescents: Dual Pathways to Adolescent Engagement Within Disaster Risk Mitigation and Adaptation

Virajitha Chimalapati and Prerna Vijaykumar Mehta

Abstract

The youth are inheriting an increasingly complex world with escalating disasters due to natural and people-made crisis. Adolescence, broadly described as the transitional period between 10 and 19 years, is a critical age where personal identity is being shaped and social skills and empathy being cultivated. Adolescent data is usually clubbed with child or youth data, and their specific needs are overlooked within disaster planning and mitigation discourse. Disconnection from schools and peer networks during disasters and extreme weather events not only causes disruptions in the academic progress of adolescents but also sets back their social and emotional development. As they are often forced to take on the more adult responsibilities in the aftermath of disasters, this disruption leaves them vulnerable to exploitation. Adolescents comprise nearly 20% of the population of India and have an important role to play within preparedness activities and building resilience within communities and families. This paper explores dual pathways through created for adolescent engagement in disaster resilience, through engaging them within co-creation of public spaces. This paper also explores how these co-creative processes facilitate the harnessing of adolescent affinity to technology to create a framework for building more resilient and prepared communities by looking at public spaces as multifunctional platforms that foster community engagement, youth development, and as essential resources during times of crisis.

Keywords

Public spaces · Adolescent development · Co-creative methodologies · Mental and physical well-being

15.1 Introduction

The Climate Change Performance Index (2023) ranks India as the 8th most vulnerable country to disasters (World Economic Forum 2023; PIB, Delhi 2022). As intensifying climate instability is leading to an increase in extreme weather events, the impacts on the younger members of the society will continue to increase (Sanson et al. 2019; UNICEF, n.d.). Today's youth are

V. Chimalapati (✉) · P. V. Mehta
WRI India, Cuttack, India
e-mail: virajithachimalapati@yahoo.co.in

V. Chimalapati
University of York, York, United Kingdom

P. V. Mehta
WRI India, Delhi, India

S. Talukdar et al. (eds.), *Communication, Science, Technology and Innovation in Disaster Risk Management*, GIScience and Geo-environmental Modelling, https://doi.org/10.1007/978-3-031-77189-7_15

thus inheriting an increasingly complex world with escalating disasters due to natural and people-made crises.

The National Youth Policy (2014) defines youth as persons between the age group of 15–29 years old, with individuals of 13–19 years being recognised as a sub-group (United Nations, n.d.; Mishra, n.d.). The general target group for involvement within disaster risk management is, however, described as "youth and adolescents between 10 and 29 years old (Central University of TamilNadu and National Institute of Disaster Management 2022; Ministry of Statistics and Programme Implementation, Government of India 2022). This age group comprises 27.5% of India's population and contributes to about 34% of India's Gross National Income (GNI) (Ramadass et al. 2017; Government of India, Ministry of Health and Family Welfare 2014).

The transitional period that spans 10–19 years of age is accompanied by rapid developmental changes in neurological, cognitive, physical, and interpersonal capacities, and is classified as adolescence (Ramadass et al. 2017; Chandramouli 2014; UNICEF 2011; Plan International n.d.; Bonnie and Backes 2019). The notions and interpretation of adolescence vary largely for different actors and within different domains. Developmental literature sub classifies adolescence into three stages: early adolescence (10–13 years), mid adolescence (14–17 years), and late adolescence or young adulthood (18–19 years). These classifications and sub classifications of adolescence overlap with those of youth and children, resulting in adolescent data being grouped either with childhood information or youth/early adult data. In India, where there is "no real concept of adolescence at all" (Bhabha 2014) this variance is blocking an understanding of the unique needs of this group (UNICEF, n.d.).

Every fifth person in India is an adolescent aged between 10 and 19 years old (UNICEF, n.d.). Adolescents from lower socio-economic communities often tend to assume adult responsibilities at a younger age while receiving meagre support from their socio-ecological environments (Newnham et al. 2019). By 2050, it is estimated that 50 per cent of India's population will be inhabiting urban areas. The past four decades has seen urban poverty rise from 18.7 to 26.8% (Save the children India 2015). Constant exposure to disasters and extreme climate events, exacerbates the challenges that are being faced by adolescents in India, especially those within India's urban slums. There is thus a huge demographic imperative to specifically look at impact of disaster on adolescents, as well as methodologies for adolescent engagement and mobilisation within disaster preparedness and climate adaptation measures (Newnham et al. 2019; de Milliano 2015). This paper accepts and works with the United Nations Population Funds' definition of an adolescent as a person aged between 10 and 19 years old (UNFPA, n.d.) and zooms in on adolescents and their role in disaster adaptation and mitigation measures in India.

Adolescence is a period where the neural systems underpinning emotional development and capacities to deal with adult challenges are developed (Ladouceur 2012; Griffin 2017). It is a critical age where incorporated behavioural patterns become established and carried forward into adulthood (Costa et al. 2021). Growing evidence links access to green spaces with mental and physical well-being, especially in adolescents (Mueller and Flouri 2021; Zhang et al. 2020). Recent research has also indicated that adolescent perceptions of their neighbourhood green spaces determine how efficiently they use and inhabit these spaces (Mueller and Flouri 2021). One of the key elements of youth development, especially adolescent development, is their capacitation as active agents, shaping their own environment and engaging in meaningful roles and leadership positions (Schusler and Krasny 2010). Programmes that foster civic participation and youth activism have also been found to be potential pathways to overall health and well-being in youth and adolescents (Arnold and Ferrari 2018). This paper thus explores the opportunities that public spaces offer to effectively integrate adolescents as active agents within disaster risk mitigation and management while outlining the pathways offered by public space development processes to adolescent empowerment and capacitation.

15.2 Impact of Disaster on Adolescents

Adolescents are greatly affected by disasters. Developmentally, they are at a phase where they are building a sense of identity and cultivating social skills and empathy (Lansford 2023). Simultaneous and ongoing disruptions of family structures, schooling patterns, health, and access to healthcare, disruptions to friendships and peer networks are the primary experience of adolescents within disasters and during extreme weather events (Fothergill 2017). Disconnection from schools, peer networks, and familiar structures not only causes disruptions in their academic progress but also sets back their social and emotional development. Family routines also have important developmental benefits that can get eroded with disruptions caused by disasters and climate events (Volkin 2020). As disasters push adolescents into taking on more adult responsibilities, these developmental disruptions leave them vulnerable to exploitation and trauma.

The recent Covid pandemic brought to the forefront the vulnerability of the adolescent community. As cities ground to a halt indefinitely, the looming sense of uncertainty, along with isolation caused by social distancing and lock downs, set back development, decreased mental health, and created psycho-social developmental challenges. It also highlighted the disproportionate impacts caused by existing societal divides, especially within lower-income communities. Lack of access to technology, and the shouldering of greater familial responsibilities during this time of crisis, resulted in increasing stress and trauma among adolescents within these communities (Volkin 2020). In fact, even within communities from mid/high income families, social and cultural setbacks caused by the pandemic, accelerated the shouldering of adult responsibilities by adolescents, transforming them into breadwinners and family heads.

The nature of the pandemic and the resulting restrictions on accessing and using public spaces directly influenced adolescent interactions with not just the urban fabric but also their peer networks and support structures (Costa et al. 2021). As more and more research started focusing on the impact of the pandemic on adolescents, it became apparent that the environment and social location of an adolescent plays a crucial role in how they experience vulnerability (Engel et al. 2021). The role of economic, social, and structural factors in adolescent/children's recovery trajectories also came to the forefront, alongside an understanding that adolescent resilience is a product of "resources and social structural support" (Fothergill 2017). Recognising the unique vulnerabilities of adolescents and designing strategies that cater to them, especially in the Indian context is becoming extremely important.

Physical environments and neighbourhood contexts play an important role in adolescent health (Engel et al. 2021; Ramadass et al. 2017). Access to public spaces including natural settings influence the "physical, cognitive, and emotional development" of adolescents by supporting memory construction processes and strengthening the immune system (Batista et al. 2018). While adolescents in urban areas experience better education and have access to better health services, due to lack of access to appropriate socio-physical infrastructure, they are also increasingly vulnerable to mental health challenges, substance use, and obesity (Mmari et al. 2014). A cross sectional school-based study of 14-year-old adolescents in urban India found evidence of distress in 15% of the respondents and depression in nearly 19% of the adolescents (Bansal et al. 2009). There is thus a pressing need to conceptualise and forefront initiatives that understand and cater to the specific needs of adolescents and work towards mitigating impacts of disaster on their psycho-social, mental, and physical development.

15.2.1 Adolescents as Active Agents Within Disaster Planning and Preparedness

Disaster preparedness could be described as the "knowledge and capacities developed to effectively anticipate, respond to, and recover from the impact of disasters" (UNDRR, n.d.). Within disaster planning and climate change management, adolescents are often looked at as vulnerable and in need of care and protection. They are rarely invited to speak about their concerns or voice their opinions (Gibbs et al. 2014; Chandramouli 2014). However, adolescents are "creative social beings and active agents", and have important roles to play in preparedness activities, and recovery processes for families and communities (Fothergill 2017). The UN Convention of the Rights of the Child (1990) enshrines the adolescents' right to be active participants and contributing to all decisions effecting their lives. In the last decade, there has been a surge of research on the role of children, adolescents, and youth in disaster risk reduction and climate change adaptation. The findings from this research emphasise the centring of adolescent voices within broader socio-economic contexts, through the disaster preparedness and adaptation cycle (Volkin 2020; Fothergill 2017; Gibbs et al. 2014). The increased understanding of the sociology of children has also led to the recognition of the potential of adolescent contribution within disaster preparedness, response, rehabilitation, and recovery (Corsaro 2014). However, adolescents need to be capacitated and informed on the methods, modalities and boundaries of age-appropriate contribution (Di Maio 2023). This section of the paper draws from this literature and secondary research to outline some of the modalities of adolescent engagement and age-appropriate contributions across the disaster preparedness cycle, i.e. disaster preparedness, response and rehabilitation phases specifically within the Indian context.

Preparedness behaviour is a broad category and includes activities such as identifying evacuation routes, performing safety drills, gathering and storing emergency supplies, being trained for disaster response, and working on disaster response activities (Fothergill 2017). Adolescent engagement through age-appropriate training programmes can motivate and improve adolescent readiness to engage in "social traumatic rescue". Regular engagement of youth/adolescent forums via the piloting and deployment of training programmes would allow the creation and mobilisation of adolescent networks into creating a "resilient task force" during disasters. Such regular engagement has been known to dilute the aftermath of disaster displacement by creating social safety nets that the adolescents can fall back on (Newnham et al. 2019). The importance of trained volunteers and coordination of preparedness activities within effective disaster mitigation and management cannot be over emphasised. Additional benefits of regular adolescent engagement within disaster mitigation and preparedness activities is the increased awareness on the shifts in life patterns necessary for transitioning into more climate sensitive modes of life (Schofield and Twigg 2019). Thus, adolescents become change-makers, bringing creativity into raising awareness and helping the community make the transitions necessary for climate change adaptation (Shahryar 2019).

The response phase within a disaster life cycle would include actions taken before, during, or immediately after a disaster and is focused towards rescue, public health, and subsistence (UNDRR, n.d.). National Service Scheme (NSS), Bharath Scouts and Guides, National Cadet Corps (NCC), Civil Defence Volunteers, Indian Red Cross Society are some of the youth forums which promote adolescent volunteerism and involve in developmental activities that can be mobilised for community building after a disaster. Adolescent contributions within these organisations and through the response phase have been known to include involvement in first aid efforts, rescue operations, food delivery and health education among others (National Institute of Disaster Management, Govt of India 2021).

The rehabilitation phase involves the restoration of basic services and facilities disrupted by the disaster, while the recovery phase includes the restoration of the economic, physical, social, cultural, and environmental systems disrupted during a disaster (National Institute of Disaster Management, Govt of India 2021). Adolescents often take on roles assigned to youth in disaster settings and are less likely to be identified as at-risk members of the community. Nevertheless, adolescents have been known to be severely impacted by disruption of their peer network and social activities (Newnham et al. 2019). Research also indicates that they are vulnerable to long-term mental health impacts and at significant risk of exploitation in the aftermath of a disaster and as a result of these disruptions (Plan International n.d.). Cultivation of a sense of agency within adolescents, especial in those from disadvantaged settings, through training and capacitation in safe rescue techniques, within disaster response efforts have been found to increase a sense of identity, develop leadership skills, and increase metal and physical health parameters (Wong et al. 2012). A focus on the provision of and access to psycho-social support by adolescents, during and in the immediate aftermath of a disaster, also becomes extremely important to ensure any developmental support needed during this period is available.

Adolescents need to be engaged in shaping their environment. Research has shown that adolescent engagement in "structured high-quality out of school activities" is connected to overall well-being and has a buffering effect on the negative impacts of insufficient socio-ecological environments (Weybright et al. 2018). Nearly 20% of the adolescents in India live in urban areas; therefore, a focus on adolescent well-being is compatible with a public health approach (Arnold and Ferrari 2018; Chandramouli 2014). The Healthy Cities for Adolescents is one such global initiative that was launched with the aim of understanding the specific needs of this age group, in India (Fondation Botnar, n.d.). This project looks at adolescents as "important contributors and equal partners in creating sustainable solutions for their growing cities" (Thygarajan et al. 2022). The Healthy Cities for Adolescents project looks at processes and methods where adolescents can be given greater agency, support, and consistency during disaster situations, through the creation of social networks and the provision of leadership opportunities within the designing and shaping of their physical environments (Fothergill 2017). By including adolescents within recovery, reconstruction, and post-disaster planning, specially within spatial planning processes, adolescent capacity to provide critical emotional support to one another and their adults can be mobilised and leveraged towards developing resilience while allowing them to create peer networks and support structures that adolescents can fall back on during disasters (Peek 2008; Fothergill 2017). Thus adolescent engagement within disaster mitigation and adaptation processes can be leveraged not only for building disaster resilience but also for the long-term psycho-social development of adolescents.

15.3 Way Forward: Dual Pathways to Adolescent Engagement with Disaster Risk Mitigation and Adaptation

Engaging adolescents within shaping urban environments and within urban development processes, especially though mobilisation in disaster risk reduction and climate change adaptation, creates potential for positive youth development. It offers them agency by allowing them to take control of and alter their physical environment, while internalising civic consciousness, disaster resilience and climate aware lifestyles and practices. This is a demographic and national imperative in the Indian context. This section of the paper proposes dual pathways to adolescent inclusion, engagement, and mobilisation within urban planning and development. They are:

Pathway 1: Co-creating public spaces with adolescents.

Pathway 2: Harnessing adolescent affinity for technology towards enhancing resilience of public spaces

These pathways are described in some detail below.

15.3.1 Pathway 1: Co-creating Public Spaces with Adolescents

Public spaces are essential components of any community, providing a place for people to come together, interact, share information, and find support during times of need as well as access resources to cope and rebuild. These spaces play a crucial role in fostering disaster resilience among communities (UN Habitat 2020). Unfortunately, many public spaces are not designed with disaster resilience in mind, leaving citizens and especially adolescents, vulnerable to the effects of disasters. Adolescents and young people are in fact the most frequent users of public spaces, as they depend on public spaces for peer interaction, and sometimes isolation (Costa et al. 2021). By involving adolescents and considering their unique needs, preferences and perspectives, cities can develop public spaces that not only facilitate their physical and emotional well-being but also offer opportunities for education, engagement, and empowerment in disaster risk reduction efforts.

A co-creative approach to public space creation not only empowers adolescents by giving them a sense of ownership and responsibility but also promotes social cohesion within communities, fostering a culture of resilience and preparedness among all age groups. Through the process of co-creation, adolescents not only become more adept at anticipating and responding to potential disasters, they also develop valuable skills in leadership, problem-solving, critical thinking, communication, and collaboration. These skills not only benefit them in disaster situations but have long-lasting positive effects on their overall quality of life and engagement in civic activities and community development. Such co-creative methods also capacitate adolescents on urban governance processes and citizen participation modalities. A vital component to building a resilient society is to shape an empowered citizen. Actively including adolescents within urban governance processes offers avenues for relationship building with local government agencies and the creation and maintenance of partnerships between local governments, grass root organisations, and youth organisations. This will not only create capacitated and informed leadership, but also create access to support systems that adolescents can leverage both during times of crisis and otherwise: thus creating a safety web of systemic support and information.

Urban planning and development are important vehicles towards integrating long-term disaster reduction and climate adaptation goals at the local level (Joshi et al. 2022). Engagement of local-level stakeholders within the urban development process has been a core strategy in bridging resource and knowledge gaps within governance and implementation of climate change adaptation and disaster preparedness (Maher and Buhmann 2019). Expanding this engagement to include the adolescent community becomes an important pathway towards harnessing this demographic dividend, especially within Indian contexts. Policy changes are the primary avenues towards the creation and promotion of healthy behaviour. Embedding equitable distribution of public spaces as medium and long-term urban policy agendas is inherent to reducing public health crises as such equitable distribution provides access to essential services, improving recreation, mobility, and provide livelihood opportunities to informal enterprises (UN Habitat 2020; Khorram-Manesh 2017). Thus, adolescent friendly co-creative public space development not only creates avenues for capacitating adolescents within co-creative policy processes but also facilitates the institutionalisation of mechanisms for adolescent participation and empowerment (Costa et al. 2021). Urban development policy that institutionalises adolescent participation thus becomes a primary pathway for positive youth development and an empowered citizenry. It also allows the leveraging of adolescent affinity to technology through decentralised urban development processes and technology led smart city interfaces (Costa et al. 2021).

15.3.2 Pathway 2: Harnessing Adolescent Affinity for Technology Towards Enhancing Resilience of Public Spaces

Technological intervention has significantly transformed the way adolescents engage with the urban public spaces. Access to smartphones and other digital devices provides them with instant connectivity to peers and information, extending social interactions beyond physical boundaries (Dienlin and Johannes 2020). Technology and social media became a source of respite during the isolation of the Covid pandemic and a zone for maintaining peer connection, especially in the case of adolescents who could afford technological interfaces. Social media platforms have become the new public squares. Games like Pokemon Go have extended the traditional notion of public spaces as physical meeting points to encompass virtual spaces and experiences. Interactive experiences with the physical spaces are documented in real time and shared on digital platforms, creating a hybrid space where digital interactions intertwine with physical ones. All this goes hand in hand with growing concerns on the impact of technology use in limiting adolescent engagement with conventional public spaces, that the allure of virtual communication may lead to fewer face-to-face interactions and reduced participation in local communities. There is thus a need to integrate technology into public spaces in a way that the virtual experiences complement and expands the physical experience.

Harnessing technology's potential offers promising possibilities for enhancing the vibrancy and perception of public spaces. Utilising augmented reality (AR) and virtual reality (VR) applications can rejuvenate underutilised areas and transform them into thriving hubs for adolescent activity (Universita Degli Studi Di Milano, n.d.). Implementing gamified elements, educational content, or interactive experiences, can make public spaces become more appealing while inspiring youth involvement in civic activities or cultural events. Integration of digital public infrastructure with physical infrastructural interventions could be applied towards enhancing physical activity and engagement. The ability of technology to allow the public realm to become more inclusive and accessible to people of diverse abilities and identities should also be harnessed. For example, public spaces can be equipped with interactive kiosks or mobile applications that provide real-time information about events, services, transportation options, walking, and cycling routes (Reach Out Australia, n.d.) that not only facilitate tourism and exploration of the urban spaces via public transportation modalities, but also allow the universal access of this information.

The New Urban Agenda underlines the role of technology within ensuring urban sustainability, creating citizen centric digital governance tools and integrating local and vulnerable populations within urban planning and development (UN-Habitat 2020). Adolescent affinity to technology, their ease with adoption of new interfaces and expansion of new technological developments offers great potential to not only centre adolescents within digital governance models, but also towards creating platforms that allow them to contribute effectively within disaster risk reduction and management. This not only strengthens their affinity towards the community but also empowers them as active contributors to civic processes, urban development and sustainable public infrastructure management. Research conducted by UN Habitat has demonstrated that engagement with technology by the younger demographic of a society provides new avenues for information gathering, collaboration and organisation, thereby increasing potential for meaningful civic engagement (Costa et al. 2020). Such contribution and engagement also becomes a platform for technological innovation within building multi-dimensional resilience in urban environments (Costa et al. 2021).

In post COVID-19 times, the application of technology within disaster management and developing community resilience has become more widespread. By leveraging technological advancements, public spaces can incorporate various systems and applications to enhance disaster planning, early warning systems, and disaster response capabilities. Adolescents' technological competency when combined with an awareness of disaster preparedness activities can become an asset in disaster risk mitigation. The integration of IoT devices, such as sensors, Wi-Fi networks, and digitalised information systems, have the potential to improve the functionality and efficiency of public spaces, while monitoring for weather events in real time, highlighting varied evacuation opportunities, enhancing disaster preparedness, and reducing vulnerabilities to extreme climate events. The integration of digital platforms within public spaces can also enable youth and adolescents to access real-time information about potential hazards, evacuation routes, and the availability of emergency services. Adolescent proficiency in digital tools, social media platforms, and the ability to access real-time information and disseminate this through their peer network is unparalleled (Benvenuti et al. 2023). Engaging adolescents in collaborative efforts with authorities and local communities, enlisting them within roles like developing educational content, tracking of relief supplies, communication and coordination during disasters, can not only provide them safe avenues to display and engage their skills but also help in meeting the particular developmental needs of this age group. Adolescent affinity and access to technology, combined with training can also be leveraged to counter misinformation, enhance public awareness and mobilise timely and effective response during times of crisis, thus minimising human and economic losses (UNICEF 2013).

In conclusion, the combination of public spaces, youth engagement, disaster resilience, and technology create a powerful framework for building more resilient and prepared communities. Public spaces serve as multifunctional platforms that foster community engagement, youth development, and act as essential resources during times of crisis. The leveraging of dual pathways of adolescent involvement in co-creative public space development processes and harnessing adolescent affinity to technology within expanding the public realm to mobilising timely and effective disaster response allows the creation of safe avenues for adolescent involvement within disaster resilience planning. The active participation of adolescents enables them to gain knowledge and skills necessary for disaster response and fosters a sense of responsibility towards their communities. Ultimately, involving adolescents as active agents in creating safe, vibrant and healthy cities especially within Indian contexts becomes a primary instrument to create a climate-resilient socio-ecological web that can help future generations successfully navigate these unprecedented challenges.

References

Arnold ME, Ferrari TM (2018) On the intersection of health, well-being and youth development. J Youth Dev

Bansal V, Goyal S, Srivastava K (2009) Study of prevalence of depression in adolescent students of a public school. Ind Psychiatry J

Batista S, Costa SC, Menezes M, Almeida I (2018) Co-creating a teenagers' sensitive public open space: a case study in Alvalade neighbourhood in Lisbon (Portugal). In: X Congresso Portuges de Sociologia

Benvenuti M, Wright M, Naslund J, Miers AC (2023) How technology use is changing adolescents' behaviors and their social, physical, and cognitive development. Curr Psychol

Bhabha J (ed) (2014) Human rights and adolescence. University of Pennsylvania Press, Philadelphia

Bonnie RJ, Backes E (eds) (2019) The promise of adolescence: realising opportunity for all youth. National Academies of Sciences, Engineering and Medicine, The National Academies Press, Washington, DC

Central University of TamilNadu and National Institute of Disaster Management (2022) Engaging youth and adolescence in Disaster Risk Management and Climate Change Adaptation

Chandramouli C (2014) A profile of adolescents and youth in India. Census of India, UNFPA

Corsaro W (2014) Children's well-being and interpretive reproduction. In: Handbook of child well-being. Springer

Costa CS, Batista JS, Menezes M (2021) What happens when teenagers reason about public open spaces? Lessons learnt from co-creation in Lisbon. Cidades, Communidades e Territorios, pp 139–155

Costa CS, Batista JS, Almeida I, Menezes M (2020) Exploring teenagers' spatial practices and needs in light of new communication technologies. Cities de Milliano C (2015) Luctor et emergo, exploring contextual variance in factors that enable adolescent resilience to flooding. Int J Disaster Risk Reduct 14(2):168–178

Di Maio MA (2023) Pre-study on the role of children and youth in building resilient societies in the Baltic region. Council of the Baltic Sea States

Dienlin T, Johannes N (2020) The impact of digital technology use on adolescent well-being. Dialog Clin Neurosci

Engel D, Chalasani S, Rakotomalala S, Shenderovich Y, Daban RA, Blight S, Kaviani Johnson A, Luna JR, Silverman A, Clarke R, Omrani OE (2021) Safety and a supportive environment—essential conditions for adolescent wellbeing. Partnership for Maternal Newborn and Child Health UHL

Fothergill A (2017) Children, youth and disaster. Oxford Research Encyclopedia, Natural Hazard Science, pp 1–28

Fondation Botnar (n.d.) Healthy city for adolescents: a fondation Botnar initiative. Retrieved from https://www.fondationbotnar.org/project/the-botnar-healthy-cities-for-adolescents-program/

Germanwatch (2023) Climate change performance index—results 2023. Germanwatch, New Climate Institute, Climate Action Network

Gibbs L, Block K, Harms L, MacDougall C, Baker E, Ireton G, Forbes D, Richardson J, Waters E (2014) Children and young people's wellbeing post-disaster: safety and stability are critical. Int J Disaster Risk Reduct 195–201

Government of India, Ministry of Health and Family Welfare (2014) Strategy handbook, Rashtriya Kishor Swasthya Karyakram. Government of India, Ministry of Health and Family Welfare, New Delhi

Griffin A (2017) Adolescent neurological development and implications for health and well-being. Healthcare

Joshi N, Wende W, Tiwari PC (2022) Urban planning as an instrument for disaster risk reduction in the Uttarakhand Himalayas. Mountain Res Dev

Khorram-Manesh A (2017) Youth are our future assets in emergency and disaster management. Bull Emerg Trauma 1–3

Ladouceur CD (2012) Neural systems supporting cognitive-affective interactions in adolescence: the role of puberty and implications for affective disorders. Front Integr Neurosci

Lansford J (2023) Adolescent development. Biswas-Diener R, Diener E (eds) Noba textbook series: psychology. DEF Publishers. Retrieved from Duke University: https://nobaproject.com/modules/adolescent-development

Maher R, Buhmann K (2019) Meaningful stakeholder engagement: bottom-up initiatives within global governance frameworks. Geoforum 231–234

Ministry of Statistics and Programme Implementation, Government of India (2022) Youth in India, 2022. National Statistical Office, Ministry of Statistics and Programme Implementation, Government of India, New Delhi

Ministry of Youth Affairs and Sports, Govt of India (2014) National Youth Policy. Ministry of Youth Affairs and Sports, Govt of India

Mishra IK (n.d.) National Youth Policy—will it help India reap the promised demographic dividend? Retrieved from Observer Research Foundation: https://www.orfonline.org/expert-speak/national-youth-policy/#:~:text=With%2066%20percent%20of%20its,the%20world%27s%20largest%20youth%20population

Mmari K, Lantos H, Blum R, Brahmbhatt H, Sangowawa A, Yu C, Delany-Meretiwa S (2014) A global study on the influence of neighborhood contextual factors on adolescent health. J Adolesc Health 13–20

Mueller MA, Flouri E (2021) Urban adolescence: the role of neighbourhood greenspace in mental well-being. Front Psychol

National Institute of Disaster Management, Govt of India (2021) Role of youth in disaster management. National Institute of Disaster Management, Ministry of Home Affairs, Government of India in collaboration with Rajiv Gandhi National Institute of Youth Development (Ministry of Youth Affairs and Sports, Govt of India)

Newnham E, Tearne J, Gao X, Guragain B, Jiao F, Ghimire L, Balsari S, Chan E, Leaning J (2019) Tailoring disaster risk reduction for adolescents: qualitative perspectives from China and Nepal. Int J Disaster Risk Reduct 337–345

Peek L (2008) Children and disasters: understanding vulnerability, developing capacities and promoting resilience—an introduction. Child Youth Environ 1–29

PIB, Delhi (2022) Climate change performance index. Ministry of Environment, Forest and Climate Change, Delhi

Plan International (n.d.) A time of transition: adolescents in humanitarian settings. Plan International UK and Plan International. https://plan-international.org/publications/

Ramadass S, Gupta S, Nongkynrih B (2017) Adolescent health in Urban India. J Fam Med Prim Care 468–476

Reach Out Australia (n.d.) Technology and teenagers. Retrieved from https://parents.au.reachout.com/skills-to-build/wellbeing/technology-and-teenagers

Sanson AV, van Hoorn J, Burke SE (2019) Responding to the impacts of the climate crisis on children and youth. Child Dev Perspect 201–207

Save the children India (2015) Forgotten voices, the world of urban children in India. Save the Children India, PwC India

Schofield H, Twigg J (2019) Making cities sustainable and resilient: lessons learned from the disaster reilience scorecard assessment and Disaster Risk Reduction (DRR) action planning

Schusler TM, Krasny ME (2010) Environmental action as context for youth development. J Environ Educ

Shahryar F (2019) Adolescents take the lead to protect their schools. Retrieved from https://www.unicef.org/rosa/stories/adolescents-taking-lead-protect-their-schools

Thygarajan S, Seemamu S, Mehta PV (2022) Youth-led urban development: a participatory approach to improving public spaces. Retrieved from WRI-India: https://wri-india.org/blog/youth-led-urban-development-participatory-approach-improving-public-spaces

UN Habitat (2020) UN-Habitat key message on Covid-19 and public space. Un Habitat

UN (2015) Transforming our world: the 2030 agenda for sustainable development. United Nations

UNCRC (1990) Conventions on the Rights of the Child. United Nations.

UNDRR (n.d.) Preparedness. Retrieved from Terminology: www.undrr.org

UNDRR (n.d.) Response. Retrieved from Sendai Framework Terminology on Disaster Risk Reduction: https://www.undrr.org/terminology/response

UNFPA (n.d.) Adolescent and youth demographics: a brief overview. United Nations Population Fund

UN-Habitat (2020) The new urban agenda. In: United Nations Conference on Housing and Sustainable Urban Development. United Nations Human Settlements Programme

Unicef Office of Research (2014) The challenges of climate change, children on the front line. Innocenti Insight, Florence

UNICEF (2011) Adolescence an age of opportunity. United Nations Chikdren's Fund (UNICEF), New York

UNICEF (2013) Actions for children and youth resilience, guide for governments. UNICEF, Protecting through Education RET, Panama City

UNICEF (n.d.) Adolescent development and participation: investing in adolescents builds strong economies, inclusive communities and vibrant societies. Retrieved from unicef.org: https://www.unicef.org/adolescence

Unicef (n.d.) Empowering adolescent girls and boys in India. Retrieved from unicef.org: https://www.unicef.org/india/what-we-do/adolescent-development-participation#:~:text=India%20has%20the%20largest%20adolescent,between%2010%20to%2019%20years

UNICEF (n.d.) The impact of climate change on children. UNICEF, Jordan

United Nations (n.d.) Global Issues, Youth. Retrieved from United Nations, Peace, dignity and equality on a healthy planet: https://www.un.org/en/global-issues/youth#:~:text=Who%20Are%20the%20Youth%3F,of%2015%20and%2024%20years.https://www.orfonline.org/expert-speak/national-youth-policy/#:~:text=With%2066%20percent%20of%20its,the%20world%27s%20largest%20youth%20population

Universita Degli Studi Di Milano (n.d.) Public space users and socio-spatial practices. Retrieved from MYC3PLACE: https://myc3place.di.unimi.it/?page_id=1051

Volkin S (2020) Hub. Retrieved from John Hopkins University: https://hub.jhu.edu/2020/05/11/covid-19-and-adolescents/

Weybright EH, Martines DA, Varrella G, Deen KM, Wright K (2018) Teens as teachers: positive outcomes and recommendations for promoting health nutrition in adolescents. J Youth Dev

Wong MC, Lau TC, Lee A (2012) The impact of leadership programme on self esteem and self efficacy in school: a randomised control trial. PLoS ONE

World Economic Forum (2023) Global Risks Report. World Economic Forum

Zhang Y, Mavoa S, Zhao J, Raphael D, Smith M (2020) The association between green space and adolescents' mental well-being: a systematic review. Int J Environ Res Public Health

Integrating Lessons and Shaping Future Pathways in Disaster and Risk Management in South Asia

16

Swapan Talukdar, Ranit Chatterjee, Somnath Bera, Atish Prashar, and Rajib Shaw

Abstract

Disaster risk management in South Asia faces complex challenges that are compounded by the region's diverse socio-economic and environmental conditions. This chapter looks at the intersections of communication, science, technology, and innovation in building resilience to disasters as explored in various case studies in this book. The main objective is to analyze and synthesize these dimensions in the context of disaster management and draw lessons from the empirical findings across South Asia to suggest forward-looking strategies. Key findings include the critical role of effective communication and the application of cutting-edge technologies in managing the differential recovery rates between urban and remote areas, as observed in the post-earthquake scenarios in Nepal, and in managing the combined crises of floods and pandemics affecting the Surjapuri community. In addition, the impact of climate variability on livelihoods and adaptation measures in agriculture underscore the urgent need for innovative solutions and disruptive technologies. The introduction of geospatial technologies and unmanned aerial vehicles (UAVs) represents a significant advance in hazard assessment and disaster response, providing detailed risk information that is essential for informed land use and policy decisions. However, the effectiveness of these technologies depends on combining them with community-oriented approaches, keeping track of technological advancements and continuous data updates. In conclusion, the chapter emphasizes the need to combine traditional disaster management practices with modern technological advances to create a robust framework for future disaster resilience and argues for a holistic approach that builds synergies among community engagement, policy

S. Talukdar
Department of Geography, Asutosh College, University of Calcutta, Kolkata 700026, India
e-mail: swapan.talukdar@asutoshcollege.in

R. Chatterjee
Graduate School of Media and Governance, Keio University, Shonan Fujisawa Campus, Fujisawa, Japan

R. Chatterjee
RIKA Institute, New Delhi, India

S. Bera (✉)
Department of Geography, Central University of South Bihar, Gaya, India
e-mail: somnath@cusb.ac.in

A. Prashar
Department of Mass Communication and Media, Central University of South Bihar, Gaya, India

R. Shaw
Graduate School of Media and Governance, Keio University, Shonan Fujisawa Campus, Fujisawa, Japan

S. Talukdar et al. (eds.), *Communication, Science, Technology and Innovation in Disaster Risk Management*, GIScience and Geo-environmental Modelling, https://doi.org/10.1007/978-3-031-77189-7_16

adaptation, and technological innovation. This synthesis not only contributes to the academic literature by outlining effective strategies but also offers practical ways to improve disaster management in South Asia.

Keywords

Disaster risk management · Community resilience · Technological innovation · Climate variability · Geospatial technology

16.1 Introduction

South Asia is one of the most disaster-prone regions in the world, with frequent natural hazards such as floods, earthquakes, cyclones, and droughts. According to the United Nations Office for Disaster Risk Reduction (UNDRR), the region has experienced more than 50 major disasters between 2000 and 2019, affecting more than 1.2 billion people and causing economic losses of more than 80 billion dollars (UNDRR 2020). The devastating 2004 tsunami in the Indian Ocean, which claimed over 230,000 lives, and the 2015 earthquake in Nepal, which killed almost 9000 people and left millions homeless, underline the urgency of effective disaster management strategies in South Asia (ADB 2019). These figures highlight the urgent need for comprehensive disaster risk reduction (DRR) and management to mitigate the impact of such events on human lives and socio-economic systems.

Disaster risk reduction (DRR) in South Asia has gained significant momentum with the adoption of the Sendai Framework for Disaster Risk Reduction (SFDRR) 2015–2030, which emphasizes disaster resilience building. The SFDRR outlines four priority areas: understanding disaster risk, strengthening disaster risk reduction, investing in disaster risk reduction to strengthen resilience, and enhancing disaster risk reduction for effective response and recovery ("Sendai Framework for Disaster Risk Reduction," 2015). Countries in South Asia, such as India, Bangladesh, and Nepal, have made significant progress in implementing these priorities while the other countries have also made necessary strides. For example, India's National Disaster Management Authority (NDMA) has developed comprehensive guidelines for disaster management plans at national, state, and district levels that emphasize community-based disaster preparedness and capacity building (NDMA 2019). These initiatives are in line with the SFDRR's goal of reducing fatalities, affected populations, and economic losses while improving early warning systems and infrastructure resilience.

Advances in communication technology have played a critical role in improving disaster management in South Asia. The integration of geographic information systems (GIS), remote sensing and early warning systems has significantly improved the region's ability to predict, monitor, and respond to disasters. For example, the South Asian Association for Regional Cooperation (SAARC) Disaster Management Center has been instrumental in promoting regional cooperation in sharing real-time data and best practices ("SAARC Disaster Management Center," 2018). The use of mobile technologies for disaster communication has also increased significantly. In Bangladesh, the Cyclone Preparedness Program (CPP) uses mobile networks to disseminate early warnings to vulnerable coastal communities to significantly reduce cyclone fatalities ("Bangladesh Cyclone Preparedness Program," 2020). This progress is in line with the SFDRR's focus on using technology to improve disaster preparedness and response.

Innovations in science and technology have further strengthened disaster risk management efforts in South Asia. UNDRR's Asia Pacific Science Technology Conference every two years takes stock of the current advancements in science technology and technology to address the Sendai priorities. The development and deployment of climate-resilient infrastructure, such as flood-resistant houses and earthquake-resistant buildings, has been instrumental in reducing the vulnerability of communities. Research and development in climate modeling and hazard mapping have provided valuable insights for risk assessment and mitigation strategies. For example, the Climate Risk and Early Warning Systems (CREWS) initiative in Nepal aims to improve the accuracy and lead time of weather forecasts to enable better preparation and timely evacuation of vulnerable populations ("CREWS Nepal Project," 2021). In addition, the use of drones for post-disaster damage assessment and delivery of relief supplies has revolutionized disaster relief while ensuring faster and efficient relief efforts ("Drones in Disaster Management," 2020). These innovations reflect the SFDRR's call to integrate scientific research and technology into all aspects of disaster risk reduction and management.

The importance of disaster and risk management in South Asia cannot be overstated, given the region's high vulnerability to natural hazards and the profound impact on people's lives and economic stability. This book presents these advances in disaster risk reduction and prevention and highlights their importance in building a more resilient future. It also aims to provide forward-looking recommendations and identify future research directions to improve resilience and disaster preparedness in the region. Continued efforts in the areas of capacity development, regional cooperation, and innovation are crucial to address changing disaster risks. By leveraging science and technology and adhering to global frameworks, South Asia can build a more resilient and sustainable future.

16.2 Insights from Disaster Risk Reduction and Capacity Development

This section provides a comparative analysis based on the findings of four chapters dealing with different aspects of disaster risk reduction and capacity building. Each case study highlights different aspects of disaster impact and community responses, providing a nuanced understanding of the complexity of disaster management in different contexts in South Asia. The chapter on disaster risk reduction and capacity development in the built environment provides a nuanced analysis of recovery processes in Nepal, highlighting the differences between urban and remote regions. The findings of this chapter are underpinned by several studies that collectively highlight the importance of robust communication and transportation systems for faster recovery in urban areas compared to remote regions. A study on capacity development for disaster risk reduction in Sri Lanka found that effective training programs significantly improve the disaster preparedness and resilience of local communities (Iizuka 2020). In Guatemala, a study emphasized the importance of improving disaster risk reduction in informal settlements through better access to resources and institutional support (Miles et al. 2012). This aligns with the case in Nepal, where remote regions faced delays due to limited access to vital information and resources, exacerbating pre-existing vulnerabilities.

By examining the combined impact of floods and the COVID-19 pandemic on the Surjapuri community, this chapter shows how dual crises affect migration patterns and disruptions to health services. Similar findings have been observed in studies looking at the resilience of communities affected by multiple hazards. For example, research in Indonesia on flood-resilient communities emphasized the need for integrated disaster risk reduction and climate change adaptation strategies to improve community resilience (Dwirahmadi et al. 2019). The impact of climate variability on the livelihoods of fisherfolk in India is consistent with studies highlighting the vulnerability of coastal communities to climate-related hazards. For example, a study in Saudi Arabia has shown that public perceptions and attitudes significantly influence disaster risk reduction efforts, highlighting the importance of awareness and preparedness (AlQahtany and Abubakar 2020).

In the drought-prone Kalahandi district of Odisha, the challenges of water scarcity mirror broader issues of agricultural risk and resilience. Drawing lessons beyond South Asia, the study of community capacity in Zimbabwe similarly identified the need for better resource management and community involvement in disaster risk reduction (Dube 2015). This chapter also highlights the critical role of community-based disaster risk reduction (CBDRR) programs. Research in the Merapi Volcano community in Indonesia has shown that people-centered disaster risk reduction programs can significantly improve community resilience (Mutiarni et al. 2022). Similarly, educational interventions in Iran have been shown to improve disaster awareness and preparedness through primary healthcare systems (Ardalan et al. 2013). The insights gleaned from these studies illustrate the multifaceted nature of disaster risk reduction and capacity development.

16.3 Advancements in Communication for Disaster Management

The role of communication in disaster management has evolved significantly due to technological advances and the integration of various digital platforms. This section explores the comparative analysis of communication strategies and their effectiveness in disaster management, drawing on findings from several studies. The chapter "Disaster Management in India" highlights the crucial role of communication and media in disaster management. The reports of the National Disaster Response Force and the National Institute of Disaster Management reveal that while infrastructural communication is often emphasized, the function of the media is rarely discussed. Integrating communication strategies, including the use of traditional and social media, can improve existing disaster management practices. Studies have shown that effective communication during disasters can lead to better prepared communities and more efficient responses. For example, Liu et al. (2016) found that social media platforms, particularly Twitter, significantly improve public engagement and information dissemination during disasters.

The potential of social media as a communication tool in disaster response is discussed in the chapter "Social media: potential communication tool in disaster response" Case studies on the Sri Lankan floods and Cyclone Ockhi show that platforms such as Twitter, when integrated with machine learning and artificial intelligence, can identify areas that require attention, enabling more effective communication. Bjerge et al. (2016) emphasize the role of technology in improving information sharing between organizations through platforms such as Virtual OSOCC, which improves coordination between emergency response agencies.

The chapter "Administrative Response to Cyclone Nilam" describes the administrative responses and challenges at different levels and highlights the barrier that local languages pose in facilitating needs-based responses. Public participation, facilitated by digital tools such as the Public Participation Geographic Information System (PPGIS), can mitigate these challenges. Mason et al. (2019) found that almost half of medical tourism providers in the Caribbean did not provide information following a disaster, highlighting the need for proactive communication strategies. Similarly, Hugelius et al. (2016) demonstrated the effectiveness of disaster radio in communicating vital news and health-related information following Typhoon Haiyan in the Philippines.

Disaster communication also includes the management of internal and external communication within organizations. Liu et al. (2018) discuss the importance of strategic risk and crisis communication in healthcare and highlight communication challenges such as media control and rumor management. The use of social media platforms such as Facebook and WhatsApp during the Chennai floods in 2015 is an example of how digital communication tools can facilitate disaster management by providing real-time updates and operational information (Bhuvana and Aram 2019).

Therefore, advances in communication for disaster management, including the use of social media, digital platforms, and traditional media, have revolutionized the way information is disseminated and coordinated during crises. This chapter emphasizes the importance of integrating communication strategies and leveraging technological advances to improve resilience and disaster preparedness in South Asia and beyond.

16.4 Science and Technology in Disaster Risk Management

The integration of science and technology into disaster risk reduction has changed the way hazards are predicted, monitored, and mitigated. This section provides a comparative analysis of several studies that look at the use of space technology to address different hazards, including landslides in mountainous regions, cyclone-related hazards in coastal areas, and floods in floodplains. These studies highlight the critical role of geospatial technologies and artificial intelligence in improving disaster preparedness and resilience. One chapter presents a landslide hazard model developed using advanced geospatial artificial intelligence (GeoAI). This model illustrates the risk dynamics of natural and landslide-induced landslides on infrastructure and agricultural land using GIS and Google Earth imagery. The application of GeoAI for landslide prediction demonstrates the potential of AI in improving the accuracy and timeliness of hazard assessments. Similarly, Thekdi et al. (2023) emphasize the synergy between AI and risk science, stating that AI methods can significantly improve risk analysis and disaster management by providing detailed and actionable insights.

The chapter on hurricane zonation uses geospatial technology to identify potentially hurricane-prone areas and their associated hazards, providing information for land use policy and planning. This aligns with the findings of Bjerge et al. (2016), who analyzed the OSOCC virtual platform and demonstrated how technological advances improve information sharing between organizations and coordination in disaster response.

Another chapter assesses the impact of flooding using the Google Earth Engine, emphasizing the need for automated mapping systems that are able to quickly integrate new data. This finding is supported by AlHinai (2020) who discusses the digital transformation of disaster management and the associated challenges for traditional systems. The importance of continuous updates and interactive web GIS platforms in improving the accessibility and usability of hazard maps is echoed by Thekdi et al. (2023), who discuss the integration of AI into disaster risk applications.

Despite their unique hazards, all these studies share a common goal: to provide risk information for better policy and planning. However, they face methodological challenges. For example, while the landslide hazard model in one chapter relies on data-driven approaches to identify hazardous areas based on past conditions, the hazard map in another chapter is created subjectively without considering historical events. Continuous updates and the development of automated mapping systems capable of quickly integrating new data are essential to maintain the relevance and accuracy of hazard maps. However, the integration of science and technology into disaster risk management through the use of space technology, AI and digital platforms greatly enhances the ability to predict, mitigate, and respond to various hazards. This chapter summarizes the key findings from several studies and provides a comprehensive understanding of the methodological challenges and opportunities for future research and policy development. By emphasizing the importance of continuous updates, stakeholder collaboration, and the integration of advanced technologies, this chapter contributes to the disaster management literature and provides a pathway for improving resilience and preparedness.

16.5 Innovation in Disaster Risk Management

Disaster risk management has evolved considerably through innovative technologies and social strategies. The various chapters in this compilation provide a comprehensive insight into these innovations and demonstrate their effectiveness in different contexts. This comparative analysis highlights key findings from the application of drones, biophilic design, collectivization, and youth engagement in public spaces and illustrates their contribution to disaster management.

The chapter "Use of Unmanned Aerial Vehicles" shows their significant role in disaster management in Nepal. Drones were used effectively for crisis mapping after the 2015 Gorkha earthquake and enabled fast and accurate data collection. This aligns with the findings of Adams and Friedland (2011), who highlight the crucial role of drones in providing detailed situational awareness in emergencies. Furthermore, the DrOTS (Drone Optimized Therapy System) project, which used drones to provide healthcare services in remote areas, reflects the potential of drones to improve medical logistics in inaccessible regions identified by Amukele et al. (2015). Furthermore, the use of drones during the Melamchi Debris Flow in 2021 for damage assessment and route finding highlights their operational efficiency in disaster scenarios, which is supported by similar findings in the disaster management literature.

The chapter "Biophilic Design" emphasizes the integration of natural elements into urban environments to improve psychological resilience. The case study in this chapter illustrates that biophilic design can mitigate the negative effects of urbanization on mental health. This supports the broader research linking green spaces and natural elements to improved emotional well-being and reduced psychological vulnerability in disaster-prone areas. These findings suggest that incorporating biophilic principles into urban planning can be a strategic measure to improve disaster resilience.

The chapter "Collectivization and Local Leadership for Resilience to Disasters" highlights the importance of social innovation in building community resilience. The findings in this chapter highlight how local groups, CSOs, NGOs and collectives play a central role in promoting community resilience to everyday risks and recurring disasters. This is in line with Aldrich (2012), who has shown that social capital and community networks are crucial for effective disaster response as they promote collective action and resource sharing. Furthermore, this chapter's focus on coping with complex disasters such as COVID-19 and cloudbursts through community trust and collaboration aligns with findings that emphasize the role of social innovation in adapting to complex crises. These parallels confirm the approach of this chapter and encourage further exploration of social innovation as a pathway to disaster resilience.

The comparative analysis of these innovative approaches in disaster management emphasizes their effectiveness and applicability in different contexts. The integration of drones, biophilic design, collectivization, and youth engagement demonstrates a multifaceted strategy that increases resilience to disasters. Together, these chapters chart a path for future disaster management and emphasize the need for technological and social innovation to overcome complex challenges.

16.6 Integrating Holistic Strategies Within the Sendai Framework for Disaster Management

Adopting a holistic approach to disaster management, as advocated in the Sendai Framework, requires the convergence of capacity building, advanced communication, and cutting-edge science and technology to effectively address the multi-faceted challenges of climate change and anthropogenic activities. This integrated strategy is essential not only for strengthening resilience, but also for ensuring sustainable development in vulnerable regions. The Sendai Framework for Disaster Risk Reduction emphasizes the importance of understanding disaster risk in all its dimensions—vulnerability, capacity, exposure of people and assets, hazard characteristics, and environment. In this context, the holistic approach goes beyond purely technical solutions and includes community participation, capacity building, and the use of local knowledge alongside scientific innovation. This approach is crucial as the frequency and intensity of disasters increases due to climate change and man-made environmental degradation.

Modern communication technologies play a crucial role in disseminating information in real time and facilitating effective disaster response and recovery. When combined with traditional communication methods, they provide greater reach, engage communities at all levels, and improve the overall resilience framework. Capacity building through education and training programs equips local populations with the necessary skills and knowledge to respond to and cope with disasters more effectively. Integrating new scientific knowledge and technologies such as remote sensing, GIS-based hazard mapping and AI for predictive analytics into disaster risk management strategies offers unprecedented opportunities to predict, monitor, and mitigate disasters. These tools can provide detailed risk assessments and enable more informed decision-making processes, which are crucial for planning and implementing effective disaster mitigation strategies. However, the effectiveness of these integrated approaches requires a strong institutional framework that supports seamless collaboration between government agencies, non-governmental organizations, and community actors. It also requires an adaptive administration that can respond to changing disaster risks influenced by climate change and anthropogenic factors.

Therefore, the holistic approach recommended by the Sendai Framework provides a solid blueprint for improving disaster resilience. At the same time, there is a need to also focus on an enabling market ecosystem for products and services in communication, science, technology, and innovation in south Asia. Such ecosystem should have a well-rounded strategy to support social enterprises and enabling policies. By combining capacity building, advanced communication and innovative technologies, this approach not only addresses current challenges but also anticipates future risks, paving the way for safer and more resilient communities.

16.7 Conclusion and Future Directions

In the face of escalating disasters exacerbated by climate change, the integration of science, technology, communication, and innovation into disaster risk management is both a glimmer of hope and a call to urgent action. As this book meticulously demonstrates, the challenges posed by disasters in South Asia are becoming increasingly complex and frequent, necessitating a paradigm shift in the understanding and management of these crises. Lessons learned from various case studies in the region underscore a critical reality: while advances in technology and communication strategies provide powerful tools for disaster management, they also highlight the significant gaps and needs faced by vulnerable communities.

Science has made great strides in the use of geospatial data, drones, and artificial intelligence to improve the accuracy and timeliness of disaster response and risk assessment. These technological innovations have proven indispensable when it comes to mapping disaster-prone areas, assessing damage, and facilitating the delivery of efficient aid. However, the effectiveness of these high-tech solutions depends heavily on their integration into well-informed, community-oriented disaster management strategies that prioritize the specific needs and capabilities of the local population. In addition, the persistent problem of communication gaps, especially in remote and underprivileged areas, is a significant obstacle to effective disaster risk management. While urban centers benefit from faster recovery due to better infrastructure and information flow, rural and remote regions remain at a severe disadvantage. This inequality not only slows the pace of recovery, but also exacerbates pre-existing vulnerabilities, making these communities more susceptible to the negative impacts of climate change. It is therefore imperative that future efforts focus not only on technological advancement, but also on building and improving communication networks that reach every corner of vulnerable populations. The convergence of climate change impacts with socio-economic vulnerabilities presents a formidable challenge, but also an opportunity to renew and redefine disaster management paradigms. Shifting to a more integrated approach that combines technology with traditional knowledge and community participation could significantly improve resilience efforts. By promoting local leadership and collectivization, as demonstrated in several case studies, communities can improve their adaptive capacity and not only survive disasters, but also successfully cope with them.

However, this optimistic outlook is clouded by the reality of increasing climate-related disasters, which pose a growing threat to stability and safety. The scientific evidence is clear and compelling and suggests that unless comprehensive and inclusive mitigation and adaptation strategies are implemented quickly, the impacts will worsen. The need for a proactive approach to policy making, education, and community engagement is more important than ever. Policies must be dynamic and evolve with climate change and new technologies to combat its effects.

The risk landscape of South Asia is becoming more complex than in the past due to rapid changes in socio-technological interactions and the environment. The mountainous, coastal, riverine, and desert regions of South Asia frequently face multiple hazards, many of which are interconnected. Sometimes, these complex interactions manifest as cascading or compound hazards. The risks associated with these hazards are not always confined to the place where they originate, nor are they limited to a specific time period, as one hazard can trigger another. Therefore, traditional risk assessment methods are insufficient for predicting and managing cascading risks. A systemic approach, which considers the interdependencies of risk components, is essential. However, methods for systemic risk assessment are still emerging and will require further development in the future.

While the road ahead is full of challenges, it is also full of opportunities for significant progress in disaster risk management. The integration of cutting-edge technologies with community-oriented strategies offers a promising path to a resilient future. However, the window of opportunity for effective action is shrinking as the impacts of climate change accelerate. So while we harness the power of innovation and technology, we must also step up our efforts to build equitable, sustainable and resilient communities to ensure that no one is left behind in the search for safety and stability in an increasingly unpredictable world. This dual approach will not only protect lives and livelihoods, but also serve as a cornerstone for sustainable development in South Asia and beyond.

References

Adams SM, Friedland CJ (2011) A survey of unmanned aerial vehicle (UAV) usage for imagery collection in disaster research and management. In: 9th international workshop on remote sensing for disaster response, vol 8, pp 1–8

ADB (2019) Disaster risk management in Asia and the Pacific. Retrieved from https://www.adb.org/publications/disaster-risk-management-asia-pacific

Aldrich DP (2012) Building resilience: social capital in post-disaster recovery. University of Chicago Press. Retrieved from https://www.press.uchicago.edu/ucp/books/book/chicago/B/bo13101652.html

AlHinai YS (2020) Disaster management digitally transformed: exploring the impact and key determinants from the UK national disaster management experience. Int J Disaster Risk Reduct 51:101851

AlQahtany AM, Abubakar IR (2020) Public perception and attitudes to disaster risks in a coastal metropolis of Saudi Arabia. Int J Disaster Risk Reduct 44:101422

Amukele TK, Sokoll LJ, Pepper D, Howard DP, Street J (2015) Can unmanned aerial systems (drones) be used for the routine transport of chemistry, hematology, and coagulation laboratory specimens? PLoS ONE 10(7):e0134020

Ardalan A, Mowafi H, Ardakani HM, Abolhasanai F, Zanganeh AM, Safizadeh H, Salari S, Zonoobi V (2013) Effectiveness of a primary health care program on urban and rural community disaster preparedness, Islamic Republic of Iran: a community intervention trial. Disaster Med Public Health Prep 7(5):481–490

Bangladesh Cyclone Preparedness Programme (2020) Annual report. Retrieved from https://www.cpp.gov.bd/

Bhuvana N, Aram IA (2019) Facebook and WhatsApp as disaster management tools during the Chennai (India) floods of 2015. Int J Disaster Risk Reduct 39:101135

Bjerge B, Clark N, Fisker P, Raju E (2016) Technology and information sharing in disaster relief. PLoS ONE 11(9):e0161783

CREWS Nepal Project (2021) Project overview. Retrieved from https://www.crews-initiative.org/en/projects/nepal

Drones in disaster management (2020) Retrieved from https://reliefweb.int/report/world/drones-disaster-response

Dube E (2015) Improving disaster risk reduction capacity of District Civil Protection Units in managing veld fires a case of Mangwe District in Matabeleland South Province, Zimbabwe. Jàmbá: J Disaster Risk Stud 7(1) 1–13

Dwirahmadi F, Rutherford S, Phung D, Chu C (2019) Understanding the operational concept of a flood-resilient urban community in Jakarta, Indonesia, from the perspectives of disaster risk reduction, climate change adaptation, and development agencies. Int J Environ Res Public Health 16(20):3993

Hugelius K, Gifford M, Örtenwall P, Adolfsson A (2016) Disaster radio for communication of vital messages and health-related information: experiences from the Haiyan Typhoon, the Philippines. Disaster Med Public Health Prep 10(4):591–597

Iizuka A (2020) Developing capacity for disaster risk reduction: lessons learned from a case of Sri Lanka. Prog Disaster Sci 6:100073

Liu BF, Fowler BM, Roberts HA, Herovic E (2018) Keeping hospitals operating during disasters through crisis communication preparedness. Public Relat Rev 44(4):585–597

Liu BF, Fraustino JD, Jin Y (2016) Social media use during disasters: how information form and source influence intended behavioral responses. Commun Res 43(5):626–646

Mason A, Flores L, Liu P, Tims K, Spencer E, Gire TG (2019) Disaster communication: an analysis of the digital communication strategies used by the medical tourism industry during the 2017 Caribbean hurricane season. J Hosp Tour Insights 2(3):241–259

Miles SB, Green RA, Svekla W (2012) Disaster risk reduction capacity assessment for precarious settlements in Guatemala City. Disasters 36(3):365–381

Mutiarni YS, Nakamura H, Bhattacharya Y (2022) The resilient community: strengthening people-centered disaster risk reduction in the Merapi volcano community, Java, Indonesia. Sustainability 14(4):2215

NDMA (2019) National disaster management plan. Retrieved from https://ndma.gov.in/en/publication.html

SAARC Disaster Management Centre (2018) Annual report. Retrieved from http://saarc-sdmc.nic.in/

Thekdi S, Tatar U, Santos J, Chatterjee S (2023) Disaster risk and artificial intelligence: a framework to characterize conceptual synergies and future opportunities. Risk Anal 43(8):1641–1656

UNDRR (2020) The human cost of disasters: an overview of the last 20 years (2000–2019). Retrieved from https://www.undrr.org/publication/human-cost-disasters-overview-last-20-years-2000-2019

Correction to: Communication, Science, Technology, and Innovation in Disaster Risk Management

Swapan Talukdar, Ranit Chatterjee, Somnath Bera,
Atish Prashar, and Rajib Shaw

Correction to:
S. Talukdar et al. (eds.), *Communication, Science, Technology and Innovation in Disaster Risk Management*, GIScience and Geo-environmental Modelling, https://doi.org/10.1007/978-3-031-77189-7

In the original version of Chapter 9, affiliation of authors A. Prashar and G. Balamurugan has been updated. The chapter has been updated with the change.

In the original version of this book, affiliation of the book editor "Swapan Talukdar" has been updated in Front mater.

The book has been updated with the change.

The updated version of this book can be found at
https://doi.org/10.1007/978-3-031-77189-7_9
https://doi.org/10.1007/978-3-031-77189-7

S. Talukdar et al. (eds.), *Communication, Science, Technology, and Innovation in Disaster Risk Management*, GIScience and Geo-environmental Modelling, https://doi.org/10.1007/978-3-031-77189-7_17

Author Index

S. Talukdar et al. (eds.), *Communication, Science, Technology and Innovation in Disaster Risk Management*, GIScience and Geo-environmental Modelling, https://doi.org/10.1007/978-3-031-77189-7

Subject Index

S. Talukdar et al. (eds.), *Communication, Science, Technology and Innovation in Disaster Risk Management*, GIScience and Geo-environmental Modelling, https://doi.org/10.1007/978-3-031-77189-7

The manufacturer's authorised representative in the EU is Springer Nature Customer Service Centre GmbH, Europaplatz 3, 69115 Heidelberg, Germany. If you have any concerns regarding our products, please contact ProductSafety@springernature.com

Printed and bound by CPI Group (UK) Ltd, Croydon, CR0 4YY
30/09/2026
02225584-0020